T0230960

**Sixth Edition**

# GMP/ISO Quality Audit Manual for Healthcare Manufacturers and Their Suppliers

**VOLUME 2**
**Regulations, Standards, and Guidelines**

## Leonard Steinborn

**CRC Press**
Taylor & Francis Group
Boca Raton  London  New York

CRC Press is an imprint of the
Taylor & Francis Group, an **informa** business

CRC Press
Taylor & Francis Group
6000 Broken Sound Parkway NW, Suite 300
Boca Raton, FL 33487-2742

First issued in paperback 2019

© 2005 by Taylor & Francis Group, LLC
CRC Press is an imprint of Taylor & Francis Group, an Informa business

No claim to original U.S. Government works

ISBN-13: 978-0-8493-1847-4 (hbk)
ISBN-13: 978-0-367-39332-8 (pbk)

This book contains information obtained from authentic and highly regarded sources. Reasonable efforts have been made to publish reliable data and information, but the author and publisher cannot assume responsibility for the validity of all materials or the consequences of their use. The authors and publishers have attempted to trace the copyright holders of all material reproduced in authors and publishers have attempted to trace the copyright holders of all material reproduced in this publication and apologize to copyright holders if permission to publish in this form has not been obtained. If any copyright material has not been acknowledged please write and let us know so we may rectify in any future reprint.

Except as permitted under U.S. Copyright Law, no part of this book may be reprinted, reproduced, transmitted, or utilized in any form by any electronic, mechanical, or other means, now known or hereafter invented, including photocopying, microfilming, and recording, or in any information storage or retrieval system, without written permission from the publishers.

For permission to photocopy or use material electronically from this work, please access www.copyright.com (http://www.copyright.com/) or contact the Copyright Clearance Center, Inc. (CCC), 222 Rosewood Drive, Danvers, MA 01923, 978-750-8400. CCC is a not-for-profit organization that provides licenses and registration for a variety of users. For organizations that have been granted a photocopy license by the CCC, a separate system of payment has been arranged.

**Trademark Notice:** Product or corporate names may be trademarks or registered trademarks, and are used only for identification and explanation without intent to infringe.

### Library of Congress Cataloging-in-Publication Data

Catalog record is available from the Library of Congress

**Visit the Taylor & Francis Web site at**
**http://www.taylorandfrancis.com**

**and the CRC Press Web site at**
**http://www.crcpress.com**

# Contents*

---

\* Although the International Pharmaceuticals Excipients Council (IPEC) Good Manufacturing Practice Guideline for Bulk Pharmaceutical Excipients is referenced in the Volume 1 checklists, we were unable to obtain distribution rights for the document. A tab is reserved for placement of the standard into this volume should you decide to obtain the document from IPEC directly. Behind the tab you will find the order information needed to procure the guideline.

# PART 11
## Electronic Records/Electronic Signatures

## CONTENTS

## SUBPART A — GENERAL PROVISIONS

### SECTION 11.1 SCOPE

(a) The regulations in this part set forth the criteria under which the agency considers electronic records, electronic signatures, and handwritten signatures executed to electronic records to be trustworthy, reliable, and generally equivalent to paper records and handwritten signatures executed on paper.

(b) This part applies to records in electronic form that are created, modified, maintained, archived, retrieved, or transmitted, under any records requirements set forth in agency regulations. This part also applies to electronic records submitted to the agency under requirements of the Federal Food, Drug, and Cosmetic Act and the Public Health Service Act, even if such records are not specifically identified in agency regulations. However, this part does not apply to paper records that are, or have been, transmitted by electronic means.

(c) Where electronic signatures and their associated electronic records meet the requirements of this part, the agency will consider the electronic signatures to be equivalent to full handwritten signatures, initials, and other general signings as required by agency regulations, unless specifically excepted by regulation(s) effective on or after August 20, 1997.

(d) Electronic records that meet the requirements of this part may be used in lieu of paper records, in accordance with §11.2, unless paper records are specifically required.

(e) Computer systems (including hardware and software), controls, and attendant documentation maintained under this part shall be readily available for, and subject to, FDA inspection.

## SECTION 11.2 IMPLEMENTATION

(a) For records required to be maintained but not submitted to the agency, persons may use electronic records in lieu of paper records or electronic signatures in lieu of traditional signatures, in whole or in part, provided that the requirements of this part are met.

(b) For records submitted to the agency, persons may use electronic records in lieu of paper records or electronic signatures in lieu of traditional signatures, in whole or in part, provided that:

(1) The requirements of this part are met; and
(2) The document or parts of a document to be submitted have been identified in public docket No. 92S-0251 as being the type of submission the agency accepts in electronic form. This docket will identify specifically what types of documents or parts of documents are acceptable for submission in electronic form without paper records and the agency receiving unit(s) (e.g., specific center, office, division, branch) to which such submissions may be made.

Documents to agency receiving unit(s) not specified in the public docket will not be considered as official if they are submitted in electronic form; paper forms of such documents will be considered as official and must accompany any electronic records.

Persons are expected to consult with the intended agency receiving unit for details on how (e.g., method of transmission, media, file formats, and technical protocols) and whether to proceed with the electronic submission.

## SECTION 11.3 DEFINITIONS

(a) The definitions and interpretations of terms contained in §201 of the act apply to those terms when used in this part.

(b) The following definitions of terms also apply to this part:

(1) Act means the Federal Food, Drug, and Cosmetic Act (§201–903 [21 U.S.C. 321–393]).
(2) Agency means the Food and Drug Administration.
(3) Biometrics means a method of verifying an individual's identity based on measurement of the individual's physical feature(s) or repeatable action(s) where those features and/or actions are both unique to that individual and measurable.
(4) Closed system means an environment in which system access is controlled by persons who are responsible for the content of electronic records that are on the system.
(5) Digital signature means an electronic signature based upon cryptographic methods of originator authentication, computed by using a set of rules and a set of parameters such that the identity of the signer and the integrity of the data can be verified.
(6) Electronic record means any combination of text, graphics, data, audio, pictorial, or other information representation in digital form that is created, modified, maintained, archived, retrieved, or distributed by a computer system.
(7) Electronic signature means a computer data compilation of any symbol or series of symbols executed, adopted, or authorized by an individual to be the legally binding equivalent of the individual's handwritten signature.
(8) Handwritten signature means the scripted name or legal mark of an individual handwritten by that individual and executed or adopted with the present intention to authenticate a writing in a permanent form. The act of signing with a writing or marking instrument such as a pen

or stylus is preserved. The scripted name or legal mark, while conventionally applied to paper, may also be applied to other devices that capture the name or mark.

(9) Open system means an environment in which system access is not controlled by persons who are responsible for the content of electronic records that are on the system.

## SUBPART B — ELECTRONIC RECORDS

### SECTION 11.10 CONTROLS FOR CLOSED SYSTEMS

Persons who use closed systems to create, modify, maintain, or transmit electronic records shall employ procedures and controls designed to ensure the authenticity, integrity, and, when appropriate, the confidentiality of electronic records, and to ensure that the signer cannot readily repudiate the signed record as not genuine. Such procedures and controls shall include the following:

(a) Validation of systems to ensure accuracy, reliability, consistent intended performance, and the ability to discern invalid or altered records.

(b) The ability to generate accurate and complete copies of records in both human readable and electronic form suitable for inspection, review, and copying by the agency. Persons should contact the agency if there are any questions regarding the ability of the agency to perform such review and copying of the electronic records.

(c) Protection of records to enable their accurate and ready retrieval throughout the records retention period.

(d) Limiting system access to authorized individuals.

(e) Use of secure, computer-generated, time-stamped audit trails to independently record the date and time of operator entries and actions that create, modify, or delete electronic records. Record changes shall not obscure previously recorded information. Such audit trail documentation shall be retained for a period at least as long as that required for the subject electronic records and shall be available for agency review and copying.

(f) Use of operational system checks to enforce permitted sequencing of steps and events, as appropriate.

(g) Use of authority checks to ensure that only authorized individuals can use the system, electronically sign a record, access the operation or computer system input or output device, alter a record, or perform the operation at hand.

(h) Use of device (e.g., terminal) checks to determine, as appropriate, the validity of the source of data input or operational instruction.

(i) Determination that persons who develop, maintain, or use electronic record/electronic signature systems have t he education, training, and experience to perform their assigned tasks.

(j) The establishment of, and adherence to, written policies that hold individuals accountable and responsible for actions initiated under their electronic signatures, in order to deter record and signature falsification.

(k) Use of appropriate controls over systems documentation including:
   (1) Adequate controls over the distribution of, access to, and use of documentation for system operation and maintenance.
   (2) Revision and change control procedures to maintain an audit trail that documents time-sequenced development and modification of systems documentation.

### SECTION 11.30 CONTROLS FOR OPEN SYSTEMS

Persons who use open systems to create, modify, maintain, or transmit electronic records shall employ procedures and controls designed to ensure the authenticity, integrity, and, as appropriate, the confidentiality of electronic records from the point of their creation to the point of their receipt. Such procedures and controls shall include those identified in §11.10, as appropriate, and additional measures such as document encryption and use of appropriate digital signature standards to ensure, as necessary under the circumstances, record authenticity, integrity, and confidentiality.

### SECTION 11.50 SIGNATURE MANIFESTATIONS

(a) Signed electronic records shall contain information associated with the signing that clearly indicates all of the following:

(1) The printed name of the signer;
(2) The date and time when the signature was executed; and
(3) The meaning (such as review, approval, responsibility, or authorship) associated with the signature.

(b) The items identified in paragraphs (a)(1), (a)(2), and (a)(3) of this section shall be subject to the same controls as for electronic records and shall be included as part of any human readable form of the electronic record (such as electronic display or printout).

### SECTION 11.70 SIGNATURE/RECORD LINKING

Electronic signatures and handwritten signatures executed to electronic records shall be linked to their respective electronic records to ensure that the signatures cannot be excised, copied, or otherwise transferred to falsify an electronic record by ordinary means.

## SUBPART C — ELECTRONIC SIGNATURES

### SECTION 11.100 GENERAL REQUIREMENTS

(a) Each electronic signature shall be unique to one individual and shall not be reused by, or reassigned to, anyone else.

(b) Before an organization establishes, assigns, certifies, or otherwise sanctions an individual's electronic signature, or any element of such electronic signature, the organization shall verify the identity of the individual.

(c) Persons using electronic signatures shall, prior to or at the time of such use, certify to the agency that the electronic signatures in their system, used on or after August 20, 1997, are intended to be the legally binding equivalent of traditional handwritten signatures.

(1) The certification shall be submitted in paper form and signed with a traditional handwritten signature, to the Office of Regional Operations (HFC-100), 5600 Fishers Lane, Rockville, MD 20857.
(2) Persons using electronic signatures shall, upon agency request, provide additional certification or testimony that a specific electronic signature is the legally binding equivalent of the signer's handwritten signature.

# SUBPART C — ELECTRONIC SIGNATURES

## SECTION 11.200 ELECTRONIC SIGNATURE COMPONENTS AND CONTROLS

(a) Electronic signatures that are not based upon biometrics shall:

(1) Employ at least two distinct identification components such as an identification code and password.
   (i) When an individual executes a series of signings during a single, continuous period of controlled system access, the first signing shall be executed using all electronic signature components; subsequent signings shall be executed using at least one electronic signature component that is executable only by, and designed to be used only by, the individual.
   (ii) When an individual executes one or more signings not performed during a single, continuous period of controlled system access, each signing shall be executed using all of the electronic signature components.
(2) Be used only by their genuine owners; and
(3) Be administered and executed to ensure that attempted use of an individual's electronic signature by anyone other than its genuine owner requires collaboration of two or more individuals.

(b) Electronic signatures based upon biometrics shall be designed to ensure that they cannot be used by anyone other than their genuine owners.

## SECTION 11.300 CONTROLS FOR IDENTIFICATION CODES/PASSWORDS

Persons who use electronic signatures based upon use of identification codes in combination with passwords shall employ controls to ensure their security and integrity. Such controls shall include:

(a) Maintaining the uniqueness of each combined identification code and password, such that no two individuals have the same combination of identification code and password.
(b) Ensuring that identification code and password issuances are periodically checked, recalled, or revised (e.g., to cover such events as password aging).
(c) Following loss management procedures to electronically deauthorize lost, stolen, missing, or otherwise potentially compromised tokens, cards, and other devices that bear or generate identification code or password information, and to issue temporary or permanent replacements using suitable, rigorous controls.
(d) Use of transaction safeguards to prevent unauthorized use of passwords and/or identification codes, and to detect and report in an immediate and urgent manner any attempts at their unauthorized use to the system security unit, and, as appropriate, to organizational management.
(e) Initial and periodic testing of devices, such as tokens or cards, that bear or generate identification code or password information to ensure that they function properly and have not been altered in an unauthorized manner.

# Part 803
## Medical Device Reporting

## CONTENTS

## SUBPART A — GENERAL PROVISIONS

### SECTION 803.17 WRITTEN MDR PROCEDURES

User facilities, importers, and manufacturers shall develop, maintain, and implement written MDR procedures for the following:

(a) Internal systems that provide for:
 (1) Timely and effective identification, communication, and evaluation of events that may be subject to medical device reporting requirements;
 (2) A standardized review process/procedure for determining when an event meets the criteria for reporting under this part; and
 (3) Timely transmission of complete medical device reports to FDA and/or manufacturers;
(b) Documentation and recordkeeping requirements for:
 (1) Information that was evaluated to determine if an event was reportable;
 (2) All medical device reports and information submitted to FDA and manufacturers;
 (3) Any information that was evaluated for the purpose of preparing the submission of annual reports; and
 (4) Systems that ensure access to information that facilitates timely follow-up and inspection by FDA.

### SECTION 803.18 FILES AND DISTRIBUTOR RECORDS

(a) User facilities, importers, and manufacturers shall establish and maintain MDR event files. All MDR event files shall be prominently identified as such and filed to facilitate timely access.
(b) (1) For purposes of this part, "MDR event files" are written or electronic files maintained by user facilities, importers, and manufacturers. MDR event files may incorporate references to other information, e.g., medical records, patient files, engineering reports, etc., in lieu of copying and maintaining duplicates in this file. MDR event files must contain:

          (i) Information in the possession of the reporting entity or references to information related to the adverse event, including all documentation of the entity's deliberations and decision-making processes used to determine if a device-related death, serious injury, or malfunction was or was not reportable under this part.

       (ii) Copies of all MDR forms, as required by this part, and other information related to the event that was submitted to FDA and other entities (e.g., an importer, distributor, or manufacturer).

    (2) User facilities, importers, and manufacturers shall permit any authorized FDA employee during all reasonable times to access, to copy, and to verify the records required by this part.

(c) User facilities shall retain an MDR event file relating to an adverse event for a period of 2 years from the date of the event. Manufacturers and importers shall retain an MDR event file relating to an adverse event for a period of 2 years from the date of the event or a period of time equivalent to the expected life of the device, whichever is greater. MDR event files must be maintained for the time periods described in this paragraph even if the device is no longer distributed.

(d) (1) A device distributor shall establish and maintain device complaint records containing any incident information, including any written, electronic, or oral communication, either received by or generated by the firm, that alleges deficiencies related to the identity (e.g., labeling), quality, durability, reliability, safety, effectiveness, or performance of a device. Information regarding the evaluation of the allegations, if any, shall also be maintained in the incident record. Device incident records shall be prominently identified as such and shall be filed by device, and may be maintained in written or electronic form. Files maintained in electronic form must be backed up.

    (2) A device distributor shall retain copies of the records required to be maintained under this section for a period of 2 years from the date of inclusion of the record in the file or for a period of time equivalent to the expected life of the device, whichever is greater, even if the distributor has ceased to distribute the device that is the subject of the record.

    (3) A device distributor shall maintain the device complaint files established under this section at the distributor's principal business establishment. A distributor that is also a manufacturer may maintain the file at the same location as the manufacturer maintains its complaint file under §820.180 and 820.198 of this chapter. A device distributor shall permit any authorized FDA employee, during all reasonable times, to have access to, and to copy and verify, the records required by this part.

(e) The manufacturer may maintain MDR event files as part of its complaint file, under §820.198 of this chapter, provided that such records are prominently identified as MDR reportable events. A report submitted under this Subpart A shall not be considered to comply with this part unless the event has been evaluated in accordance with the requirements of §820.162 and 820.198 of this chapter. MDR files shall contain an explanation of why any information required by this part was not submitted or could not be obtained. The results of the evaluation of each event are to be documented and maintained in the manufacturer's MDR event file.

# Part 806
## Medical Devices;
## Reports of Corrections and Removals

## CONTENTS

## SUBPART B — REPORTS AND RECORDS

### SECTION 806.10 REPORTS OF CORRECTIONS AND REMOVALS

(a) Each device manufacturer or importer shall submit a written report to FDA of any correction or removal of a device initiated by such manufacturer or importer if the correction or removal was initiated:

    (1) To reduce a risk to health posed by the device; or

    (2) To remedy a violation of the Act caused by the device which may present a risk to health unless the information has already been provided as set forth in paragraph (f) of this section or the corrective or removal action is exempt from the reporting requirements under §806.1(b).

(b) The manufacturer or importer shall submit any report required by paragraph (a) of this section within 10 working days of initiating such correction or removal.

(c) The manufacturer or importer shall include the following information in the report:

    (1) The seven-digit registration number of the entity responsible for submission of the report of corrective or removal action (if applicable); the month, day, and year that the report is made; and a sequence number (i.e., 001 for the first report, 002 for the second report, 003 etc.); and the report type designation "C" or "R." For example, the complete number for the first correction report submitted on June 1, 1997, will appear as follows for a firm with the registration number 1234567: 1234567-6/1/97-001-C. The second correction report number submitted by the same firm on July 1, 1997, would be 1234567-7/1/97-002-C etc. For removals, the number will appear as follows: 1234567-6/1/97-001-R and 1234567-7/1/97-002-R, etc. Firms that do not have a seven-digit registration number may use seven zeros followed by the month, day, year, and sequence number (i.e., 0000000-6/1/97-001-C for corrections and 0000000-7/1/97-001-R for removals). Reports received without a seven-digit registration number will be assigned a seven-digit central file number by the district office reviewing the reports.

(2) The name, address, and telephone number of the manufacturer or importer, and the name, title, address, and telephone number of the manufacturer or importer representative responsible for conducting the device correction or removal.

(3) The brand name and the common name, classification name, or usual name of the device and the intended use of the device.

(4) Marketing status of the device, i.e., any applicable premarket notification number, premarket approval number, or indication that the device is a preamendments device, and the device listing number. A manufacturer or importer that does not have an FDA establishment registration number shall indicate in the report whether it has ever registered with FDA.

(5) The model, catalog, or code number of the device and the manufacturing lot or serial number of the device or other identification number.

(6) The manufacturer's name, address, telephone number, and contact person if different from that of the person submitting the report.

(7) A description of the event(s) giving rise to the information reported and the corrective or removal actions that have been, and are expected to be taken.

(8) Any illness or injuries that have occurred with use of the device. If applicable, include the medical device report numbers.

(9) The total number of devices manufactured or distributed subject to the correction or removal and the number in the same batch, lot, or equivalent unit of production subject to the correction or removal.

(10) The date of manufacture or distribution and the device's expiration date or expected life.

(11) The names, addresses, and telephone numbers of all domestic and foreign consignees of the device and the dates and number of devices distributed to each such consignee.

(12) A copy of all communications regarding the correction or removal and the names and addresses of all recipients of the communications not provided in accordance with paragraph (c)(11) of this section.

(13) If any required information is not immediately available, a statement as to why it is not available and when it will be submitted.

(d) If, after submitting a report under this part, a manufacturer or importer determines that the same correction or removal should be extended to additional lots or batches of the same device, the manufacturer or importer shall within 10 working days of initiating the extension of the correction or removal, amend the report by submitting an amendment citing the original report number assigned according to paragraph (c)(1) of this section, all of the information required by paragraph (c)(2), and any information required by paragraphs (c)(3) through (c)(12) of this section that is different from the information submitted in the original report. The manufacturer or importer shall also provide a statement in accordance with paragraph (c)(13) of this section for any required information that is not readily available.

(e) A report submitted by a manufacturer or importer under this section (and any release by FDA of that report or information) does not necessarily reflect a conclusion by the manufacturer, importer, or FDA that the report or information constitutes an admission that the device caused or contributed to a death or serious injury. A manufacturer or importer need not admit, and may deny, that the report or information submitted under this section constitutes an admission that the device caused or contributed to a death or serious injury.

(f) No report of a correction or removal is required under this part, if a report of the correction or removal is required and has been submitted under parts 803, 804, or 1004 of this chapter.

# Part 821
## Medical Device Tracking Requirements

## CONTENTS

## SUBPART B — TRACKING REQUIREMENTS

### SECTION 821.25 DEVICE TRACKING SYSTEM AND CONTENT REQUIREMENTS: MANUFACTURER REQUIREMENTS

(a) A manufacturer of a tracked device shall adopt a method of tracking for each such type of device it distributes that enables a manufacturer to provide FDA with the following information in writing for each tracked device distributed:

  (1) Except as required by order under §518(e) of the Act, within 3 working days of a request from FDA, prior to the distribution of a tracked device to a patient, the name, address, and telephone number of the distributor, multiple distributor, or final distributor holding the device for distribution and the location of the device;

  (2) Within 10 working days of a request from FDA for tracked devices that are intended for use by a single patient over the life of the device, after distribution to or implantation in a patient:

    (i) The lot number, batch number, model number, or serial number of the device or other identifier necessary to provide for effective tracking of the devices;

    (ii) The date the device was shipped by the manufacturer;

    (iii) The name, address, telephone number, and social security number (if available) of the patient receiving the device, unless not released by the patient under §821.55(a);

    (iv) The date the device was provided to the patient;

    (v) The name, mailing address, and telephone number of the prescribing physician;

    (vi) The name, mailing address, and telephone number of the physician regularly following the patient if different than the prescribing physician; and

    (vii) If applicable, the date the device was explanted and the name, mailing address, and telephone number of the explanting physician; the date of the patient's death; or the date the device was returned to the manufacturer, permanently retired from use, or otherwise permanently disposed of.

(3) Except as required by order under §518(e) of the Act, within 10 working days of a request from FDA for tracked devices that are intended for use by more than one patient, after the distribution of the device to the multiple distributor:

    (i) The lot model number, batch number, serial number of the device or other identifier necessary to provide for effective tracking of the device;

    (ii) The date the device was shipped by the manufacturer;

    (iii) The name, address, and telephone number of the multiple distributor;

    (iv) The name, address, telephone number, and social security number (if available) of the patient using the device, unless not released by the patient under §821.55(a);

    (v) The location of the device;

    (vi) The date the device was provided for use by the patient;

    (vii) The name, address, and telephone number of the prescribing physician; and

    (viii) If and when applicable, the date the device was returned to the manufacturer, permanently retired from use, or otherwise permanently disposed of.

(b) A manufacturer of a tracked device shall keep current records in accordance with its standard operating procedure of the information identified in paragraphs (a)(1), (a)(2) and (a)(3)(i) through (a)(3)(iii) of this section on each tracked device released for distribution for as long as such device is in use or in distribution for use.

(c) A manufacturer of a tracked device shall establish a written standard operating procedure for the collection, maintenance, and auditing of the data specified in paragraphs (a) and (b) of this section. A manufacturer shall make this standard operating procedure available to FDA upon request. A manufacturer shall incorporate the following into the standard operating procedure:

    (1) Data collection and recording procedures, which shall include a procedure for recording when data that is required under this part is missing and could not be collected and the reason why such required data is missing and could not be collected;

    (2) A method for recording all modifications or changes to the tracking system or to the data collected and maintained under the tracking system, reasons for any modification or change, and dates of any modification or change. Modification and changes included under this requirement include modifications to the data (including termination of tracking), the data format, the recording system, and the file maintenance procedures system; and

    (3) A quality assurance program that includes an audit procedure to be run for each device product subject to tracking, at not less than 6-month intervals for the first 3 years of distribution and at least once a year thereafter. This audit procedure shall provide for statistically relevant sampling of the data collected to ensure the accuracy of data and performance testing of the functioning of the tracking system.

(d) When a manufacturer becomes aware that a distributor, final distributor, or multiple distributor has not collected, maintained, or furnished any record or information required by this part, the manufacturer shall notify the FDA district office responsible for the area in which the distributor, final distributor, or multiple distributor is located of the failure of such persons to comply with the requirements of this part. Manufacturers shall have taken reasonable steps to obtain compliance by the distributor, multiple distributor, or final distributor in question before notifying FDA.

(e) A manufacturer may petition for an exemption or variance from one or more requirements of this part according to the procedures in §821.2 of this chapter.

# Part 820
## Quality System Regulation

**CONTENTS**

## SUBPART A — GENERAL PROVISIONS

### SECTION 820.1 SCOPE

#### (a) Applicability

(1) Current good manufacturing practice (CGMP) requirements are set forth in this quality system regulation. The requirements in this part govern the methods used in, and the facilities and controls used for, the design, manufacture, packaging, labeling, storage, installation, and servicing of all finished devices intended for human use. The requirements in this part are intended to ensure that finished devices will be safe and effective and otherwise in compliance with the Federal Food, Drug, and Cosmetic Act (the act). This part establishes basic requirements applicable to manufacturers of finished medical devices. If a manufacturer engages in only some operations subject to the requirements in this part, and not in others, that manufacturer need only comply with those requirements applicable to the operations in which it is engaged. With respect to class I devices, design controls apply only to those devices listed in §820.30(a)(2).

This regulation does not apply to manufacturers of components or parts of finished devices, but such manufacturers are encouraged to use appropriate provisions of this regulation as guidance. Manufacturers of human blood and blood components are not subject to this part, but are subject to part 606 of this chapter.

(2) The provisions of this part shall be applicable to any finished device as defined in this part, intended for human use, that is manufactured, imported, or offered for import in any State or Territory of the United States, the District of Columbia, or the Commonwealth of Puerto Rico.

(3) In this regulation the term "where appropriate" is used several times. When a requirement is qualified by "where appropriate," it is deemed to be "appropriate" unless the manufacturer can document justification otherwise. A requirement is "appropriate" if nonimplementation could reasonably be expected to result in the product not meeting its specified requirements or the manufacturer not being able to carry out any necessary corrective action.

#### (b) Limitations

The quality system regulation in this part supplements regulations in other parts of this chapter except where explicitly stated otherwise. In the event that it is impossible to comply with all applicable regulations, both in this part and in other parts of this chapter, the regulations specifically applicable to the device in question shall supersede any other generally applicable requirements.

### (c) Authority

Part 820 is established and issued under authority of §501, 502, 510, 513, 514, 515, 518, 519, 520, 522, 701, 704, 801, 803 of the act (21 U.S.C. 351, 352, 360, 360c, 360d, 360e, 360h, 360i, 360j, 360l, 371, 374, 381, 383). The failure to comply with any applicable provision in this part renders a device adulterated under §501(h) of the Act. Such a device, as well as any person responsible for the failure to comply, is subject to regulatory action.

### (d) Foreign manufacturers

If a manufacturer who offers devices for import into the United States refuses to permit or allow the completion of a Food and Drug Administration (FDA) inspection of the foreign facility for the purpose of determining compliance with this part, it shall appear for purposes of §801(a) of the Act, that the methods used in, and the facilities and controls used for, the design, manufacture, packaging, labeling, storage, installation, or servicing of any devices produced at such facility that are offered for import into the United States do not conform to the requirements of §520(f) of the Act and this part and that the devices manufactured at that facility are adulterated under §501(h) of the Act.

### (e) Exemptions or variances

(1) Any person who wishes to petition for an exemption or variance from any device quality system requirement is subject to the requirements of §520(f)(2) of the Act.

Petitions for an exemption or variance shall be submitted according to the procedures set forth in §10.30 of this chapter, the FDA's administrative procedures. Guidance is available from the Center for Devices and Radiological Health, Division of Small Manufacturers Assistance (HFZ-220), 1350 Piccard Dr., Rockville, MD 20850, U.S.A., telephone 1-800-638-2041 or 1-301-443-6597, FAX 301-443-8818.

(2) FDA may initiate and grant a variance from any device quality system requirement when the agency determines that such variance is in the best interest of the public health. Such variance will remain in effect only so long as there remains a public health need for the device and the device would not likely be made sufficiently available without the variance.

## SECTION 820.3 DEFINITIONS

(a) Act means the Federal Food, Drug, and Cosmetic Act, as amended (§201-903, 52 Stat. 1040 *et seq.*, as amended (21 U.S.C. 321-394)). All definitions in §201 of the Act shall apply to the regulations in this part.

(b) Complaint means any written, electronic, or oral communication that alleges deficiencies related to the identity, quality, durability, reliability, safety, effectiveness, or performance of a device after it is released for distribution.

(c) Component means any raw material, substance, piece, part, software, firmware, labeling, or assembly which is intended to be included as part of the finished, packaged, and labeled device.

(d) Control number means any distinctive symbols, such as a distinctive combination of letters or numbers, or both, from which the history of the manufacturing, packaging, labeling, and distribution of a unit, lot, or batch of finished devices can be determined.

(e) Design history file (DHF) means a compilation of records which describes the design history of a finished device.

(f) Design input means the physical and performance requirements of a device that are used as a basis for device design.

(g) Design output means the results of a design effort at each design phase and at the end of the total design effort. The finished design output is the basis for the device master record. The total finished design output consists of the device, its packaging and labeling, and the device master record.

(h) Design review means a documented, comprehensive, systematic examination of a design to evaluate the adequacy of the design requirements, to evaluate the capability of the design to meet these requirements, and to identify problems.

(i) Device history record (DHR) means a compilation of records containing the production history of a finished device.

(j) Device master record (DMR) means a compilation of records containing the procedures and specifications for a finished device.

(k) Establish means define, document (in writing or electronically), and implement.

(l) Finished device means any device or accessory to any device that is suitable for use or capable of functioning, whether or not it is packaged, labeled, or sterilized.

(m) Lot or batch means one or more components or finished devices that consist of a single type, model, class, size, composition, or software version that are manufactured under essentially the same conditions and that are intended to have uniform characteristics and quality within specified limits.

(n) Management with executive responsibility means those senior employees of a manufacturer who have the authority to establish or make changes to the manufacturer's quality policy and quality system.

(o) Manufacturer means any person who designs, manufactures, fabricates, assembles, or processes a finished device. Manufacturer includes but is not limited to those who perform the functions of contract sterilization, installation, relabeling, remanufacturing, repacking, or specification development, and initial distributors of foreign entities performing these functions.

(p) Manufacturing material means any material or substance used in or used to facilitate the manufacturing process, a concomitant constituent, or a byproduct constituent produced during the manufacturing process, which is present in or on the finished device as a residue or impurity not by design or intent of the manufacturer.

(q) Nonconformity means the nonfulfillment of a specified requirement.

(r) Product means components, manufacturing materials, in-process devices, finished devices, and returned devices.

(s) Quality means the totality of features and characteristics that bear on the ability of a device to satisfy fitness-for-use, including safety and performance.

(t) Quality audit means a systematic, independent examination of a manufacturer's quality system that is performed at defined intervals and at sufficient frequency to determine whether both quality system activities and the results of such activities comply with quality system procedures, that these procedures are implemented effectively, and that these procedures are suitable to achieve quality system objectives.

(u) Quality policy means the overall intentions and direction of an organization with respect to quality, as established by management with executive responsibility.

(v) Quality system means the organizational structure, responsibilities, procedures, processes, and resources for implementing quality management.

(w) Remanufacturer means any person who processes, conditions, renovates, repackages, restores, or does any other act to a finished device that significantly changes the finished device's performance or safety specifications, or intended use.

(x) Rework means action taken on a nonconforming product so that it will fulfill the specified DMR requirements before it is released for distribution.

(y) Specification means any requirement with which a product, process, service, or other activity must conform.

(z) Validation means confirmation by examination and provision of objective evidence that the particular requirements for a specific intended use can be consistently fulfilled.

   (1) Process validation means establishing by objective evidence that a process consistently produces a result or product meeting its predetermined specifications.

   (2) Design validation means establishing by objective evidence that device specifications conform with user needs and intended use(s).

(aa) Verification means confirmation by examination and provision of objective evidence that specified requirements have been fulfilled.

## SECTION 820.5 QUALITY SYSTEM

Each manufacturer shall establish and maintain a quality system that is appropriate for the specific medical device(s) designed or manufactured, and that meets the requirements of this part.

## SUBPART B — QUALITY SYSTEM REQUIREMENTS

## SECTION 820.20 MANAGEMENT RESPONSIBILITY

(a) Quality policy. Management with executive responsibility shall establish its policy and objectives for, and commitment to, quality. Management with executive responsibility shall ensure that the quality policy is understood, implemented, and maintained at all levels of the organization.

(b) Organization. Each manufacturer shall establish and maintain an adequate organizational structure to ensure that devices are designed and produced in accordance with the requirements of this part.

   (1) Responsibility and authority. Each manufacturer shall establish the appropriate responsibility, authority, and interrelation of all personnel who manage, perform, and assess work affecting quality, and provide the independence and authority necessary to perform these tasks.

   (2) Resources. Each manufacturer shall provide adequate resources, including the assignment of trained personnel, for management, performance of work, and assessment activities, including internal quality audits, to meet the requirements of this part.

   (3) Management representative. Management with executive responsibility shall appoint, and document such appointment of, a member of management who, irrespective of other responsibilities, shall have established authority over and responsibility for:

      (i) Ensuring that quality system requirements are effectively established and effectively maintained in accordance with this part; and

      (ii) Reporting on the performance of the quality system to management with executive responsibility for review.

(c) Management review. Management with executive responsibility shall review the suitability and effectiveness of the quality system at defined intervals and with sufficient frequency according to established procedures to ensure that the quality system satisfies the requirements of this part and the manufacturer's established quality policy and objectives. The dates and results of quality system reviews shall be documented.

(d) Quality planning. Each manufacturer shall establish a quality plan which defines the quality practices, resources, and activities relevant to devices that are designed and manufactured. The manufacturer shall establish how the requirements for quality will be met.

(e) Quality system procedures. Each manufacturer shall establish quality system procedures and instructions. An outline of the structure of the documentation used in the quality system shall be established where appropriate.

## Section 820.22 Quality Audit

Each manufacturer shall establish procedures for quality audits and conduct such audits to assure that the quality system is in compliance with the established quality system requirements and to determine the effectiveness of the quality system. Quality audits shall be conducted by individuals who do not have direct responsibility for the matters being audited. Corrective action(s), including a reaudit of deficient matters, shall be taken when necessary. A report of the results of each quality audit, and reaudit(s) where taken, shall be made and such reports shall be reviewed by management having responsibility for the matters audited. The dates and results of quality audits and reaudits shall be documented.

## Section 820.25 Personnel

(a) General. Each manufacturer shall have sufficient personnel with the necessary education, background, training, and experience to assure that all activities required by this part are correctly performed.

(b) Training. Each manufacturer shall establish procedures for identifying training needs and ensure that all personnel are trained to adequately perform their assigned responsibilities. Training shall be documented.
   (1) As part of their training, personnel shall be made aware of device defects which may occur from the improper performance of their specific jobs.
   (2) Personnel who perform verification and validation activities shall be made aware of defects and errors that may be encountered as part of their job functions.

## SUBPART C — DESIGN CONTROLS

## Section 820.30 Design Controls

(a) General
   (1) Each manufacturer of any class III or class II device, and the class I devices listed in paragraph (a)(2) of this section, shall establish and maintain procedures to control the design of the device in order to ensure that specified design requirements are met.
   (2) The following class I devices are subject to design controls:
      (i) Devices automated with computer software; and

(ii) The devices listed in the following chart.

| Section | Device |
|---|---|
| 868.6810 | Catheter, Tracheobronchial Suction. |
| 878.4460 | Glove, Surgeon's. |
| 880.676 | Restraint, Protective. |
| 892.5650 | System, Applicator, Radionuclide Manual. |
| 892.5740 | Source, Radionuclide Teletherapy. |

(b) Design and development planning. Each manufacturer shall establish and maintain plans that describe or reference the design and development activities and define responsibility for implementation. The plans shall identify and describe the interfaces with different groups or activities that provide, or result in, input to the design and development process. The plans shall be reviewed, updated, and approved as design and development evolves.

(c) Design input. Each manufacturer shall establish and maintain procedures to ensure that the design requirements relating to a device are appropriate and address the intended use of the device, including the needs of the user and patient. The procedures shall include a mechanism for addressing incomplete, ambiguous, or conflicting requirements. The design input requirements shall be documented and shall be reviewed and approved by a designated individual(s). The approval, including the date and signature of the individual(s) approving the requirements, shall be documented.

(d) Design output. Each manufacturer shall establish and maintain procedures for defining and documenting design output in terms that allow an adequate evaluation of conformance to design input requirements. Design output procedures shall contain or make reference to acceptance criteria and shall ensure that those design outputs that are essential for the proper functioning of the device are identified. Design output shall be documented, reviewed, and approved before release. The approval, including the date and signature of the individual(s) approving the output, shall be documented.

(e) Design review. Each manufacturer shall establish and maintain procedures to ensure that formal documented reviews of the design results are planned and conducted at appropriate stages of the device's design development. The procedures shall ensure that participants at each design review include representatives of all functions concerned with the design stage being reviewed and an individual(s) who does not have direct responsibility for the design stage being reviewed, as well as any specialists needed. The results of a design review, including identification of the design, the date, and the individual(s) performing the review, shall be documented in the design history file (the DHF).

(f) Design verification. Each manufacturer shall establish and maintain procedures for verifying the device design. Design verification shall confirm that the design output meets the design-input requirements. The results of the design verification, including identification of the design, method(s), the date, and the individual(s) performing the verification, shall be documented in the DHF.

(g) Design validation. Each manufacturer shall establish and maintain procedures for validating the device design. Design validation shall be performed under defined operating conditions on initial production units, lots, or batches, or their equivalents. Design validation shall ensure that devices conform to defined user needs and intended uses and shall include testing of production units under actual or simulated use conditions. Design validation shall include software validation and risk analysis, where appropriate. The results of the design validation, including identification of the design, method(s), the date, and the individual(s) performing the validation, shall be documented in the DHF.

(h) Design transfer. Each manufacturer shall establish and maintain procedures to ensure that the device design is correctly translated into production specifications.

(i) Design changes. Each manufacturer shall establish and maintain procedures for the identification, documentation, validation or, where appropriate verification, review, and approval of design changes before their implementation.

(j) Design history file. Each manufacturer shall establish and maintain a DHF for each type of device. The DHF shall contain or reference the records necessary to demonstrate that the design was developed in accordance with the approved design plan and the requirements of this part.

## SUBPART D — DOCUMENT CONTROLS

### SECTION 820.40 DOCUMENT CONTROLS

Each manufacturer shall establish and maintain procedures to control all documents that are required by this part. The procedures shall provide for the following:

(a) Document approval and distribution. Each manufacturer shall designate an individual(s) to review for adequacy and approve prior to issuance all documents established to meet the requirements of this part. The approval, including the date and signature of the individual(s) approving the document, shall be documented. Documents established to meet the requirements of this part shall be available at all locations for which they are designated, used, or otherwise necessary, and all obsolete documents shall be promptly removed from all points of use or otherwise prevented from unintended use.

(b) Document changes. Changes to documents shall be reviewed and approved by an individual(s) in the same function or organization that performed the original review and approval, unless specifically designated otherwise. Approved changes shall be communicated to the appropriate personnel in a timely manner. Each manufacturer shall maintain records of changes to documents. Change records shall include a description of the change, identification of the affected documents, the signature of the approving individual(s), the approval date, and when the change becomes effective.

## SUBPART E — PURCHASING CONTROLS

### SECTION 820.50 PURCHASING CONTROLS

Each manufacturer shall establish and maintain procedures to ensure that all purchased or otherwise received product and services conform to specified requirements.

(a) Evaluation of suppliers, contractors, and consultants. Each manufacturer shall establish and maintain the requirements, including quality requirements, that must be met by suppliers, contractors, and consultants. Each manufacturer shall:

(1) Evaluate and select potential suppliers, contractors, and consultants on the basis of their ability to meet specified requirements, including quality requirements. The evaluation shall be documented.

(2) Define the type and extent of control to be exercised over the product, services, suppliers, contractors, and consultants, based on the evaluation results.

(3) Establish and maintain records of acceptable suppliers, contractors, and consultants.

(b) Purchasing data. Each manufacturer shall establish and maintain data that clearly describe or reference the specified requirements, including quality requirements, for purchased or otherwise received product and services. Purchasing documents shall include, where possible, an agreement that the suppliers, contractors, and consultants agree to notify the manufacturer of changes in the product or service so that manufacturers may determine whether the changes may affect the quality of a finished device. Purchasing data shall be approved in accordance with §820.40.

## SUBPART F — IDENTIFICATION AND TRACEABILITY

### Section 820.60 Identification

Each manufacturer shall establish and maintain procedures for identifying product during all stages of receipt, production, distribution, and installation to prevent mixups.

### Section 820.65 Traceability

Each manufacturer of a device that is intended for surgical implant into the body or to support or sustain life and whose failure to perform when properly used in accordance with instructions for use provided in the labeling can be reasonably expected to result in a significant injury to the user shall establish and maintain procedures for identifying with a control number each unit, lot, or batch of finished devices and where appropriate components. The procedures shall facilitate corrective action. Such identification shall be documented in the DHR.

## SUBPART G — PRODUCTION AND PROCESS CONTROLS

### Section 820.70 Production and Process Controls

(a) General. Each manufacturer shall develop, conduct, control, and monitor production processes to ensure that a device conforms to its specifications. Where deviations from device specifications could occur as a result of the manufacturing process, the manufacturer shall establish and maintain process control procedures that describe any process controls necessary to ensure conformance to specifications. Where process controls are needed they shall include:

(1) Documented instructions, standard operating procedures (SOP's), and methods that define and control the manner of production;

(2) Monitoring and control of process parameters and component and device characteristics during production;

(3) Compliance with specified reference standards or codes;

(4) The approval of processes and process equipment; and

(5) Criteria for workmanship which shall be expressed in documented standards or by means of identified and approved representative samples.

(b) Production and process changes. Each manufacturer shall establish and maintain procedures for changes to a specification, method, process, or procedure. Such changes shall be verified or, where appropriate, validated according to §820.75, before implementation and these activities shall be documented. Changes shall be approved in accordance with §820.40.

(c) Environmental control. Where environmental conditions could reasonably be expected to have an adverse effect on product quality, the manufacturer shall establish and maintain procedures to adequately control these environmental conditions. Environmental control system(s) shall be periodically inspected to verify that the system, including necessary equipment, is adequate and functioning properly. These activities shall be documented and reviewed.

(d) Personnel. Each manufacturer shall establish and maintain requirements for the health, cleanliness, personal practices, and clothing of personnel if contact between such personnel and product or environment could reasonably be expected to have an adverse effect on product quality. The manufacturer shall ensure that maintenance and other personnel who are required to work temporarily under special environmental conditions are appropriately trained or supervised by a trained individual.

(e) Contamination control. Each manufacturer shall establish and maintain procedures to prevent contamination of equipment or product by substances that could reasonably be expected to have an adverse effect on product quality.

(f) Buildings. Buildings shall be of suitable design and contain sufficient space to perform necessary operations, prevent mixups, and assure orderly handling.

(g) Equipment. Each manufacturer shall ensure that all equipment used in the manufacturing process meets specified requirements and is appropriately designed, constructed, placed, and installed to facilitate maintenance, adjustment, cleaning, and use.

(1) Maintenance schedule. Each manufacturer shall establish and maintain schedules for the adjustment, cleaning, and other maintenance of equipment to ensure that manufacturing specifications are met. Maintenance activities, including the date and individual(s) performing the maintenance activities, shall be documented.

(2) Inspection. Each manufacturer shall conduct periodic inspections in accordance with established procedures to ensure adherence to applicable equipment maintenance schedules. The inspections, including the date and individual(s) conducting the inspections, shall be documented.

(3) Adjustment. Each manufacturer shall ensure that any inherent limitations or allowable tolerances are visibly posted on or near equipment requiring periodic adjustments or are readily available to personnel performing these adjustments.

(h) Manufacturing material. Where a manufacturing material could reasonably be expected to have an adverse effect on product quality, the manufacturer shall establish and maintain procedures for the use and removal of such manufacturing material to ensure that it is removed or limited to an amount that does not adversely affect the device's quality. The removal or reduction of such manufacturing material shall be documented.

(i) Automated processes. When computers or automated data processing systems are used as part of production or the quality system, the manufacturer shall validate computer software for its intended use according to an established protocol. All software changes shall be validated before approval and issuance. These validation activities and results shall be documented.

### Section 820.72 Inspection, Measuring, and Test Equipment

(a) Control of inspection, measuring, and test equipment. Each manufacturer shall ensure that all inspection, measuring, and test equipment, including mechanical, automated, or electronic inspection and test equipment, is suitable for its intended purposes and is capable of producing valid results. Each manufacturer shall establish and maintain procedures to ensure that equipment is routinely calibrated, inspected, checked, and maintained. The procedures shall include provisions for handling, preservation, and storage of equipment, so that its accuracy and fitness for use are maintained. These activities shall be documented.

(b) Calibration. Calibration procedures shall include specific directions and limits for accuracy and precision. When accuracy and precision limits are not met, there shall be provisions for remedial action to re-establish the limits and to evaluate whether there was any adverse effect on the device's quality. These activities shall be documented.

   (1) Calibration standards. Calibration standards used for inspection, measuring, and test equipment shall be traceable to national or international standards. If national or international standards are not practical or available, the manufacturer shall use an independent reproducible standard. If no applicable standard exists, the manufacturer shall establish and maintain an in-house standard.

   (2) Calibration records. The equipment identification, calibration dates, the individual performing each calibration, and the next calibration date shall be documented. These records shall be displayed on or near each piece of equipment or shall be readily available to the personnel using such equipment and to the individuals responsible for calibrating the equipment.

### Section 820.75 Process Validation

(a) Where the results of a process cannot be fully verified by subsequent inspection and test, the process shall be validated with a high degree of assurance and approved according to established procedures. The validation activities and results, including the date and signature of the individual(s) approving the validation and where appropriate the major equipment validated, shall be documented.

(b) Each manufacturer shall establish and maintain procedures for monitoring and control of process parameters for validated processes to ensure that the specified requirements continue to be met.

   (1) Each manufacturer shall ensure that validated processes are performed by qualified individual(s).

   (2) For validated processes, the monitoring and control methods and data, the date performed, and, where appropriate, the individual(s) performing the process or the major equipment used shall be documented.

(c) When changes or process deviations occur, the manufacturer shall review and evaluate the process and perform revalidation where appropriate. These activities shall be documented.

## SUBPART H — ACCEPTANCE ACTIVITIES

### Section 820.86 Acceptance Status

Each manufacturer shall identify by suitable means the acceptance status of product, to indicate the conformance or nonconformance of product with acceptance criteria. The identification of acceptance status shall be maintained throughout manufacturing, packaging, labeling, installation, and servicing of the product to ensure that only product which has passed the required acceptance activities is distributed, used, or installed.

## SUBPART I — NONCONFORMING PRODUCT

### SECTION 820.90 NONCONFORMING PRODUCT

(a) Control of nonconforming product. Each manufacturer shall establish and maintain procedures to control product that does not conform to specified requirements. The procedures shall address the identification, documentation, evaluation, segregation, and disposition of nonconforming product. The evaluation of nonconformance shall include a determination of the need for an investigation and notification of the persons or organizations responsible for the nonconformance. The evaluation and any investigation shall be documented.

(b) Nonconformity review and disposition.

(1) Each manufacturer shall establish and maintain procedures that define the responsibility for review and the authority for the disposition of nonconforming product. The procedures shall set forth the review and disposition process. Disposition of nonconforming product shall be documented. Documentation shall include the justification for use of nonconforming product and the signature of the individual(s) authorizing the use.

(2) Each manufacturer shall establish and maintain procedures for rework, to include retesting and re-evaluation of the nonconforming product after rework, to ensure that the product meets its current approved specifications. Rework and re-evaluation activities, including a determination of any adverse effect from the rework upon the product, shall be documented in the DHR.

## SUBPART J — CORRECTIVE AND PREVENTIVE ACTION

### SECTION 820.100 CORRECTIVE AND PREVENTIVE ACTION

(a) Each manufacturer shall establish and maintain procedures for implementing corrective and preventive action. The procedures shall include requirements for:

(1) Analyzing processes, work operations, concessions, quality audit reports, quality records, service records, complaints, returned product, and other sources of quality data to identify existing and potential causes of nonconforming product, or other quality problems. Appropriate statistical methodology shall be employed where necessary to detect recurring quality problems;

(2) Investigating the cause of nonconformities relating to product, processes, and the quality system;

(3) Identifying the action(s) needed to correct and prevent recurrence of nonconforming product and other quality problems;

(4) Verifying or validating the corrective and preventive action to ensure that such action is effective and does not adversely affect the finished device;

(5) Implementing and recording changes in methods and procedures needed to correct and prevent identified quality problems;

(6) Ensuring that information related to quality problems or nonconforming product is disseminated to those directly responsible for assuring the quality of such product or the prevention of such problems; and

(7) Submitting relevant information on identified quality problems, as well as corrective and preventive actions, for management review.

(b) All activities required under this section, and their results, shall be documented.

## SUBPART K — LABELING AND PACKAGING CONTROL

### SECTION 820.120 DEVICE LABELING

Each manufacturer shall establish and maintain procedures to control labeling activities.

(a) Label integrity. Labels shall be printed and applied so as to remain legible and affixed during the customary conditions of processing, storage, handling, distribution, and where appropriate, use.

(b) Labeling inspection. Labeling shall not be released for storage or use until a designated individual(s) has examined the labeling for accuracy including, where applicable, the correct expiration date, control number, storage instructions, handling instructions, and any additional processing instructions. The release, including the date and signature of the individual(s) performing the examination, shall be documented in the DHR.

(c) Labeling storage. Each manufacturer shall store labeling in a manner that provides proper identification and is designed to prevent mixups.

(d) Labeling operations. Each manufacturer shall control labeling and packaging operations to prevent labeling mixups. The label and labeling used for each production unit, lot, or batch shall be documented in the DHR.

(e) Control number. Where a control number is required by §820.65, that control number shall be on or shall accompany the device through distribution.

### SECTION 820.130 DEVICE PACKAGING

Each manufacturer shall ensure that device packaging and shipping containers are designed and constructed to protect the device from alteration or damage during the customary conditions of processing, storage, handling, and distribution.

### SECTION 820.140 HANDLING

Each manufacturer shall establish and maintain procedures to ensure that mixups, damage, deterioration, contamination, or other adverse effects to product do not occur during handling.

## SUBPART L — HANDLING, STORAGE, DISTRIBUTION, AND INSTALLATION

### SECTION 820.150 STORAGE

(a) Each manufacturer shall establish and maintain procedures for the control of storage areas and stock rooms for product to prevent mixups, damage, deterioration, contamination, or other adverse effects pending use or distribution and to ensure that no obsolete, rejected, or deteriorated product is used or distributed. When the quality of product deteriorates over time, it shall be stored in a manner to facilitate proper stock rotation, and its condition shall be assessed as appropriate.

(b) Each manufacturer shall establish and maintain procedures that describe the methods for authorizing receipt from and dispatch to storage areas and stock rooms.

## SECTION 820.160 DISTRIBUTION

(a) Each manufacturer shall establish and maintain procedures for control and distribution of finished devices to ensure that only those devices approved for release are distributed and that purchase orders are reviewed to ensure that ambiguities and errors are resolved before devices are released for distribution. Where a device's fitness for use or quality deteriorates over time, the procedures shall ensure that expired devices or devices deteriorated beyond acceptable fitness for use are not distributed.

(b) Each manufacturer shall maintain distribution records which include or refer to the location of:
    (1) The name and address of the initial consignee;
    (2) The identification and quantity of devices shipped;
    (3) The date shipped; and
    (4) Any control number(s) used.

## SECTION 820.170 INSTALLATION

(a) Each manufacturer of a device requiring installation shall establish and maintain adequate installation and inspection instructions, and, where appropriate, test procedures. Instructions and procedures shall include directions for ensuring proper installation so that the device will perform as intended after installation. The manufacturer shall distribute the instructions and procedures with the device or otherwise make them available to the person(s) installing the device.

(b) The person installing the device shall ensure that the installation, inspection, and any required testing are performed in accordance with the manufacturer's instructions and procedures and shall document the inspection and any test results to demonstrate proper installation.

# SUBPART M — RECORDS

## SECTION 820.180 GENERAL REQUIREMENTS

All records required by this part shall be maintained at the manufacturing establishment or other location that is reasonably accessible to responsible officials of the manufacturer and to employees of FDA designated to perform inspections. Such records, including those not stored at the inspected establishment, shall be made readily available for review and copying by FDA employee(s). Such records shall be legible and shall be stored to minimize deterioration and to prevent loss. Those records stored in automated data processing systems shall be backed up.

(a) Confidentiality. Records deemed confidential by the manufacturer may be marked to aid FDA in determining whether information may be disclosed under the public information regulation in part 20 of this chapter.

(b) Record retention period. All records required by this part shall be retained for a period of time equivalent to the design and expected life of the device, but in no case less than 2 years from the date of release for commercial distribution by the manufacturer.

(c) Exceptions. This section does not apply to the reports required by §820.20(c) Management review, §820.22 Quality audits, and supplier audit reports used to meet the requirements of §820.50(a) Evaluation of suppliers, contractors, and consultants, but does apply to procedures established under these provisions. Upon request of a designated employee of FDA, an employee in management with executive responsibility shall certify in writing that the management reviews and quality audits required under this part, and supplier audits where applicable, have been performed and documented, the dates on which they were performed, and that any required corrective action has been undertaken.

## Section 820.181 Device Master Record

Each manufacturer shall maintain device master records (DMRs). Each manufacturer shall ensure that each DMR is prepared and approved in accordance with §820.40. The DMR for each type of device shall include, or refer to the location of, the following information:

(a) Device specifications including appropriate drawings, composition, formulation, component specifications, and software specifications;

(b) Production process specifications including the appropriate equipment specifications, production methods, production procedures, and production environment specifications;

(c) Quality assurance procedures and specifications including acceptance criteria and the quality assurance equipment to be used;

(d) Packaging and labeling specifications, including methods and processes used; and

(e) Installation, maintenance, and servicing procedures and methods.

## Section 820.184 Device History Record

Each manufacturer shall maintain device history records (DHRs). Each manufacturer shall establish and maintain procedures to ensure that DHR's for each batch, lot, or unit are maintained to demonstrate that the device is manufactured in accordance with the DMR and the requirements of this part. The DHR shall include, or refer to the location of, the following information:

(a) The dates of manufacture;

(b) The quantity manufactured;

(c) The quantity released for distribution;

(d) The acceptance records which demonstrate the device is manufactured in accordance with the DMR;

(e) The primary identification label and labeling used for each production unit; and

(f) Any device identification(s) and control number(s) used.

## Section 820.186 Quality System Record

Each manufacturer shall maintain a quality system record (QSR). The QSR shall include, or refer to the location of, procedures and the documentation of activities required by this part that are not specific to a particular type of device(s), including, but not limited to, the records required by §820.20. Each manufacturer shall ensure that the QSR is prepared and approved in accordance with §820.40.

## Section 820.198 Complaint Files

(a) Each manufacturer shall maintain complaint files. Each manufacturer shall establish and maintain procedures for receiving, reviewing, and evaluating complaints by a formally designated unit. Such procedures shall ensure that:

   (1) All complaints are processed in a uniform and timely manner;

   (2) Oral complaints are documented upon receipt; and

   (3) Complaints are evaluated to determine whether the complaint represents an event which is required to be reported to FDA under part 803 or 804 of this chapter, Medical Device Reporting.

(b) Each manufacturer shall review and evaluate all complaints to determine whether an investigation is necessary. When no investigation is made, the manufacturer shall maintain a record that includes the reason no investigation was made and the name of the individual responsible for the decision not to investigate.

(c) Any complaint involving the possible failure of a device, labeling, or packaging to meet any of its specifications shall be reviewed, evaluated, and investigated, unless such investigation has already been performed for a similar complaint and another investigation is not necessary.

(d) Any complaint that represents an event which must be reported to FDA under part 803 or 804 of this chapter shall be promptly reviewed, evaluated, and investigated by a designated individual(s) and shall be maintained in a separate portion of the complaint files or otherwise clearly identified. In addition to the information required by §820.198(e), records of investigation under this paragraph shall include a determination of:

   (1) Whether the device failed to meet specifications;

   (2) Whether the device was being used for treatment or diagnosis; and

   (3) The relationship, if any, of the device to the reported incident or adverse event.

(e) When an investigation is made under this section, a record of the investigation shall be maintained by the formally designated unit identified in paragraph (a) of this section. The record of investigation shall include:

   (1) The name of the device;

   (2) The date the complaint was received;

   (3) Any device identification(s) and control number(s) used;

   (4) The name, address, and phone number of the complainant;

   (5) The nature and details of the complaint;

   (6) The dates and results of the investigation;

   (7) Any corrective action taken; and

   (8) Any reply to the complainant.

(f) When the manufacturer's formally designated complaint unit is located at a site separate from the manufacturing establishment, the investigated complaint(s) and the record(s) of investigation shall be reasonably accessible to the manufacturing establishment.

(g) If a manufacturer's formally designated complaint unit is located outside of the United States, records required by this section shall be reasonably accessible in the United States at either:

   (1) A location in the United States where the manufacturer's records are regularly kept; or

   (2) The location of the initial distributor.

# SUBPART N — SERVICING

## SECTION 820.200 SERVICING

(a) Where servicing is a specified requirement, each manufacturer shall establish and maintain instructions and procedures for performing and verifying that the servicing meets the specified requirements.

(b) Each manufacturer shall analyze service reports with appropriate statistical methodology in accordance with §820.100.

(c) Each manufacturer who receives a service report that represents an event which must be reported to FDA under part 803 or 804 of this chapter shall automatically consider the report a complaint and shall process it in accordance with the requirements of §820.198.

(d) Service reports shall be documented and shall include:

    (1) The name of the device serviced;

    (2) Any device identification(s) and control number(s) used;

    (3) The date of service;

    (4) The individual(s) servicing the device;

    (5) The service performed; and

    (6) The test and inspection data.

# SUBPART O — STATISTICAL TECHNIQUES

## SECTION 820.250 STATISTICAL TECHNIQUES

(a) Where appropriate, each manufacturer shall establish and maintain procedures for identifying valid statistical techniques required for establishing, controlling, and verifying the acceptability of process capability and product characteristics.

(b) Sampling plans, when used, shall be written and based on a valid statistical rationale. Each manufacturer shall establish and maintain procedures to ensure that sampling methods are adequate for their intended use and to ensure that when changes occur the sampling plans are reviewed. These activities shall be documented.

# Part 211
## Current Good Manufacturing Practice for Finished Pharmaceuticals

**CONTENTS**

## SUBPART A — GENERAL PROVISIONS

### Section 211.1 Scope

(a) The regulations in this part contain the minimum current good manufacturing practice for preparation of drug products for administration to humans or animals.

(b) The current good manufacturing practice regulations in this chapter, as they pertain to drug products, and in parts 600 through 680 of this chapter, as they pertain to biological products for human use, shall be considered to supplement, not supersede, the regulations in this part unless the regulations explicitly provide otherwise. In the event it is impossible to comply with applicable regulations both in this part and in other parts of this chapter or in parts 600 through 680 of this chapter, the regulation specifically applicable to the drug product in question shall supersede the regulation in this part.

(c) Pending consideration of a proposed exemption, published in the Federal Register of September 29, 1978, the requirements in this part shall not be enforced for OTC drug products if the products and all their ingredients are ordinarily marketed and consumed as human foods, and which products may also fall within the legal definition of drugs by virtue of their intended use. Therefore, until further notice, regulations under part 110 of this chapter, and where applicable, parts 113 to 129 of this chapter, shall be applied in determining whether these OTC drug products that are also foods are manufactured, processed, packed, or held under current good manufacturing practice.

## Section 211.3 Definitions

The definitions set forth in §210.3 of this chapter apply in this part.

# SUBPART B — ORGANIZATION AND PERSONNEL

## Section 211.22 Responsibilities of Quality Control Unit

(a) There shall be a quality control unit that shall have the responsibility and authority to approve or reject all components, drug product containers, closures, in-process materials, packaging material, labeling, and drug products, and the authority to review production records to assure that no errors have occurred or, if errors have occurred, that they have been fully investigated. The quality control unit shall be responsible for approving or rejecting drug products manufactured, processed, packed, or held under contract by another company.

(b) Adequate laboratory facilities for the testing and approval (or rejection) of components, drug product containers, closures, packaging materials, in-process materials, and drug products shall be available to the quality control unit.

(c) The quality control unit shall have the responsibility for approving or rejecting all procedures or specifications impacting on the identity, strength, quality, and purity of the drug product.

(d) The responsibilities and procedures applicable to the quality control unit shall be in writing; such written procedures shall be followed.

## Section 211.25 Personnel Qualifications

(a) Each person engaged in the manufacture, processing, packing, or holding of a drug product shall have education, training, and experience, or any combination thereof, to enable that person to perform the assigned functions. Training shall be in the particular operations that the employee performs and in current good manufacturing practice (including the current good manufacturing practice regulations in this chapter and written procedures required by these regulations) as they relate to the employee's functions. Training in current good manufacturing practice shall be conducted by qualified individuals on a continuing basis and with sufficient frequency to assure that employees remain familiar with CGMP requirements applicable to them.

(b) Each person responsible for supervising the manufacture, processing, packing, or holding of a drug product shall have the education, training, and experience, or any combination thereof, to perform assigned functions in such a manner as to provide assurance that the drug product has the safety, identity, strength, quality, and purity that it purports or is represented to possess.

(c) There shall be an adequate number of qualified personnel to perform and supervise the manufacture, processing, packing, or holding of each drug product.

### SECTION 211.28 PERSONNEL RESPONSIBILITIES

(a) Personnel engaged in the manufacture, processing, packing, or holding of a drug product shall wear clean clothing appropriate for the duties they perform. Protective apparel, such as head, face, hand, and arm coverings, shall be worn as necessary to protect drug products from contamination.

(b) Personnel shall practice good sanitation and health habits.

(c) Only personnel authorized by supervisory personnel shall enter those areas of the buildings and facilities designated as limited-access areas.

(d) Any person shown at any time (either by medical examination or supervisory observation) to have an apparent illness or open lesions that may adversely affect the safety or quality of drug products shall be excluded from direct contact with components, drug product containers, closures, in-process materials, and drug products until the condition is corrected or determined by competent medical personnel not to jeopardize the safety or quality of drug products. All personnel shall be instructed to report to supervisory personnel any health conditions that may have an adverse effect on drug products.

### SECTION 211.34 CONSULTANTS

Consultants advising on the manufacture, processing, packing, or holding of drug products shall have sufficient education, training, and experience, or any combination thereof, to advise on the subject for which they are retained. Records shall be maintained stating the name, address, and qualifications of any consultants and the type of service they provide.

## SUBPART C — BUILDINGS AND FACILITIES

### SECTION 211.42 DESIGN AND CONSTRUCTION FEATURES

(a) Any building or buildings used in the manufacture, processing, packing, or holding of a drug product shall be of suitable size, construction and location to facilitate cleaning, maintenance, and proper operations.

(b) Any such building shall have adequate space for the orderly placement of equipment and materials to prevent mixups between different components, drug product containers, closures, labeling, in-process materials, or drug products, and to prevent contamination. The flow of components, drug product containers, closures, labeling, in-process materials, and drug products through the building or buildings shall be designed to prevent contamination.

(c) Operations shall be performed within specifically defined areas of adequate size. There shall be separate or defined areas or such other control systems for the firm's operations as are necessary to prevent contamination or mixups during the course of the following procedures:

(1) Receipt, identification, storage, and withholding from use of components, drug product containers, closures, and labeling, pending the appropriate sampling, testing, or examination by the quality control unit before release for manufacturing or packaging;

(2) Holding rejected components, drug product containers, closures, and labeling before disposition;

(3) Storage of released components, drug product containers, closures, and labeling;

(4) Storage of in-process materials;

(5) Manufacturing and processing operations;

(6) Packaging and labeling operations;

(7) Quarantine storage before release of drug products;

(8) Storage of drug products after release;

(9) Control and laboratory operations;

(10) Aseptic processing, which includes as appropriate:

(i) Floors, walls, and ceilings of smooth, hard surfaces that are easily cleanable;

(ii) Temperature and humidity controls;

(iii) An air supply filtered through high-efficiency particulate air filters under positive pressure, regardless of whether flow is laminar or nonlaminar;

(iv) A system for monitoring environmental conditions;

(v) A system for cleaning and disinfecting the room and equipment to produce aseptic conditions;

(vi) A system for maintaining any equipment used to control the aseptic conditions.

(d) Operations relating to the manufacture, processing, and packing of penicillin shall be performed in facilities separate from those used for other drug products for human use.

## SECTION 211.44 LIGHTING

Adequate lighting shall be provided in all areas.

## SECTION 211.46 VENTILATION, AIR FILTRATION, AIR HEATING AND COOLING

(a) Adequate ventilation shall be provided.

(b) Equipment for adequate control over air pressure, micro-organisms, dust, humidity, and temperature shall be provided when appropriate for the manufacture, processing, packing, or holding of a drug product.

(c) Air filtration systems, including prefilters and particulate matter air filters, shall be used when appropriate on air supplies to production areas. If air is recirculated to production areas, measures shall be taken to control recirculation of dust from production. In areas where air contamination occurs during production, there shall be adequate exhaust systems or other systems adequate to control contaminants.

(d) Air-handling systems for the manufacture, processing, and packing of penicillin shall be completely separate from those for other drug products for human use.

## SECTION 211.48 PLUMBING

(a) Potable water shall be supplied under continuous positive pressure in a plumbing system free of defects that could contribute contamination to any drug product. Potable water shall meet the standards prescribed in the Environmental Protection Agency's Primary Drinking Water Regulations set forth in 40 CFR part 141. Water not meeting such standards shall not be permitted in the potable water system.

(b) Drains shall be of adequate size and, where connected directly to a sewer, shall be provided with an air break or other mechanical device to prevent back-siphonage.

## SECTION 211.50 SEWAGE AND REFUSE

Sewage, trash, and other refuse in and from the building and immediate premises shall be disposed of in a safe and sanitary manner.

## SECTION 211.52 WASHING AND TOILET FACILITIES

Adequate washing facilities shall be provided, including hot and cold water, soap or detergent, air driers or single-service towels, and clean toilet facilities easily accesible to working areas.

## SECTION 211.56 SANITATION

(a) Any building used in the manufacture, processing, packing, or holding of a drug product shall be maintained in a clean and sanitary condition. Any such building shall be free of infestation by rodents, birds, insects, and other vermin (other than laboratory animals). Trash and organic waste matter shall be held and disposed of in a timely and sanitary manner.

(b) There shall be written procedures assigning responsibility for sanitation and describing in sufficient detail the cleaning schedules, methods, equipment, and materials to be used in cleaning the buildings and facilities; such written procedures shall be followed.

(c) There shall be written procedures for use of suitable rodenticides, insecticides, fungicides, fumigating agents, and cleaning and sanitizing agents. Such written procedures shall be designed to prevent the contamination of equipment, components, drug product containers, closures, packaging, labeling materials, or drug products and shall be followed. Rodenticides, insecticides, and fungicides shall not be used unless registered and used in accordance with the Federal Insecticide, Fungicide, and Rodenticide Act (7 U.S.C. 135).

(d) Sanitation procedures shall apply to work performed by contractors or temporary employees as well as work performed by full-time employees during the ordinary course of operations.

## SECTION 211.58 MAINTENANCE

Any building used in the manufacture, processing, packing, or holding of a drug product shall be maintained in a good state of repair.

# SUBPART D — EQUIPMENT

## SECTION 211.63 EQUIPMENT DESIGN, SIZE, AND LOCATION

Equipment used in the manufacture, processing, packing, or holding of a drug product shall be of appropriate design, adequate size, and suitably located to facilitate operations for its intended use and for its cleaning and maintenance.

## SECTION 211.65 EQUIPMENT CONSTRUCTION

(a) Equipment shall be constructed so that surfaces that contact components, in-process materials, or drug products shall not be reactive, additive, or absorptive so as to alter the safety, identity, strength, quality, or purity of the drug product beyond the official or other established requirements.

(b) Any substances required for operation, such as lubricants or coolants, shall not come into contact with components, drug product containers, closures, in-process materials, or drug products so as to alter the safety, identity, strength, quality, or purity of the drug product beyond the official or other established requirements.

## SECTION 211.67 EQUIPMENT CLEANING AND MAINTENANCE

(a) Equipment and utensils shall be cleaned, maintained, and sanitized at appropriate intervals to prevent malfunctions or contamination that would alter the safety, identity, strength, quality, or purity of the drug product beyond the official or other established requirements.

(b) Written procedures shall be established and followed for cleaning and maintenance of equipment, including utensils, used in the manufacture, processing, packing, or holding of a drug product. These procedures shall include, but are not necessarily limited to, the following:

(1) Assignment of responsibility for cleaning and maintaining equipment;

(2) Maintenance and cleaning schedules, including, where appropriate, sanitizing schedules;

(3) A description in sufficient detail of the methods, equipment, and materials used in cleaning and maintenance operations, and the methods of disassembling and reassembling equipment as necessary to assure proper cleaning and maintenance;

(4) Removal or obliteration of previous batch identification;

(5) Protection of clean equipment from contamination prior to use;

(6) Inspection of equipment for cleanliness immediately before use.

(c) Records shall be kept of maintenance, cleaning, sanitizing, and inspection as specified in §211.180 and 211.182.

## SECTION 211.68 AUTOMATIC, MECHANICAL, AND ELECTRONIC EQUIPMENT

(a) Automatic, mechanical, or electronic equipment or other types of equipment, including computers, or related systems that will perform a function satisfactorily, may be used in the manufacture, processing, packing, and holding of a drug product. If such equipment is so used, it shall be routinely calibrated, inspected, or checked according to a written program designed to assure proper performance. Written records of those calibration checks and inspections shall be maintained.

(b) Appropriate controls shall be exercised over computer or related systems to assure that changes in master production and control records or other records are instituted only by authorized personnel. Input to and output from the computer or related system of formulas or other records or data shall be checked for accuracy. The degree and frequency of input/output verification shall be based on the complexity and reliability of the computer or related system. A backup file of data entered into the computer or related system shall be maintained except where certain data, such as calculations performed in connection with laboratory analysis, are eliminated by computerization or other automated processes. In such instances a written record of the program shall be maintained along with appropriate validation data. Hard copy or alternative systems, such as duplicates, tapes, or microfilm, designed to assure that backup data are exact and complete and that it is secure from alteration, inadvertent erasures, or loss shall be maintained.

## SECTION 211.72 FILTERS

Filters for liquid filtration used in the manufacture, processing, or packing of injectable drug products intended for human use shall not release fibers into such products. Fiber-releasing filters may not be used in the manufacture, processing, or packing of these injectable drug products unless it is not possible to manufacture such drug products without the use of such filters. If use of a fiber-releasing filter is necessary, an additional non-fiber-releasing filter of 0.22 micron maximum mean porosity (0.45 micron if the manufacturing conditions so dictate) shall subsequently be used to reduce the content of particles in the injectable drug product. Use of an asbestos-containing filter, with or without subsequent use of a specific non-fiber-releasing filter, is permissible only upon submission of proof to the appropriate bureau of the Food and Drug Administration that use of a non-fiber-releasing filter will, or is likely to, compromise the safety or effectiveness of the injectable drug product.

## SUBPART E — CONTROL OF COMPONENTS AND DRUG PRODUCT CONTAINERS AND CLOSURES

### SECTION 211.80 GENERAL REQUIREMENTS

(a) There shall be written procedures describing in sufficient detail the receipt, identification, storage, handling, sampling, testing, and approval or rejection of components and drug product containers and closures; such written procedures shall be followed.
(b) Components and drug product containers and closures shall at all times be handled and stored in a manner to prevent contamination.
(c) Bagged or boxed components of drug product containers, or closures shall be stored off the floor and suitably spaced to permit cleaning and inspection.
(d) Each container or grouping of containers for components or drug product containers, or closures shall be identified with a distinctive code for each lot in each shipment received. This code shall be used in recording the disposition of each lot. Each lot shall be appropriately identified as to its status (i.e., quarantined, approved, or rejected).

### SECTION 211.82 RECEIPT AND STORAGE OF UNTESTED COMPONENTS, DRUG PRODUCT CONTAINERS, AND CLOSURES

(a) Upon receipt and before acceptance, each container or grouping of containers of components, drug product containers, and closures shall be examined visually for appropriate labeling as to contents, container damage or broken seals, and contamination.
(b) Components, drug product containers, and closures shall be stored under quarantine until they have been tested or examined, as appropriate, and released. Storage within the area shall conform to the requirements of §211.80.

## SECTION 211.84 TESTING AND APPROVAL OR REJECTION OF COMPONENTS, DRUG PRODUCT CONTAINERS, AND CLOSURES

(a) Each lot of components, drug product containers, and closures shall be withheld from use until the lot has been sampled, tested, or examined, as appropriate, and released for use by the quality control unit.

(b) Representative samples of each shipment of each lot shall be collected for testing or examination. The number of containers to be sampled, and the amount of material to be taken from each container, shall be based upon appropriate criteria such as statistical criteria for component variability, confidence levels, and degree of precision desired, the past quality history of the supplier, and the quantity needed for analysis and reserve where required by §211.170.

(c) Samples shall be collected in accordance with the following procedures:

   (1) The containers of components selected shall be cleaned where necessary, by appropriate means.

   (2) The containers shall be opened, sampled, and resealed in a manner designed to prevent contamination of their contents and contamination of other components, drug product containers, or closures.

   (3) Sterile equipment and aseptic sampling techniques shall be used when necessary.

   (4) If it is necessary to sample a component from the top, middle, and bottom of its container, such sample subdivisions shall not be composited for testing.

   (5) Sample containers shall be identified so that the following information can be determined: name of the material sampled, the lot number, the container from which the sample was taken, the date on which the sample was taken, and the name of the person who collected the sample.

   (6) Containers from which samples have been taken shall be marked to show that samples have been removed from them.

(d) Samples shall be examined and tested as follows:

   (1) At least one test shall be conducted to verify the identity of each component of a drug product. Specific identity tests, if they exist, shall be used.

   (2) Each component shall be tested for conformity with all appropriate written specifications for purity, strength, and quality. In lieu of such testing by the manufacturer, a report of analysis may be accepted from the supplier of a component, provided that at least one specific identity test is conducted on such component by the manufacturer, and provided that the manufacturer establishes the reliability of the supplier's analyses through appropriate validation of the supplier's test results at appropriate intervals.

   (3) Containers and closures shall be tested for conformance with all appropriate written procedures. In lieu of such testing by the manufacturer, a certificate of testing may be accepted from the supplier, provided that at least a visual identification is conducted on such containers/closures by the manufacturer and provided that the manufacturer establishes the reliability of the supplier's test results through appropriate validation of the supplier's test results at appropriate intervals.

   (4) When appropriate, components shall be microscopically examined.

   (5) Each lot of a component, drug product container, or closure that is liable to contamination with filth, insect infestation, or other extraneous adulterant shall be examined against established specifications for such contamination.

   (6) Each lot of a component, drug product container, or closure that is liable to microbiological contamination that is objectionable in view of its intended use shall be subjected to microbiological tests before use.

(e) Any lot of components, drug product containers, or closures that meets the appropriate written specifications of identity, strength, quality, and purity and related tests under paragraph (d) of this section may be approved and released for use. Any lot of such material that does not meet such specifications shall be rejected.

## Section 211.86 Use of Approved Components, Drug Product Containers, and Closures

Components, drug product containers, and closures approved for use shall be rotated so that the oldest approved stock is used first.

Deviation from this requirement is permitted if such deviation is temporary and appropriate.

## Section 211.87 Retesting of Approved Components, Drug Product Containers, and Closures

Components, drug product containers, and closures shall be retested or re-examined, as appropriate, for identity, strength, quality, and purity and approved or rejected by the quality control unit in accordance with §211.84 as necessary, e.g., after storage for long periods or after exposure to air, heat or other conditions that might adversely affect the component, drug product container, or closure.

## Section 211.89 Rejected Components, Drug Product Containers, and Closures

Rejected components, drug product containers, and closures shall be identified and controlled under a quarantine system designed to prevent their use in manufacturing or processing operations for which they are unsuitable.

## Section 211.94 Drug Product Containers and Closures

(a) Drug product containers and closures shall not be reactive, additive, or absorptive so as to alter the safety, identity, strength, quality, or purity of the drug beyond the official or established requirements.
(b) Container closure systems shall provide adequate protection against foreseeable external factors in storage and use that can cause deterioration or contamination of the drug product.
(c) Drug product containers and closures shall be clean and, where indicated by the nature of the drug, sterilized and processed to remove pyrogenic properties to assure that they are suitable for their intended use.
(d) Standards or specifications, methods of testing, and, where indicated, methods of cleaning, sterilizing, and processing to remove pyrogenic properties shall be written and followed for drug product containers and closures.

## SUBPART F — PRODUCTION AND PROCESS CONTROLS

## Section 211.100 Written Procedures; Deviations

(a) There shall be written procedures for production and process control designed to assure that the drug products have the identity, strength, quality, and purity they purport or are represented to possess. Such procedures shall include all requirements in this subpart. These written procedures, including any changes, shall be drafted, reviewed, and approved by the appropriate organizational units and reviewed and approved by the quality control unit.

(b) Written production and process control procedures shall be followed in the execution of the various production and process control functions and shall be documented at the time of performance. Any deviation from the written procedures shall be recorded and justified.

## SECTION 211.101 CHARGE-IN OF COMPONENTS

Written production and control procedures shall include the following, which are designed to assure that the drug products produced have the identity, strength, quality, and purity they purport or are represented to possess:

(a) The batch shall be formulated with the intent to provide not less than 100 percent of the labeled or established amount of active ingredient.

(b) Components for drug product manufacturing shall be weighed, measured, or subdivided as appropriate. If a component is removed from the original container to another, the new container shall be identified with the following information:
(1) Component name or item code;
(2) Receiving or control number;
(3) Weight or measure in new container;
(4) Batch for which component was dispensed, including its product name, strength, and lot number.

(c) Weighing, measuring, or subdividing operations for components shall be adequately supervised. Each container of component dispensed to manufacturing shall be examined by a second person to assure that:
(1) The component was released by the quality control unit;
(2) The weight or measure is correct as stated in the batch production records;
(3) The containers are properly identified.

(d) Each component shall be added to the batch by one person and verified by a second person.

## SECTION 211.103 CALCULATION OF YIELD

Actual yields and percentages of theoretical yield shall be determined at the conclusion of each appropriate phase of manufacturing, processing, packaging, or holding of the drug product. Such calculations shall be performed by one person and independently verified by a second person.

## SECTION 211.105 EQUIPMENT IDENTIFICATION

(a) All compounding and storage containers, processing lines, and major equipment used during the production of a batch of a drug product shall be properly identified at all times to indicate their contents and, when necessary, the phase of processing of the batch.

(b) Major equipment shall be identified by a distinctive identification number or code that shall be recorded in the batch production record to show the specific equipment used in the manufacture of each batch of a drug product. In cases where only one of a particular type of equipment exists in a manufacturing facility, the name of the equipment may be used in lieu of a distinctive identification number or code.

## SECTION 211.110 SAMPLING AND TESTING OF IN-PROCESS MATERIALS AND DRUG PRODUCTS

(a) To assure batch uniformity and integrity of drug products, written procedures shall be established and followed that describe the in-process controls, and tests, or examinations to be conducted on appropriate samples of in-process materials of each batch. Such control procedures shall be established to monitor the output and to validate the performance of those manufacturing processes that may be responsible for causing variability in the characteristics of in-process material and the drug product. Such control procedures shall include, but are not limited to, the following, where appropriate:
  (1) Tablet or capsule weight variation;
  (2) Disintegration time;
  (3) Adequacy of mixing to assure uniformity and homogeneity;
  (4) Dissolution time and rate;
  (5) Clarity, completeness, or pH of solutions.

(b) Valid in-process specifications for such characteristics shall be consistent with drug product final specifications and shall be derived from previous acceptable process average and process variability estimates where possible and determined by the application of suitable statistical procedures where appropriate. Examination and testing of samples shall assure that the drug product and in-process material conform to specifications.

(c) In-process materials shall be tested for identity, strength, quality, and purity as appropriate, and approved or rejected by the quality control unit, during the production process, e.g., at commencement or completion of significant phases or after storage for long periods.

(d) Rejected in-process materials shall be identified and controlled under a quarantine system designed to prevent their use in manufacturing or processing operations for which they are unsuitable.

## SECTION 211.111 TIME LIMITATIONS ON PRODUCTION

When appropriate, time limits for the completion of each phase of production shall be established to assure the quality of the drug product. Deviation from established time limits may be acceptable if such deviation does not compromise the quality of the drug product. Such deviation shall be justified and documented.

## SECTION 211.113 CONTROL OF MICROBIOLOGICAL CONTAMINATION

(a) Appropriate written procedures, designed to prevent objectionable microorganisms in drug products not required to be sterile, shall be established and followed.

(b) Appropriate written procedures, designed to prevent microbiological contamination of drug products purporting to be sterile, shall be established and followed. Such procedures shall include validation of any sterilization process.

## SECTION 211.115 REPROCESSING

(a) Written procedures shall be established and followed prescribing a system for reprocessing batches that do not conform to standards or specifications and the steps to be taken to insure that the reprocessed batches will conform with all established standards, specifications, and characteristics.

(b) Reprocessing shall not be performed without the review and approval of the quality control unit.

# SUBPART G — PACKAGING AND LABELING CONTROL

## SECTION 211.122 MATERIALS EXAMINATION AND USAGE CRITERIA

(a) There shall be written procedures describing in sufficient detail the receipt, identification, storage, handling, sampling, examination, and/or testing of labeling and packaging materials; such written procedures shall be followed. Labeling and packaging materials shall be representatively sampled, and examined or tested upon receipt and before use in packaging or labeling of a drug product.

(b) Any labeling or packaging materials meeting appropriate written specifications may be approved and released for use. Any labeling or packaging materials that do not meet such specifications shall be rejected to prevent their use in operations for which they are unsuitable.

(c) Records shall be maintained for each shipment received of each different labeling and packaging material indicating receipt, examination or testing, and whether accepted or rejected.

(d) Labels and other labeling materials for each different drug product, strength, dosage form, or quantity of contents shall be stored separately with suitable identification. Access to the storage area shall be limited to authorized personnel.

(e) Obsolete and outdated labels, labeling, and other packaging materials shall be destroyed.

(f) Use of gang-printed labeling for different drug products, or different strengths or net contents of the same drug product, is prohibited unless the labeling from gang-printed sheets is adequately differentiated by size, shape, or color.

(g) If cut labeling is used, packaging and labeling operations shall include one of the following special control procedures:
  (1) Dedication of labeling and packaging lines to each different strength of each different drug product;
  (2) Use of appropriate electronic or electromechanical equipment to conduct a 100-percent examination for correct labeling during or after completion of finishing operations; or
  (3) Use of visual inspection to conduct a 100-percent examination for correct labeling during or after completion of finishing operations for hand-applied labeling. Such examination shall be performed by one person and independently verified by a second person.

(h) Printing devices on, or associated with, manufacturing lines used to imprint labeling upon the drug product unit label or case shall be monitored to assure that all imprinting conforms to the print specified in the batch production record.

## SECTION 211.125 LABELING ISSUANCE

(a) Strict control shall be exercised over labeling issued for use in drug product labeling operations.

(b) Labeling materials issued for a batch shall be carefully examined for identity and conformity to the labeling specified in the master or batch production records.

(c) Procedures shall be used to reconcile the quantities of labeling issued, used, and returned, and shall require evaluation of discrepancies found between the quantity of drug product finished and the quantity of labeling issued when such discrepancies are outside narrow preset limits based on historical operating data. Such discrepancies shall be investigated in accordance with §211.192. Labeling reconciliation is waived for cut or roll labeling if a 100-percent examination for correct labeling is performed in accordance with §211.122(g)(2).

(d) All excess labeling bearing lot or control numbers shall be destroyed.

(e) Returned labeling shall be maintained and stored in a manner to prevent mixups and provide proper identification.

(f) Procedures shall be written describing in sufficient detail the control procedures employed for the issuance of labeling; such written procedures shall be followed.

## SECTION 211.130 PACKAGING AND LABELING OPERATIONS

There shall be written procedures designed to assure that correct labels, labeling, and packaging materials are used for drug products; such written procedures shall be followed. These procedures shall incorporate the following features:

(a) Prevention of mixups and cross-contamination by physical or spatial separation from operations on other drug products.

(b) Identification and handling of filled drug product containers that are set aside and held in unlabeled condition for future labeling operations to preclude mislabeling of individual containers, lots, or portions of lots. Identification need not be applied to each individual container but shall be sufficient to determine name, strength, quantity of contents, and lot or control number of each container.

(c) Identification of the drug product with a lot or control number that permits determination of the history of the manufacture and control of the batch.

(d) Examination of packaging and labeling materials for suitability and correctness before packaging operations, and documentation of such examination in the batch production record.

(e) Inspection of the packaging and labeling facilities immediately before use to assure that all drug products have been removed from previous operations. Inspection shall also be made to assure that packaging and labeling materials not suitable for subsequent operations have been removed. Results of inspection shall be documented in the batch production records.

## SECTION 211.132 TAMPER-EVIDENT PACKAGING REQUIREMENTS FOR OVER-THE-COUNTER (OTC) HUMAN DRUG PRODUCTS

(a) General. The Food and Drug Administration has the authority under the Federal Food, Drug, and Cosmetic Act (the act) to establish a uniform national requirement for tamper-evident packaging of OTC drug products that will improve the security of OTC drug packaging and help assure the safety and effectiveness of OTC drug products. An OTC drug product (except a dermatological, dentifrice, insulin, or lozenge product) for retail sale that is not packaged in a tamper-resistant package or that is not properly labeled under this section is adulterated under §501 of the act or misbranded under §502 of the act, or both.

(b) Requirements for tamper-evident package.

(1) Each manufacturer and packer who packages an OTC drug product (except a dermatological, dentifrice, insulin, or lozenge product) for retail sale shall package the product in a tamper-evident package, if this product is accessible to the public while held for sale. A tamper-evident package is one having one or more indicators or barriers to entry which, if breached or missing, can reasonably be expected to provide visible evidence to consumers that tampering has occurred. To reduce the likelihood of successful tampering and to increase the likelihood that consumers will discover if a product has

been tampered with, the package is required to be distinctive by design or by the use of one or more indicators or barriers to entry that employ an identifying characteristic (e.g., a pattern, name, registered trademark, logo, or picture). For purposes of this section, the term "distinctive by design" means the packaging cannot be duplicated with commonly available materials or through commonly available processes. A tamper-evident package may involve an immediate-container and closure system or secondary-container or carton system or any combination of systems intended to provide a visual indication of package integrity. The tamper-evident feature shall be designed to and shall remain intact when handled in a reasonable manner during manufacture, distribution, and retail display.

(2) In addition to the tamper-evident packaging feature described in paragraph (b)(1) of this section, any two-piece, hard gelatin capsule covered by this section must be sealed using an acceptable tamper-evident technology.

(c) Labeling.

(1) In order to alert consumers to the specific tamper-evident feature(s) used, each retail package of an OTC drug product covered by this section (except ammonia inhalant in crushable glass ampules, containers of compressed medical oxygen, or aerosol products that depend upon the power of a liquefied or compressed gas to expel the contents from the container) is required to bear a statement that:

   (i) Identifies all tamper-evident feature(s) and any capsule sealing technologies used to comply with paragraph (b) of this section;

  (ii) Is prominently placed on the package; and

 (iii) Is so placed that it will be unaffected if the tamper-evident feature of the package is breached or missing.

(2) If the tamper-evident feature chosen to meet the requirements in paragraph (b) of this section uses an identifying characteristic, that characteristic is required to be referred to in the labeling statement. For example, the labeling statement on a bottle with a shrink band could say "For your protection, this bottle has an imprinted seal around the neck."

(d) Request for exemptions from packaging and labeling requirements. A manufacturer or packer may request an exemption from the packaging and labeling requirements of this section. A request for an exemption is required to be submitted in the form of a citizen petition under §10.30 of this chapter and should be clearly identified on the envelope as a "Request for Exemption from the Tamper-Evident Packaging Rule." The petition is required to contain the following:

(1) The name of the drug product or, if the petition seeks an exemption for a drug class, the name of the drug class, and a list of products within that class.

(2) The reasons that the drug product's compliance with the tamper-evident packaging or labeling requirements of this section is unnecessary or cannot be achieved.

(3) A description of alternative steps that are available, or that the petitioner has already taken, to reduce the likelihood that the product or drug class will be the subject of malicious adulteration.

(4) Other information justifying an exemption.

(e) OTC drug products subject to approved new drug applications. Holders of approved new drug applications for OTC drug products are required under §314.70 of this chapter to provide the agency with notification of changes in packaging and labeling to comply with the requirements of this section. Changes in packaging and labeling required by this regulation

may be made before FDA approval, as provided under §314.70(c) of this chapter. Manufacturing changes by which capsules are to be sealed require prior FDA approval under
§314.70(b) of this chapter.

(f) Poison Prevention Packaging Act of 1970. This section does not affect any requirements for
"special packaging" as defined under §310.3(l) of this chapter and required under the Poison
Prevention Packaging Act of 1970.

## Section 211.134 Drug Product Inspection

(a) Packaged and labeled products shall be examined during finishing operations to provide
assurance that containers and packages in the lot have the correct label.

(b) A representative sample of units shall be collected at the completion of finishing operations
and shall be visually examined for correct labeling.

(c) Results of these examinations shall be recorded in the batch production or control records.

## Section 211.137 Expiration Dating

(a) To assure that a drug product meets applicable standards of identity, strength, quality, and
purity at the time of use, it shall bear an expiration date determined by appropriate stability
testing described in §211.166.

(b) Expiration dates shall be related to any storage conditions stated on the labeling, as determined by stability studies described in §211.166.

(c) If the drug product is to be reconstituted at the time of dispensing, its labeling shall bear
expiration information for both the reconstituted and unreconstituted drug products.

(d) Expiration dates shall appear on labeling in accordance with the requirements of §201.17
of this chapter.

(e) Homeopathic drug products shall be exempt from the requirements of this section.

(f) Allergenic extracts that are labeled "No U.S. Standard of Potency" are exempt from the
requirements of this section.

(g) New drug products for investigational use are exempt from the requirements of this section,
provided that they meet appropriate standards or specifications as demonstrated by stability
studies during their use in clinical investigations. Where new drug products for investigational use are to be reconstituted at the time of dispensing, their labeling shall bear expiration
information for the reconstituted drug product.

(h) Pending consideration of a proposed exemption, published in the Federal Register of September 29, 1978, the requirements in this section shall not be enforced for human OTC drug
products if their labeling does not bear dosage limitations and they are stable for at least 3
years as supported by appropriate stability data.

## SUBPART H — HOLDING AND DISTRIBUTION

## Section 211.142 Warehousing Procedures

Written procedures describing the warehousing of drug products shall be established and followed. They shall include:

(a) Quarantine of drug products before release by the quality control unit.

(b) Storage of drug products under appropriate conditions of temperature, humidity, and light so that the identity, strength, quality, and purity of the drug products are not affected.

## SECTION 211.150 DISTRIBUTION PROCEDURES

Written procedures shall be established, and followed, describing the distribution of drug products. They shall include:

(a) A procedure whereby the oldest approved stock of a drug product is distributed first. Deviation from this requirement is permitted if such deviation is temporary and appropriate.

(b) A system by which the distribution of each lot of drug product can be readily determined to facilitate its recall if necessary.

# SUBPART I — LABORATORY CONTROLS

## SECTION 211.160 GENERAL REQUIREMENTS

(a) The establishment of any specifications, standards, sampling plans, test procedures, or other laboratory control mechanisms required by this subpart, including any change in such specifications, standards, sampling plans, test procedures, or other laboratory control mechanisms, shall be drafted by the appropriate organizational unit and reviewed and approved by the quality control unit. The requirements in this subpart shall be followed and shall be documented at the time of performance. Any deviation from the written specifications, standards, sampling plans, test procedures, or other laboratory control mechanisms shall be recorded and justified.

(b) Laboratory controls shall include the establishment of scientifically sound and appropriate specifications, standards, sampling plans, and test procedures designed to assure that components, drug product containers, closures, in-process materials, labeling, and drug products conform to appropriate standards of identity, strength, quality, and purity. Laboratory controls shall include:

   (1) Determination of conformance to appropriate written specifications for the acceptance of each lot within each shipment of components, drug product containers, closures, and labeling used in the manufacture, processing, packing, or holding of drug products. The specifications shall include a description of the sampling and testing procedures used. Samples shall be representative and adequately identified. Such procedures shall also require appropriate retesting of any component, drug product container, or closure that is subject to deterioration.

   (2) Determination of conformance to written specifications and a description of sampling and testing procedures for in-process materials. Such samples shall be representative and properly identified.

   (3) Determination of conformance to written descriptions of sampling procedures and appropriate specifications for drug products. Such samples shall be representative and properly identified.

   (4) The calibration of instruments, apparatus, gauges, and recording devices at suitable intervals in accordance with an established written program containing specific directions,

schedules, limits for accuracy and precision, and provisions for remedial action in the event accuracy and/or precision limits are not met. Instruments, apparatus, gauges, and recording devices not meeting established specifications shall not be used.

## SECTION 211.165 TESTING AND RELEASE FOR DISTRIBUTION

(a) For each batch of drug product, there shall be appropriate laboratory determination of satisfactory conformance to final specifications for the drug product, including the identity and strength of each active ingredient, prior to release. Where sterility and/or pyrogen testing are conducted on specific batches of shortlived radiopharmaceuticals, such batches may be released prior to completion of sterility and/or pyrogen testing, provided such testing is completed as soon as possible.

(b) There shall be appropriate laboratory testing, as necessary, of each batch of drug product required to be free of objectionable microorganisms.

(c) Any sampling and testing plans shall be described in written procedures that shall include the method of sampling and the number of units per batch to be tested; such written procedure shall be followed.

(d) Acceptance criteria for the sampling and testing conducted by the quality control unit shall be adequate to assure that batches of drug products meet each appropriate specification and appropriate statistical quality control criteria as a condition for their approval and release. The statistical quality control criteria shall include appropriate acceptance levels and/or appropriate rejection levels.

(e) The accuracy, sensitivity, specificity, and reproducibility of test methods employed by the firm shall be established and documented. Such validation and documentation may be accomplished in accordance with §211.194(a)(2).

(f) Drug products failing to meet established standards or specifications and any other relevant quality control criteria shall be rejected. Reprocessing may be performed. Prior to acceptance and use, reprocessed material must meet appropriate standards, specifications, and any other relevant critieria.

## SECTION 211.166 STABILITY TESTING

(a) There shall be a written testing program designed to assess the stability characteristics of drug products. The results of such stability testing shall be used in determining appropriate storage conditions and expiration dates. The written program shall be followed and shall include:
  (1) Sample size and test intervals based on statistical criteria for each attribute examined to assure valid estimates of stability;
  (2) Storage conditions for samples retained for testing;
  (3) Reliable, meaningful, and specific test methods;
  (4) Testing of the drug product in the same container-closure system as that in which the drug product is marketed;
  (5) Testing of drug products for reconstitution at the time of dispensing (as directed in the labeling) as well as after they are reconstituted.

(b) An adequate number of batches of each drug product shall be tested to determine an appropriate expiration date and a record of such data shall be maintained. Accelerated studies, combined with basic stability information on the components, drug products, and container-closure system, may be used to support tentative expiration dates provided full shelf life studies are not available and are being conducted. Where data from accelerated studies are used to project a tentative expiration date that is beyond a date supported by actual shelf life studies, there must be stability studies conducted, including drug product testing at appropriate intervals, until the tentative expiration date is verified or the appropriate expiration date determined.

(c) For homeopathic drug products, the requirements of this section are as follows:

    (1) There shall be a written assessment of stability based at least on testing or examination of the drug product for compatibility of the ingredients, and based on marketing experience with the drug product to indicate that there is no degradation of the product for the normal or expected period of use.

    (2) Evaluation of stability shall be based on the same container-closure system in which the drug product is being marketed.

(d) Allergenic extracts that are labeled "No U.S. Standard of Potency" are exempt from the requirements of this section.

## Section 211.167 Special Testing Requirements

(a) For each batch of drug product purporting to be sterile and/or pyrogen-free, there shall be appropriate laboratory testing to determine conformance to such requirements. The test procedures shall be in writing and shall be followed.

(b) For each batch of ophthalmic ointment, there shall be appropriate testing to determine conformance to specifications regarding the presence of foreign particles and harsh or abrasive substances. The test procedures shall be in writing and shall be followed.

(c) For each batch of controlled-release dosage form, there shall be appropriate laboratory testing to determine conformance to the specifications for the rate of release of each active ingredient. The test procedures shall be in writing and shall be followed.

## Section 211.170 Reserve Samples

(a) An appropriately identified reserve sample that is representative of each lot in each shipment of each active ingredient shall be retained. The reserve sample consists of at least twice the quantity necessary for all tests required to determine whether the active ingredient meets its established specifications, except for sterility and pyrogen testing. The retention time is as follows:

    (1) For an active ingredient in a drug product other than those described in paragraphs (a)(2) and (3) of this section, the reserve sample shall be retained for 1 year after the expiration date of the last lot of the drug product containing the active ingredient.

    (2) For an active ingredient in a radioactive drug product, except for nonradioactive reagent kits, the reserve sample shall be retained for:

        (i) Three months after the expiration date of the last lot of the drug product containing the active ingredient if the expiration dating period of the drug product is 30 days or less; or

        (ii) Six months after the expiration date of the last lot of the drug product containing the active ingredient if the expiration dating period of the drug product is more than 30 days.

(3) For an active ingredient in an OTC drug product that is exempt from bearing an expiration date under §211.137, the reserve sample shall be retained for 3 years after distribution of the last lot of the drug product containing the active ingredient.

(b) An appropriately identified reserve sample that is representative of each lot or batch of drug product shall be retained and stored under conditions consistent with product labeling. The reserve sample shall be stored in the same immediate container-closure system in which the drug product is marketed or in one that has essentially the same characteristics. The reserve sample consists of at least twice the quantity necessary to perform all the required tests, except those for sterility and pyrogens. Except for those for drug products described in paragraph (b)(2) of this section, reserve samples from representative sample lots or batches selected by acceptable statistical procedures shall be examined visually at least once a year for evidence of deterioration unless visual examination would affect the integrity of the reserve sample. Any evidence of reserve sample deterioration shall be investigated in accordance with §211.192. The results of the examination shall be recorded and maintained with other stability data on the drug product. Reserve samples of compressed medical gases need not be retained. The retention time is as follows:

(1) For a drug product other than those described in paragraphs (b)(2) and (3) of this section, the reserve sample shall be retained for 1 year after the expiration date of the drug product.

(2) For a radioactive drug product, except for nonradioactive reagent kits, the reserve sample shall be retained for:

(i) Three months after the expiration date of the drug product if the expiration dating period of the drug product is 30 days or less; or

(ii) Six months after the expiration date of the drug product if the expiration dating period of the drug product is more than 30 days.

(3) For an OTC drug product that is exempt from bearing an expiration date under §211.137, the reserve sample must be retained for 3 years after the lot or batch of drug product is distributed.

## SECTION 211.173 LABORATORY ANIMALS

Animals used in testing components, in-process materials, or drug products for compliance with established specifications shall be maintained and controlled in a manner that assures their suitability for their intended use. They shall be identified, and adequate records shall be maintained showing the history of their use.

## SECTION 211.176 PENICILLIN CONTAMINATION

If a reasonable possibility exists that a non-penicillin drug product has been exposed to cross-contamination with penicillin, the non-penicillin drug product shall be tested for the presence of penicillin. Such drug product shall not be marketed if detectable levels are found when tested according to procedures specified in "Procedures for Detecting and Measuring Penicillin Contamination in Drugs," which is incorporated by reference. Copies are available from the Division of Research and Testing (HFD-470), Center for Drug Evaluation and Research, Food and Drug Administration, 5100 Paint Branch Pkwy., College Park, MD 20740, or available for inspection at the Office of the Federal Register, 800 North Capitol Street, NW., Suite 700, Washington, D.C. 20408.

# SUBPART J — RECORDS AND REPORTS

## SECTION 211.180 GENERAL REQUIREMENTS

(a) Any production, control, or distribution record that is required to be maintained in compliance with this part and is specifically associated with a batch of a drug product shall be retained for at least 1 year after the expiration date of the batch or, in the case of certain OTC drug products lacking expiration dating because they meet the criteria for exemption under §211.137, 3 years after distribution of the batch.

(b) Records shall be maintained for all components, drug product containers, closures, and labeling for at least 1 year after the expiration date or, in the case of certain OTC drug products lacking expiration dating because they meet the criteria for exemption under §211.137, 3 years after distribution of the last lot of drug product incorporating the component or using the container, closure, or labeling.

(c) All records required under this part, or copies of such records, shall be readily available for authorized inspection during the retention period at the establishment where the activities described in such records occurred. These records or copies thereof shall be subject to photocopying or other means of reproduction as part of such inspection. Records that can be immediately retrieved from another location by computer or other electronic means shall be considered as meeting the requirements of this paragraph.

(d) Records required under this part may be retained either as original records or as true copies such as photocopies, microfilm, microfiche, or other accurate reproductions of the original records. Where reduction techniques, such as microfilming, are used, suitable reader and photocopying equipment shall be readily available.

(e) Written records required by this part shall be maintained so that data therein can be used for evaluating, at least annually, the quality standards of each drug product to determine the need for changes in drug product specifications or manufacturing or control procedures. Written procedures shall be established and followed for such evaluations and shall include provisions for:

   (1) A review of a representative number of batches, whether approved or rejected, and, where applicable, records associated with the batch.

   (2) A review of complaints, recalls, returned or salvaged drug products, and investigations conducted under §211.192 for each drug product.

(f) Procedures shall be established to assure that the responsible officials of the firm, if they are not personally involved in or immediately aware of such actions, are notified in writing of any investigations conducted under §211.198, 211.204, or 211.208 of these regulations, any recalls, reports of inspectional observations issued by the Food and Drug Administration, or any regulatory actions relating to good manufacturing practices brought by the Food and Drug Administration.

## SECTION 211.182 EQUIPMENT CLEANING AND USE LOG

A written record of major equipment cleaning, maintenance (except routine maintenance such as lubrication and adjustments), and use shall be included in individual equipment logs that show the date, time, product, and lot number of each batch processed. If equipment is dedicated to manufacture of one product, then individual equipment logs are not required, provided that lots or batches of such product follow in numerical order and are manufactured in numerical sequence. In cases where dedicated equipment is employed, the records of cleaning, maintenance, and use shall be part of the batch record. The persons performing and double-checking the cleaning and maintenance shall date and sign or initial the log indicating that the work was performed. Entries in the log shall be in chronological order.

## Section 211.184 Component, Drug Product Container, Closure, and Labeling Records

These records shall include the following:

(a) The identity and quantity of each shipment of each lot of components, drug product containers, closures, and labeling; the name of the supplier; the supplier's lot number(s) if known; the receiving code as specified in §211.80; and the date of receipt. The name and location of the prime manufacturer, if different from the supplier, shall be listed if known.

(b) The results of any test or examination performed (including those performed as required by §211.82(a), §211.84(d), or §211.122(a)) and the conclusions derived therefrom.

(c) An individual inventory record of each component, drug product container, and closure and, for each component, a reconciliation of the use of each lot of such component. The inventory record shall contain sufficient information to allow determination of any batch or lot of drug product associated with the use of each component, drug product container, and closure.

(d) Documentation of the examination and review of labels and labeling for conformity with established specifications in accord with §211.122(c) and 211.130(c).

(e) The disposition of rejected components, drug product containers, closure, and labeling.

## Section 211.186 Master Production and Control Records

(a) To assure uniformity from batch to batch, master production and control records for each drug product, including each batch size thereof, shall be prepared, dated, and signed (full signature, handwritten) by one person and independently checked, dated, and signed by a second person. The preparation of master production and control records shall be described in a written procedure and such written procedure shall be followed.

(b) Master production and control records shall include:

   (1) The name and strength of the product and a description of the dosage form;

   (2) The name and weight or measure of each active ingredient per dosage unit or per unit of weight or measure of the drug product, and a statement of the total weight or measure of any dosage unit;

   (3) A complete list of components designated by names or codes sufficiently specific to indicate any special quality characteristic;

   (4) An accurate statement of the weight or measure of each component, using the same weight system (metric, avoirdupois, or apothecary) for each component. Reasonable variations may be permitted, however, in the amount of components necessary for the preparation in the dosage form, provided they are justified in the master production and control records;

   (5) A statement concerning any calculated excess of component;

   (6) A statement of theoretical weight or measure at appropriate phases of processing;

   (7) A statement of theoretical yield, including the maximum and minimum percentages of theoretical yield beyond which investigation according to §211.192 is required;

(8) A description of the drug product containers, closures, and packaging materials, including a specimen or copy of each label and all other labeling signed and dated by the person or persons responsible for approval of such labeling;

(9) Complete manufacturing and control instructions, sampling and testing procedures, specifications, special notations, and precautions to be followed.

## SECTION 211.188 BATCH PRODUCTION AND CONTROL RECORDS

Batch production and control records shall be prepared for each batch of drug product produced and shall include complete information relating to the production and control of each batch. These records shall include:

(a) An accurate reproduction of the appropriate master production or control record, checked for accuracy, dated, and signed;

(b) Documentation that each significant step in the manufacture, processing, packing, or holding of the batch was accomplished, including:

(1) Dates;

(2) Identity of individual major equipment and lines used;

(3) Specific identification of each batch of component or in-process material used;

(4) Weights and measures of components used in the course of processing;

(5) In-process and laboratory control results;

(6) Inspection of the packaging and labeling area before and after use;

(7) A statement of the actual yield and a statement of the percentage of theoretical yield at appropriate phases of processing;

(8) Complete labeling control records, including specimens or copies of all labeling used;

(9) Description of drug product containers and closures;

(10) Any sampling performed;

(11) Identification of the persons performing and directly supervising or checking each significant step in the operation;

(12) Any investigation made according to §211.192.

(13) Results of examinations made in accordance with §211.134.

## SECTION 211.192 PRODUCTION RECORD REVIEW

All drug product production and control records, including those for packaging and labeling, shall be reviewed and approved by the quality control unit to determine compliance with all established, approved written procedures before a batch is released or distributed. Any unexplained discrepancy (including a percentage of theoretical yield exceeding the maximum or minimum percentages established in master production and control records) or the failure of a batch or any of its components to meet any of its specifications shall be thoroughly investigated, whether or not the batch has already been distributed. The investigation shall extend to other batches of the same drug product and other drug products that may have been associated with the specific failure or discrepancy. A written record of the investigation shall be made and shall include the conclusions and follow-up.

## SECTION 211.194 LABORATORY RECORDS

(a) Laboratory records shall include complete data derived from all tests necessary to assure compliance with established specifications and standards, including examinations and assays, as follows:

(1) A description of the sample received for testing with identification of source (that is, location from where sample was obtained), quantity, lot number or other distinctive code, date sample was taken, and date sample was received for testing.

(2) A statement of each method used in the testing of the sample. The statement shall indicate the location of data that establish that the methods used in the testing of the sample meet proper standards of accuracy and reliability as applied to the product tested. (If the method employed is in the current revision of the United States Pharmacopeia, National Formulary, Association of Official Analytical Chemists, Book of Methods,* or in other recognized standard references, or is detailed in an approved new drug application and the referenced method is not modified, a statement indicating the method and reference will suffice). The suitability of all testing methods used shall be verified under actual conditions of use.

(3) A statement of the weight or measure of sample used for each test, where appropriate.

(4) A complete record of all data secured in the course of each test, including all graphs, charts, and spectra from laboratory instrumentation, properly identified to show the specific component, drug product container, closure, in-process material, or drug product, and lot tested.

(5) A record of all calculations performed in connection with the test, including units of measure, conversion factors, and equivalency factors.

(6) A statement of the results of tests and how the results compare with established standards of identity, strength, quality, and purity for the component, drug product container, closure, in-process material, or drug product tested.

(7) The initials or signature of the person who performs each test and the date(s) the tests were performed.

(8) The initials or signature of a second person showing that the original records have been reviewed for accuracy, completeness, and compliance with established standards.

(b) Complete records shall be maintained of any modification of an established method employed in testing. Such records shall include the reason for the modification and data to verify that the modification produced results that are at least as accurate and reliable for the material being tested as the established method.

(c) Complete records shall be maintained of any testing and standardization of laboratory reference standards, reagents, and standard solutions.

(d) Complete records shall be maintained of the periodic calibration of laboratory instruments, apparatus, gauges, and recording devices required by §211.160(b)(4).

(e) Complete records shall be maintained of all stability testing performed in accordance with §211.166.

## Section 211.196 Distribution Records

Distribution records shall contain the name and strength of the product and description of the dosage form, name and address of the consignee, date and quantity shipped, and lot or control number of the drug product. For compressed medical gas products, distribution records are not required to contain lot or control numbers.

---

* Copies may be obtained from: Association of Official Analytical Chemists, 2200 Wilson Blvd., Suite 400, Arlington, VA 22201-3301.

## Section 211.198 Complaint Files

(a) Written procedures describing the handling of all written and oral complaints regarding a drug product shall be established and followed. Such procedures shall include provisions for review by the quality control unit, of any complaint involving the possible failure of a drug product to meet any of its specifications and, for such drug products, a determination as to the need for an investigation in accordance with §211.192. Such procedures shall include provisions for review to determine whether the complaint represents a serious and unexpected adverse drug experience which is required to be reported to the Food and Drug Administration in accordance with §310.305 of this chapter.

(b) A written record of each complaint shall be maintained in a file designated for drug product complaints. The file regarding such drug product complaints shall be maintained at the establishment where the drug product involved was manufactured, processed, or packed, or such file may be maintained at another facility if the written records in such files are readily available for inspection at that other facility. Written records involving a drug product shall be maintained until at least 1 year after the expiration date of the drug product, or 1 year after the date that the complaint was received, whichever is longer. In the case of certain OTC drug products lacking expiration dating because they meet the criteria for exemption under §211.137, such written records shall be maintained for 3 years after distribution of the drug product.

(1) The written record shall include the following information, where known: the name and strength of the drug product, lot number, name of complainant, nature of complaint, and reply to complainant.

(2) Where an investigation under §211.192 is conducted, the written record shall include the findings of the investigation and follow-up. The record or copy of the record of the investigation shall be maintained at the establishment where the investigation occurred in accordance with §211.180(c).

(3) Where an investigation under §211.192 is not conducted, the written record shall include the reason that an investigation was found not to be necessary and the name of the responsible person making such a determination.

## SUBPART K — RETURNED AND SALVAGED DRUG PRODUCTS

### Section 211.204 Returned Drug Products

Returned drug products shall be identified as such and held. If the conditions under which returned drug products have been held, stored, or shipped before or during their return, or if the condition of the drug product, its container, carton, or labeling, as a result of storage or shipping, casts doubt on the safety, identity, strength, quality or purity of the drug product, the returned drug product shall be destroyed unless examination, testing, or other investigations prove the drug product meets appropriate standards of safety, identity, strength, quality, or purity. A drug product may be reprocessed provided the subsequent drug product meets appropriate standards, specifications, and characteristics. Records of returned drug products shall be maintained and shall include the name and label potency of the drug product dosage form, lot number (or control number or batch number), reason for the return, quantity returned, date of disposition, and ultimate disposition of the returned drug product. If the reason for a drug product being returned implicates associated batches, an appropriate investigation shall be conducted in accordance with the requirements of §211.192. Procedures for the holding, testing, and reprocessing of returned drug products shall be in writing and shall be followed.

## Section 211.208 Drug Product Salvaging

Drug products that have been subjected to improper storage conditions including extremes in temperature, humidity, smoke, fumes, pressure, age, or radiation due to natural disasters, fires, accidents, or equipment failures shall not be salvaged and returned to the marketplace. Whenever there is a question whether drug products have been subjected to such conditions, salvaging operations may be conducted only if there is (a) evidence from laboratory tests and assays (including animal feeding studies where applicable) that the drug products meet all applicable standards of identity, strength, quality, and purity and (b) evidence from inspection of the premises that the drug products and their associated packaging were not subjected to improper storage conditions as a result of the disaster or accident. Organoleptic examinations shall be acceptable only as supplemental evidence that the drug products meet appropriate standards of identity, strength, quality, and purity. Records including name, lot number, and disposition shall be maintained for drug products subject to this section.

# Guidance for Industry* Q7A Good Manufacturing Practice Guidance for Active Pharmaceutical Ingredients

U.S. Department of Health and Human Services
Food and Drug Administration
Center for Drug Evaluation and Research (CDER)
Center for Biologics Evaluation and Research (CBER)
August 2001 ICH

## CONTENTS

---

\* This guidance was developed within the Expert Working Group (Q7A) of the International Conference on Harmonisation of Technical Requirements for Registration of Pharmaceuticals for Human Use (ICH) and has been subject to consultation by the regulatory parties, in accordance with the ICH process. This document has been endorsed by the ICH Steering Committee at *Step 4* of the ICH process, November 2000. At *Step 4* of the process, the final draft is recommended for adoption to the regulatory bodies of the European Union, Japan, and the United States.

Arabic numbers in subheadings reflect the organizational breakdown in the document endorsed by the ICH Steering Committee at Step 4 of the ICH process, November 2000.

This guidance represents the Food and Drug Administration's (FDA's) current thinking on this topic. It does not create or confer any rights for or on any person and does not operate to bind FDA or the public. An alternative approach may be used if such approach satisfies the requirements of the applicable statutes and regulations.

# I. INTRODUCTION (1)

## A. Objective (1.1)

This document is intended to provide guidance regarding good manufacturing practice (GMP) for the manufacturing of active pharmaceutical ingredients (APIs) under an appropriate system for managing quality. It is also intended to help ensure that APIs meet the quality and purity characteristics that they purport, or are represented, to possess.

In this guidance, the term *manufacturing* is defined to include all operations of receipt of materials, production, packaging, repackaging, labeling, relabeling, quality control, release, storage and distribution of APIs and the related controls. In this guidance, the term *should* identifies recommendations that, when followed, will ensure compliance with CGMPs. An alternative approach may be used if such approach satisfies the requirements of the applicable statutes. For the purposes of this guidance, the terms *current good manufacturing practices* and *good manufacturing practices* are equivalent.

The guidance as a whole does not cover safety aspects for the personnel engaged in manufacturing, nor aspects related to protecting the environment. These controls are inherent responsibilities of the manufacturer and are governed by national laws.

This guidance is not intended to define registration and/or filing requirements or modify pharmacopoeial requirements. This guidance does not affect the ability of the responsible regulatory agency to establish specific registration/filing requirements regarding APIs within the context of marketing/manufacturing authorizations or drug applications. All commitments in registration/filing documents should be met.

## B. Regulatory Applicability (1.2)

Within the world community, materials may vary as to their legal classification as an API. When a material is classified as an API in the region or country in which it is manufactured or used in a drug product, it should be manufactured according to this guidance.

## C. Scope (1.3)

This guidance applies to the manufacture of APIs for use in human drug (medicinal) products. It applies to the manufacture of sterile APIs only up to the point immediately prior to the APIs being rendered sterile. The sterilization and aseptic processing of sterile APIs are not covered by this guidance, but should be performed in accordance with GMP guidances for drug (medicinal) products as defined by local authorities.

This guidance covers APIs that are manufactured by chemical synthesis, extraction, cell culture/fermentation, recovery from natural sources, or any combination of these processes. Specific guidance for APIs manufactured by cell culture/fermentation is described in Section XVIII (18).

This guidance excludes all vaccines, whole cells, whole blood and plasma, blood and plasma derivatives (plasma fractionation), and gene therapy APIs. However, it does include APIs that are produced using blood or plasma as raw materials. Note that cell substrates (mammalian, plant, insect or microbial cells, tissue or animal sources including transgenic animals) and early process steps may be subject to GMP but are not covered by this guidance. In addition, the guidance does not apply to medical gases, bulk-packaged drug (medicinal) products (e.g., tablets or capsules in bulk containers), or radiopharmaceuticals.

**TABLE 1**
**Application of this Guidance to API Manufacturing**

| Type of Manufacturing | Application of this Guidance to Steps (Shown in Gray) Used in this Type of Manufacturing | | | | |
|---|---|---|---|---|---|
| Chemical manufacturing | Production of the API starting material | Introduction of the API starting material into process | Production of intermediate(s) | Isolation and purification | Physical processing, and packaging |
| API derived from animal sources | Collection of organ, fluid, or tissue | Cutting, mixing, and/or initial processing | Introduction of the API starting material into process | Isolation and purification | Physical processing, and packaging |
| API extracted from plant sources | Collection of plant | Cutting and initial extraction(s) | Introduction of the API starting material into process | Isolation and purification | Physical processing, and packaging |
| Herbal extracts used as API | Collection of plants | Cutting and initial extraction | | Further extraction | Physical processing, and packaging |
| API consisting of comminuted or powdered herbs | Collection of plants and/or cultivation and harvesting | Cutting/ comminuting | | | Physical processing, and packaging |
| Biotechnology: fermentation/cell culture | Establishment of master cell bank and working cell bank | Maintenance of working cell bank | Cell culture and/or fermentation | Isolation and purification | Physical processing, and packaging |
| "Classical" Fermentation to produce an API | Establishment of cell bank | Maintenance of the cell bank | Introduction of the cells into fermentation | Isolation and purification | Physical processing, and packaging |

**Increasing GMP requirements** ⟹

Section XIX (19) contains guidance that applies only to the manufacture of APIs used in the production of drug (medicinal) products specifically for clinical trials (investigational medicinal products).

An *API starting material* is a raw material, an intermediate, or an API that is used in the production of an API and that is incorporated as a significant structural fragment into the structure of the API. An API starting material can be an article of commerce, a material purchased from one or more suppliers under contract or commercial agreement, or produced in-house. API starting materials normally have defined chemical properties and structure.

The company should designate and document the rationale for the point at which production of the API begins. For synthetic processes, this is known as the point at which API starting materials are entered into the process. For other processes (e.g., fermentation, extraction, purification), this rationale should be established on a case-by-case basis. Table 1 gives guidance on the point at which the API starting material is normally introduced into the process.

From this point on, appropriate GMP as defined in this guidance should be applied to these intermediate and/or API manufacturing steps. This would include the validation of critical process steps determined to impact the quality of the API. However, it should be noted that the fact that a company chooses to validate a process step does not necessarily define that step as critical.

The guidance in this document would normally be applied to the steps shown in gray in Table 1. However, all steps shown may not need to be completed. The stringency of GMP in API manufacturing should increase as the process proceeds from early API steps to final steps, purification, and packaging. Physical processing of APIs, such as granulation, coating or physical manipulation of particle size (e.g., milling, micronizing) should be conducted according to this guidance.

This GMP guidance does not apply to steps prior to the introduction of the defined API starting material.

## II. QUALITY MANAGEMENT (2)

### A. PRINCIPLES (2.1)

Quality should be the responsibility of all persons involved in manufacturing.

Each manufacturer should establish, document, and implement an effective system for managing quality that involves the active participation of management and appropriate manufacturing personnel.

The system for managing quality should encompass the organizational structure, procedures, processes and resources, as well as activities to ensure confidence that the API will meet its intended specifications for quality and purity. All quality-related activities should be defined and documented.

There should be a quality unit(s) that is independent of production and that fulfills both quality assurance (QA) and quality control (QC) responsibilities. The quality unit can be in the form of separate QA and QC units or a single individual or group, depending upon the size and structure of the organization.

The persons authorized to release intermediates and APIs should be specified.

All quality-related activities should be recorded at the time they are performed.

Any deviation from established procedures should be documented and explained. Critical deviations should be investigated, and the investigation and its conclusions should be documented.

No materials should be released or used before the satisfactory completion of evaluation by the quality unit(s) unless there are appropriate systems in place to allow for such use (e.g., release under quarantine as described in Section X (10) or the use of raw materials or intermediates pending completion of evaluation).

Procedures should exist for notifying responsible management in a timely manner of regulatory inspections, serious GMP deficiencies, product defects and related actions (e.g., quality-related complaints, recalls, and regulatory actions).

### B. RESPONSIBILITIES OF THE QUALITY UNIT(S) (2.2)

The quality unit(s) should be involved in all quality-related matters.

The quality unit(s) should review and approve all appropriate quality-related documents.

The main responsibilities of the independent quality unit(s) should not be delegated. These responsibilities should be described in writing and should include, but not necessarily be limited to:

1. Releasing or rejecting all APIs. Releasing or rejecting intermediates for use outside the control of the manufacturing company
2. Establishing a system to release or reject raw materials, intermediates, packaging, and labeling materials

3. Reviewing completed batch production and laboratory control records of critical process steps before release of the API for distribution
4. Making sure that critical deviations are investigated and resolved
5. Approving all specifications and master production instructions
6. Approving all procedures affecting the quality of intermediates or APIs
7. Making sure that internal audits (self-inspections) are performed
8. Approving intermediate and AN contract manufacturers
9. Approving changes that potentially affect intermediate or API quality
10. Reviewing and approving validation protocols and reports
11. Making sure that quality-related complaints are investigated and resolved
12. Making sure that effective systems are used for maintaining and calibrating critical equipment
13. Making sure that materials are appropriately tested and the results are reported
14. Making sure that there is stability data to support retest or expiry dates and storage conditions on APIs and/or intermediates, where appropriate
15. Performing product quality reviews (as defined in Section 2.5)

## C. Responsibility for Production Activities (2.3)

The responsibility for production activities should be described in writing and should include, but not necessarily be limited to:

1. Preparing, reviewing, approving, and distributing the instructions for the production of intermediates or APIs according to written procedures
2. Producing APIs and, when appropriate, intermediates according to pre-approved instructions
3. Reviewing all production batch records and ensuring that these are completed and signed
4. Making sure that all production deviations are reported and evaluated and that critical deviations are investigated and the conclusions are recorded
5. Making sure that production facilities are clean and, when appropriate, disinfected
6. Making sure that the necessary calibrations are performed and records kept
7. Making sure that the premises and equipment are maintained and records kept
8. Making sure that validation protocols and reports are reviewed and approved
9. Evaluating proposed changes in product, process or equipment
10. Making sure that new and, when appropriate, modified facilities and equipment are qualified

## D. Internal Audits (Self Inspection) (2.4)

To verify compliance with the principles of GMP for APIs, regular internal audits should be performed in accordance with an approved schedule.

Audit findings and corrective actions should be documented and brought to the attention of responsible management of the firm. Agreed corrective actions should be completed in a timely and effective manner.

## E. Product Quality Review (2.5)

Regular quality-reviews of APIs should be conducted with the objective of verifying the consistency of the process. Such reviews should normally be conducted and documented annually and should include at least:

- A review of critical in-process control and critical API test results
- A review of all batches that failed to meet established specification(s)
- A review of all critical deviations or nonconformances and related investigations
- A review of any changes carried out to the processes or analytical methods
- A review of results of the stability monitoring program
- A review of all quality-related returns, complaints and recalls
- A review of adequacy of corrective actions

The results of this review should be evaluated and an assessment made of whether corrective action or any revalidation should be undertaken. Reasons for such corrective action should be documented. Agreed corrective actions should be completed in a timely and effective manner.

# III. PERSONNEL (3)

## A. Personnel Qualifications (3.1)

There should be an adequate number of personnel qualified by appropriate education, training, and/or experience to perform and supervise the manufacture of intermediates and APIs.

The responsibilities of all personnel engaged in the manufacture of intermediates and APIs should be specified in writing.

Training should be regularly conducted by qualified individuals and should cover, at a minimum, the particular operations that the employee performs and GMP as it relates to the employee's functions. Records of training should be maintained. Training should be periodically assessed.

## B. Personnel Hygiene (3.2)

Personnel should practice good sanitation and health habits.

Personnel should wear clean clothing suitable for the manufacturing activity with which they are involved and this clothing should be changed, when appropriate. Additional protective apparel, such as head, face, hand, and arm coverings, should be worn, when necessary, to protect intermediates and APIs from contamination.

Personnel should avoid direct contact with intermediates or APIs.

Smoking, eating, drinking, chewing and the storage of food should be restricted to certain designated areas separate from the manufacturing areas.

Personnel suffering from an infectious disease or having open lesions on the exposed surface of the body should not engage in activities that could result in compromising the quality of APIs. Any person shown at any time (either by medical examination or supervisory observation) to have an apparent illness or open lesions should be excluded from activities where the health condition could adversely affect the quality of the APIs until the condition is corrected or qualified medical personnel determine that the person's inclusion would not jeopardize the safety or quality of the APIs.

## C. Consultants (3.3)

Consultants advising on the manufacture and control of intermediates or APIs should have sufficient education, training, and experience, or any combination thereof, to advise on the subject for which they are retained.

Records should be maintained stating the name, address, qualifications, and type of service provided by these consultants.

# IV. BUILDINGS AND FACILITIES (4)

## A. Design and Construction (4.1)

Buildings and facilities used in the manufacture of intermediates and APIs should be located, designed, and constructed to facilitate cleaning, maintenance, and operations as appropriate to the type and stage of manufacture. Facilities should also be designed to minimize potential contamination. Where microbiological specifications have been established for the intermediate or API, facilities should also be designed to limit exposure to objectionable microbiological contaminants, as appropriate.

Buildings and facilities should have adequate space for the orderly placement of equipment and materials to prevent mix-ups and contamination.

Where the equipment itself (e.g., closed or contained systems) provides adequate protection of the material, such equipment can be located outdoors.

The flow of materials and personnel through the building or facilities should be designed to prevent mix-ups or contamination.

There should be defined areas or other control systems for the following activities:

- Receipt, identification, sampling, and quarantine of incoming materials, pending release or rejection
- Quarantine before release or rejection of intermediates and APIs
- Sampling of intermediates and APIs
- Holding rejected materials before further disposition (e.g., return, reprocessing or destruction)
- Storage of released materials
- Production operations
- Packaging and labeling operations
- Laboratory operations

Adequate and clean washing and toilet facilities should be provided for personnel. These facilities should be equipped with hot and cold water, as appropriate, soap or detergent, air dryers, or single service towels. The washing and toilet facilities should be separate from, but easily accessible to, manufacturing areas. Adequate facilities for showering and/or changing clothes should be provided, when appropriate.

Laboratory areas/operations should normally be separated from production areas. Some laboratory areas, in particular those used for in-process controls, can be located in production areas, provided the operations of the production process do not adversely affect the accuracy of the laboratory measurements, and the laboratory and its operations do not adversely affect the production process, intermediate, or API.

## B. Utilities (4.2)

All utilities that could affect product quality (e.g., steam, gas, compressed air, heating, ventilation, and air conditioning) should be qualified and appropriately monitored and action should be taken when limits are exceeded. Drawings for these utility systems should be available.

Adequate ventilation, air filtration and exhaust systems should be provided, where appropriate. These systems should be designed and constructed to minimize risks of contamination and cross-contamination and should include equipment for control of air pressure, microorganisms (if appropriate), dust, humidity, and temperature, as appropriate to the stage of manufacture. Particular attention should be given to areas where APIs are exposed to the environment.

If air is recirculated to production areas, appropriate measures should be taken to control risks of contamination and cross-contamination.

Permanently installed pipework should be appropriately identified. This can be accomplished by identifying individual lines, documentation, computer control systems, or alternative means. Pipework should be located to avoid risks of contamination of the intermediate or API.

Drains should be of adequate size and should be provided with an air break or a suitable device to prevent back-siphonage, when appropriate.

## C. WATER (4.3)

Water used in the manufacture of APIs should be demonstrated to be suitable for its intended use.

Unless otherwise justified, process water should, at a minimum, meet World Health Organization (WHO) guidelines for drinking (potable) water quality.

If drinking (potable) water is insufficient to ensure API quality and tighter chemical and/or microbiological water quality specifications are called for, appropriate specifications for physical/chemical attributes, total microbial counts, objectionable organisms, and/or endotoxins should be established.

Where water used in the process is treated by the manufacturer to achieve a defined quality, the treatment process should be validated and monitored with appropriate action limits.

Where the manufacturer of a nonsterile API either intends or claims that it is suitable for use in further processing to produce a sterile drug (medicinal) product, water used in the final isolation and purification steps should be monitored and controlled for total microbial counts, objectionable organisms, and endotoxins.

## D. CONTAINMENT (4.4)

Dedicated production areas, which can include facilities, air handling equipment and/or process equipment, should be employed in the production of highly sensitizing materials, such as penicillins or cephalosporins.

The use of dedicated production areas should also be considered when material of an infectious nature or high pharmacological activity or toxicity is involved (e.g., certain steroids or cytotoxic anticancer agents) unless validated inactivation and/or cleaning procedures are established and maintained.

Appropriate measures should be established and implemented to prevent cross-contamination from personnel and materials moving from one dedicated area to another.

Any production activities (including weighing, milling, or packaging) of highly toxic nonpharmaceutical materials, such as herbicides and pesticides, should not be conducted using the buildings and/or equipment being used for the production of APIs. Handling and storage of these highly toxic nonpharmaceutical materials should be separate from APIs.

## E. LIGHTING (4.5)

Adequate lighting should be provided in all areas to facilitate cleaning, maintenance, and proper operations.

## F. Sewage and Refuse (4.6)

Sewage, refuse, and other waste (e.g., solids, liquids, or gaseous by-products from manufacturing) in and from buildings and the immediate surrounding area should be disposed of in a safe, timely, and sanitary manner. Containers and/or pipes for waste material should be clearly identified.

## G. Sanitation and Maintenance (4.7)

Buildings used in the manufacture of intermediates and APIs should be properly maintained and repaired and kept in a clean condition.

Written procedures should be established assigning responsibility for sanitation and describing the cleaning schedules, methods, equipment, and materials to be used in cleaning buildings and facilities.

When necessary, written procedures should also be established for the use of suitable rodenticides, insecticides, fungicides, fumigating agents, and cleaning and sanitizing agents to prevent the contamination of equipment, raw materials, packaging/labeling materials, intermediates, and APIs.

# V. PROCESS EQUIPMENT (5)

## A. Design and Construction (5.1)

Equipment used in the manufacture of intermediates and APIs should be of appropriate design and adequate size, and suitably located for its intended use, cleaning, sanitation (where appropriate), and maintenance.

Equipment should be constructed so that surfaces that contact raw materials, intermediates, or APIs do not alter the quality of the intermediates and APIs beyond the official or other established specifications.

Production equipment should be used only within its qualified operating range.

Major equipment (e.g., reactors, storage containers) and permanently installed processing lines used during the production of an intermediate or API should be appropriately identified.

Any substances associated with the operation of equipment, such as lubricants, heating fluids or coolants, should not contact intermediates or APIs so as to alter the quality of APIs or intermediates beyond the official or other established specifications. Any deviations from this practice should be evaluated to ensure that there are no detrimental effects on the material's fitness for use. Wherever possible, food grade lubricants and oils should be used.

Closed or contained equipment should be used whenever appropriate. Where open equipment is used, or equipment is opened, appropriate precautions should be taken to minimize the risk of contamination.

A set of current drawings should be maintained for equipment and critical installations (e.g., instrumentation and utility systems).

## B. Equipment Maintenance and Cleaning (5.2)

Schedules and procedures (including assignment of responsibility) should be established for the preventative maintenance of equipment.

Written procedures should be established for cleaning equipment and its subsequent release for use in the manufacture of intermediates and APIs. Cleaning procedures should contain sufficient details to enable operators to clean each type of equipment in a reproducible and effective manner. These procedures should include:

- Assignment of responsibility for cleaning of equipment
- Cleaning schedules, including, where appropriate, sanitizing schedules
- A complete description of the methods and materials, including dilution of cleaning agents used to clean equipment
- When appropriate, instructions for disassembling and reassembling each article of equipment to ensure proper cleaning
- Instructions for the removal or obliteration of previous batch identification
- Instructions for the protection of clean equipment from contamination prior to use
- Inspection of equipment for cleanliness immediately before use, if practical
- Establishing the maximum time that may elapse between the completion of processing and equipment cleaning, when appropriate

Equipment and utensils should be cleaned, stored, and, where appropriate, sanitized or sterilized to prevent contamination or carry-over of a material that would alter the quality of the intermediate or API beyond the official or other established specifications.

Where equipment is assigned to continuous production or campaign production of successive batches of the same intermediate or API, equipment should be cleaned at appropriate intervals to prevent build-up and carry-over of contaminants (e.g., degradants or objectionable levels of micro-organisms).

Nondedicated equipment should be cleaned between production of different materials to prevent cross-contamination.

Acceptance criteria for residues and the choice of cleaning procedures and cleaning agents should be defined and justified.

Equipment should be identified as to its contents and its cleanliness status by appropriate means.

## C. Calibration (5.3)

Control, weighing, measuring, monitoring, and testing equipment critical for ensuring the quality of intermediates or APIs should be calibrated according to written procedures and an established schedule.

Equipment calibrations should be performed using standards traceable to certified standards, if they exist.

Records of these calibrations should be maintained.

The current calibration status of critical equipment should be known and verifiable.

Instruments that do not meet calibration criteria should not be used.

Deviations from approved standards of calibration on critical instruments should be investigated to determine if these could have had an effect on the quality of the intermediate(s) or API(s) manufactured using this equipment since the last successful calibration.

## D. Computerized Systems (5.4)

GMP-related computerized systems should be validated. The depth and scope of validation depends on the diversity, complexity, and criticality of the computerized application.

Appropriate installation and operational qualifications should demonstrate the suitability of computer hardware and software to perform assigned tasks.

Commercially available software that has been qualified does not require the same level of testing. If an existing system was not validated at time of installation, a retrospective validation could be conducted if appropriate documentation is available.

Computerized systems should have sufficient controls to prevent unauthorized access or changes to data. There should be controls to prevent omissions in data (e.g., system turned off and data not captured). There should be a record of any data change made, the previous entry, who made the change, and when the change was made.

Written procedures should be available for the operation and maintenance of computerized systems.

Where critical data are being entered manually, there should be an additional check on the accuracy of the entry. This can be done by a second operator or by the system itself.

Incidents related to computerized systems that could affect the quality of intermediates or APIs or the reliability of records or test results should be recorded and investigated.

Changes to computerized systems should be made according to a change procedure and should be formally authorized, documented, and tested. Records should be kept of all changes, including modifications and enhancements made to the hardware, software, and any other critical component of the system. These records should demonstrate that the system is maintained in a validated state.

If system breakdowns or failures would result in the permanent loss of records, a back-up system should be provided. A means of ensuring data protection should be established for all computerized systems.

Data can be recorded by a second means in addition to the computer system.

## VI. DOCUMENTATION AND RECORDS (6)

### A. DOCUMENTATION SYSTEM AND SPECIFICATIONS (6.1)

All documents related to the manufacture of intermediates or APIs should be prepared, reviewed, approved, and distributed according to written procedures. Such documents can be in paper or electronic form.

The issuance, revision, superseding, and withdrawal of all documents should be controlled by maintaining revision histories.

A procedure should be established for retaining all appropriate documents (e.g., development history reports, scale-up reports, technical transfer reports, process validation reports, training records, production records, control records, and distribution records). The retention periods for these documents should be specified.

All production, control, and distribution records should be retained for at least 1 year after the expiry date of the batch. For APIs with retest dates, records should be retained for at least 3 years after the batch is completely distributed.

When entries are made in records, these should be made indelibly in spaces provided for such entries, directly after performing the activities, and should identify the person making the entry. Corrections to entries should be dated and signed and leave the original entry still legible.

During the retention period, originals or copies of records should be readily available at the establishment where the activities described in such records occurred. Records that can be promptly retrieved from another location by electronic or other means are acceptable.

Specifications, instructions, procedures, and records can be retained either as originals or as true copies such as photocopies, microfilm, microfiche, or other accurate reproductions of the original records. Where reduction techniques such as microfilming or electronic records are used, suitable retrieval equipment and a means to produce a hard copy should be readily available.

Specifications should be established and documented for raw materials, intermediates where necessary, APIs, and labeling and packaging materials. In addition, specifications may be appropriate for certain other materials, such as process aids, gaskets, or other materials used during the production of intermediates or APIs that could critically affect quality. Acceptance criteria should be established and documented for in-process controls.

If electronic signatures are used on documents, they should be authenticated and secure.

## B. Equipment Cleaning and Use Record (6.2)

Records of major equipment use, cleaning, sanitation, and/or sterilization and maintenance should show the date, time (if appropriate), product, and batch number of each batch processed in the equipment and the person who performed the cleaning and maintenance.

If equipment is dedicated to manufacturing one intermediate or API, individual equipment records are not necessary if batches of the intermediate or API follow in traceable sequence. In cases where dedicated equipment is employed, the records of cleaning, maintenance, and use can be part of the batch record or maintained separately.

## C. Records of Raw Materials, Intermediates, API Labeling and Packaging Materials (6.3)

Records should be maintained including:

- The name of the manufacturer, identity, and quantity of each shipment of each batch of raw materials, intermediates, or labeling and packaging materials for API's; the name of the supplier; the supplier's control number(s), if known, or other identification number; the number allocated on receipt; and the date of receipt
- The results of any test or examination performed and the conclusions derived from this
- Records tracing the use of materials
- Documentation of the examination and review of API labeling and packaging materials for conformity with established specifications
- The final decision regarding rejected raw materials, intermediates, or API labeling and packaging materials

Master (approved) labels should be maintained for comparison to issued labels.

## D. Master Production Instructions (Master Production and Control Records) (6.4)

To ensure uniformity from batch to batch, master production instructions for each intermediate and API should be prepared, dated, and signed by one person and independently checked, dated, and signed by a person in the quality unit(s).

Master production instructions should include:

- The name of the intermediate or API being manufactured and an identifying document reference code, if applicable
- A complete list of raw materials and intermediates designated by names or codes sufficiently specific to identify any special quality characteristics

- An accurate statement of the quantity or ratio of each raw material or intermediate to be used, including the unit of measure. Where the quantity is not fixed, the calculation for each batch size or rate of production should be included. Variations to quantities should be included where they are justified
- The production location and major production equipment to be used
- Detailed production instructions, including the:
    sequences to be followed
    ranges of process parameters to be used
    sampling instructions and in-process controls with their acceptance criteria, where appropriate
    time limits for completion of individual processing steps and/or the total process, where appropriate
    expected yield ranges at appropriate phases of processing or time
- Where appropriate, special notations and precautions to be followed, or cross-references to these
- The instructions for storage of the intermediate or API to ensure its suitability for use, including the labelling and packaging materials and special storage conditions with time limits, where appropriate.

## E. BATCH PRODUCTION RECORDS (BATCH PRODUCTION AND CONTROL RECORDS) (6.5)

Batch production records should be prepared for each intermediate and API and should include complete information relating to the production and control of each batch. The batch production record should be checked before issuance to ensure that it is the correct version and a legible accurate reproduction of the appropriate master production instruction. If the batch production record is produced from a separate part of the master document, that document should include a reference to the current master production instruction being used.

These records should be numbered with a unique batch or identification number, dated and signed when issued. In continuous production, the product code together with the date and time can serve as the unique identifier until the final number is allocated.

Documentation of completion of each significant step in the batch production records (batch production and control records) should include:

- Dates and, when appropriate, times
- Identity of major equipment (e.g., reactors, driers, mills, etc.) used
- Specific identification of each batch, including weights, measures, and batch numbers of raw materials, intermediates, or any reprocessed materials used during manufacturing
- Actual results recorded for critical process parameters
- Any sampling performed
- Signatures of the persons performing and directly supervising or checking each critical step in the operation
- In-process and laboratory test results
- Actual yield at appropriate phases or times
- Description of packaging and label for intermediate or API
- Representative label of API or intermediate if made commercially available
- Any deviation noted, its evaluation, investigation conducted (if appropriate) or reference to that investigation if stored separately
- Results of release testing

Written procedures should be established and followed for investigating critical deviations or the failure of a batch of intermediate or API to meet specifications. The investigation should extend to other batches that may have been associated with the specific failure or deviation.

## F. Laboratory Control Records (6.6)

Laboratory control records should include complete data derived from all tests conducted to ensure compliance with established specifications and standards, including examinations and assays, as follows:

- A description of samples received for testing, including the material name or source, batch number or other distinctive code, date sample was taken, and, where appropriate, the quantity and date the sample was received for testing
- A statement of or reference to each test method used
- A statement of the weight or measure of sample used for each test as described by the method; data on or cross-reference to the preparation and testing of reference standards, reagents and standard solutions
- A complete record of all raw data generated during each test, in addition to graphs, charts and spectra from laboratory instrumentation, properly identified to show the specific material and batch tested
- A record of all calculations performed in connection with the test, including, for example, units of measure, conversion factors, and equivalency factors
- A statement of the test results and how they compare with established acceptance criteria
- The signature of the person who performed each test and the date(s) the tests were performed
- The date and signature of a second person showing that the original records have been reviewed for accuracy, completeness, and compliance with established standards

Complete records should also be maintained for:

- Any modifications to an established analytical method
- Periodic calibration of laboratory instruments, apparatus, gauges, and recording devices
- All stability testing performed on APIs
- Out-of-specification (OOS) investigations

## G. Batch Production Record Review (6.7)

Written procedures should be established and followed for the review and approval of batch production and laboratory control records, including packaging and labeling, to determine compliance of the intermediate or API with established specifications before a batch is released or distributed.

Batch production and laboratory control records of critical process steps should be reviewed and approved by the quality unit(s) before an API batch is released or distributed. Production and laboratory control records of noncritical process steps can be reviewed by qualified production personnel or other units following procedures approved by the quality unit(s).

All deviation, investigation, and OOS reports should be reviewed as part of the batch record review before the batch is released.

The quality unit(s) can delegate to the production unit the responsibility and authority for release of intermediates, except for those shipped outside the control of the manufacturing company.

# VII. MATERIALS MANAGEMENT (7)

## A. GENERAL CONTROLS (7.1)

There should be written procedures describing the receipt, identification, quarantine, storage, handling, sampling, testing, and approval or rejection of materials.

Manufacturers of intermediates and/or APIs should have a system for evaluating the suppliers of critical materials.

Materials should be purchased against an agreed specification, from a supplier, or suppliers, approved by the quality unit(s).

If the supplier of a critical material is not the manufacturer of that material, the name and address of that manufacturer should be known by the intermediate and/or API manufacturer.

Changing the source of supply of critical raw materials should be treated according to Section 13, Change Control.

## B. RECEIPT AND QUARANTINE (7.2)

Upon receipt and before acceptance, each container or grouping of containers of materials should be examined visually for correct labeling (including correlation between the name used by the supplier and the in-house name, if these are different), container damage, broken seals and evidence of tampering or contamination. Materials should be held under quarantine until they have been sampled, examined, or tested, as appropriate, and released for use.

Before incoming materials are mixed with existing stocks (e.g., solvents or stocks in silos), they should be identified as correct, tested, if appropriate, and released. Procedures should be available to prevent discharging incoming materials wrongly into the existing stock.

If bulk deliveries are made in nondedicated tankers, there should be assurance of no cross-contamination from the tanker. Means of providing this assurance could include one or more of the following:

- certificate of cleaning
- testing for trace impurities
- audit of the supplier

Large storage containers and their attendant manifolds, filling, and discharge lines should be appropriately identified.

Each container or grouping of containers (batches) of materials should be assigned and identified with a distinctive code, batch, or receipt number. This number should be used in recording the disposition of each batch. A system should be in place to identify the status of each batch.

## C. SAMPLING AND TESTING OF INCOMING PRODUCTION MATERIALS (7.3)

At least one test to verify the identity of each batch of material should be conducted, with the exception of the materials described below. A supplier's certificate of analysis can be used in place of performing other tests, provided that the manufacturer has a system in place to evaluate suppliers.

Supplier approval should include an evaluation that provides adequate evidence (e.g., past quality history) that the manufacturer can consistently provide material meeting specifications. Complete analyses should be conducted on at least three batches before reducing in-house testing. However, as a minimum, a complete analysis should be performed at appropriate intervals and compared with the certificates of analysis. Reliability of certificates of analysis should be checked at regular intervals.

Processing aids, hazardous or highly toxic raw materials, other special materials, or materials transferred to another unit within the company's control do not need to be tested if the manufacturer's certificate of analysis is obtained, showing that these raw materials conform to established specifications. Visual examination of containers, labels, and recording of batch numbers should help in establishing the identity of these materials. The lack of on-site testing for these materials should be justified and documented.

Samples should be representative of the batch of material from which they are taken. Sampling methods should specify the number of containers to be sampled, which part of the container to sample, and the amount of material to be taken from each container. The number of containers to sample and the sample size should be based on a sampling plan that takes into consideration the criticality of the material, material variability, past quality history of the supplier, and the quantity needed for analysis.

Sampling should be conducted at defined locations and by procedures designed to prevent contamination of the material sampled and contamination of other materials.

Containers from which samples are withdrawn should be opened carefully and subsequently reclosed. They should be marked to indicate that a sample has been taken.

## D. Storage (7.4)

Materials should be handled and stored in a manner to prevent degradation, contamination, and cross-contamination.

Materials stored in fiber drums, bags, or boxes should be stored off the floor and, when appropriate, suitably spaced to permit cleaning and inspection.

Materials should be stored under conditions and for a period that have no adverse effect on their quality, and should normally be controlled so that the oldest stock is used first.

Certain materials in suitable containers can be stored outdoors, provided identifying labels remain legible and containers are appropriately cleaned before opening and use.

Rejected materials should be identified and controlled under a quarantine system designed to prevent their unauthorized use in manufacturing.

## E. Re-evaluation (7.5)

Materials should be re-evaluated, as appropriate, to determine their suitability for use (e.g., after prolonged storage or exposure to heat or humidity).

## VIII. PRODUCTION AND IN-PROCESS CONTROLS (8)

### A. Production Operations (8.1)

Raw materials for intermediate and API manufacturing should be weighed or measured under appropriate conditions that do not affect their suitability for use. Weighing and measuring devices should be of suitable accuracy for the intended use.

If a material is subdivided for later use in production operations, the container receiving the material should be suitable and should be so identified that the following information is available:

- Material name and/or item code
- Receiving or control number
- Weight or measure of material in the new container
- Re-evaluation or retest date if appropriate

Critical weighing, measuring, or subdividing operations should be witnessed or subjected to an equivalent control. Prior to use, production personnel should verify that the materials are those specified in the batch record for the intended intermediate or API.

Other critical activities should be witnessed or subjected to an equivalent control.

Actual yields should be compared with expected yields at designated steps in the production process. Expected yields with appropriate ranges should be established based on previous laboratory, pilot scale, or manufacturing data. Deviations in yield associated with critical process steps should be investigated to determine their impact or potential impact on the resulting quality of affected batches.

Any deviation should be documented and explained. Any critical deviation should be investigated.

The processing status of major units of equipment should be indicated either on the individual units of equipment or by appropriate documentation, computer control systems, or alternative means.

Materials to be reprocessed or reworked should be appropriately controlled to prevent unauthorized use.

## B. Time Limits (8.2)

If time limits are specified in the master production instruction (see 6.40), these time limits should be met to ensure the quality of intermediates and APIs. Deviations should be documented and evaluated. Time limits may be inappropriate when processing to a target value (e.g., pH adjustment, hydrogenation, drying to predetermined specification) because completion of reactions or processing steps are determined by in-process sampling and testing.

Intermediates held for further processing should be stored under appropriate conditions to ensure their suitability for use.

## C. In-Process Sampling and Controls (8.3)

Written procedures should be established to monitor the progress and control the performance of processing steps that cause variability in the quality characteristics of intermediates and APIs. In-process controls and their acceptance criteria should be defined based on the information gained during the developmental stage or from historical data.

The acceptance criteria and type and extent of testing can depend on the nature of the intermediate or API being manufactured, the reaction or process step being conducted, and the degree to which the process introduces variability in the product's quality. Less stringent in-process controls may be appropriate in early processing steps, whereas tighter controls may be appropriate for later processing steps (e.g., isolation and purification steps).

Critical in-process controls (and critical process monitoring), including control points and methods, should be stated in writing and approved by the quality unit(s).

In-process controls can be performed by qualified production department personnel and the process adjusted without prior quality unit(s) approval if the adjustments are made within pre-established limits approved by the quality unit(s). All tests and results should be fully documented as part of the batch record.

Written procedures should describe the sampling methods for in-process materials, intermediates, and APIs. Sampling plans and procedures should be based on scientifically sound sampling practices.

In-process sampling should be conducted using procedures designed to prevent contamination of the sampled material and other intermediates or APIs. Procedures should be established to ensure the integrity of samples after collection.

Out-of-specification (OOS) investigations are not normally needed for in-process tests that are performed for the purpose of monitoring and/or adjusting the process.

### D. BLENDING BATCHES OF INTERMEDIATES OR APIs (8.4)

For the purpose of this document, blending is defined as the process of combining materials within the same specification to produce a homogeneous intermediate or API. In-process mixing of fractions from single batches (e.g., collecting several centrifuge loads from a single crystallization batch) or combining fractions from several batches for further processing is considered to be part of the production process and is not considered to be blending.

Out-of-specification batches should not be blended with other batches for the purpose of meeting specifications. Each batch incorporated into the blend should have been manufactured using an established process and should have been individually tested and found to meet appropriate specifications prior to blending.

Acceptable blending operations include, but are not limited to:

- Blending of small batches to increase batch size
- Blending of tailings (i.e., relatively small quantities of isolated material) from batches of the same intermediate or API to form a single batch

Blending processes should be adequately controlled and documented, and the blended batch should be tested for conformance to established specifications, where appropriate.

The batch record of the blending process should allow traceability back to the individual batches that make up the blend.

Where physical attributes of the API are critical (e.g., APIs intended for use in solid oral dosage forms or suspensions), blending operations should be validated to show homogeneity of the combined batch. Validation should include testing of critical attributes (e.g., particle size distribution, bulk density, and tap density) that may be affected by the blending process.

If the blending could adversely affect stability, stability testing of the final blended batches should be performed.

The expiry or retest date of the blended batch should be based on the manufacturing date of the oldest tailings or batch in the blend.

### E. CONTAMINATION CONTROL (8.5)

Residual materials can be carried over into successive batches of the same intermediate or API if there is adequate control. Examples include residue adhering to the wall of a micronizer, residual layer of damp crystals remaining in a centrifuge bowl after discharge, and incomplete discharge of fluids or crystals from a processing vessel upon transfer of the material to the next step in the process. Such carryover should not result in the carryover of degradants or microbial contamination that may adversely alter the established API impurity profile.

Production operations should be conducted in a manner that prevents contamination of intermediates or APIs by other materials.

Precautions to avoid contamination should be taken when APIs are handled after purification.

# IX. PACKAGING AND IDENTIFICATION LABELING OF APIS AND INTERMEDIATES (9)

## A. GENERAL (9.1)

There should be written procedures describing the receipt, identification, quarantine, sampling, examination, and/or testing, release, and handling of packaging and labeling materials.

Packaging and labeling materials should conform to established specifications. Those that do not comply with such specifications should be rejected to prevent their use in operations for which they are unsuitable.

Records should be maintained for each shipment of labels and packaging materials showing receipt, examination, or testing, and whether accepted or rejected.

## B. PACKAGING MATERIALS (9.2)

Containers should provide adequate protection against deterioration or contamination of the intermediate or API that may occur during transportation and recommended storage.

Containers should be clean and, where indicated by the nature of the intermediate or API, sanitized to ensure that they are suitable for their intended use. These containers should not be reactive, additive, or absorptive so as to alter the quality of the intermediate or API beyond the specified limits.

If containers are reused, they should be cleaned in accordance with documented procedures, and all previous labels should be removed or defaced

## C. LABEL ISSUANCE AND CONTROL (9.3)

Access to the label storage areas should be limited to authorized personnel.

Procedures should be established to reconcile the quantities of labels issued, used, and returned and to evaluate discrepancies found between the number of containers labeled and the number of labels issued. Such discrepancies should be investigated, and the investigation should be approved by the quality unit(s).

All excess labels bearing batch numbers or other batch-related printing should be destroyed. Returned labels should be maintained and stored in a manner that prevents mix-ups and provides proper identification.

Obsolete and out-dated labels should be destroyed.

Printing devices used to print labels for packaging operations should be controlled to ensure that all imprinting conforms to the print specified in the batch production record.

Printed labels issued for a batch should be carefully examined for proper identity and conformity to specifications in the master production record. The results of this examination should be documented.

A printed label representative of those used should be included in the batch production record.

## D. PACKAGING AND LABELING OPERATIONS (9.4)

There should be documented procedures designed to ensure that correct packaging materials and labels are used.

Labeling operations should be designed to prevent mix-ups. There should be physical or spatial separation from operations involving other intermediates or APIs.

Labels used on containers of intermediates or APIs should indicate the name or identifying code, batch number, and storage conditions when such information is critical to ensure the quality of intermediate or API.

If the intermediate or API is intended to be transferred outside the control of the manufacturer's material management system, the name and address of the manufacturer, quantity of contents, special transport conditions, and any special legal requirements should also be included on the label. For intermediates or APIs with an expiry date, the expiry date should be indicated on the label and certificate of analysis. For intermediates or APIs with a retest date, the retest date should be indicated on the label and/or certificate of analysis.

Packaging and labeling facilities should be inspected immediately before use to ensure that all materials not needed for the next packaging operation have been removed. This examination should be documented in the batch production records, the facility log, or other documentation system.

Packaged and labeled intermediates or APIs should be examined to ensure that containers and packages in the batch have the correct label. This examination should be part of the packaging operation. Results of these examinations should be recorded in the batch production or control records.

Intermediate or API containers that are transported outside of the manufacturer's control should be sealed in a manner such that, if the seal is breached or missing, the recipient will be alerted to the possibility that the contents may have been altered.

# X. STORAGE AND DISTRIBUTION (10)

## A. WAREHOUSING PROCEDURES (10.1)

Facilities should be available for the storage of all materials under appropriate conditions (e.g., controlled temperature and humidity when necessary). Records should be maintained of these conditions if they are critical for the maintenance of material characteristics.

Unless there is an alternative system to prevent the unintentional or unauthorized use of quarantined, rejected, returned, or recalled materials, separate storage areas should be assigned for their temporary storage until the decision as to their future use has been made.

## B. DISTRIBUTION PROCEDURES (10.2)

APIs and intermediates should be released only for distribution to third parties after they have been released by the quality unit(s). APIs and intermediates can be transferred under quarantine to another unit under the company's control when authorized by the quality unit(s) and if appropriate controls and documentation are in place.

APIs and intermediates should be transported in a manner that does not adversely affect their quality.

Special transport or storage conditions for an API or intermediate should be stated on the label.

The manufacturer should ensure that the contract acceptor (contractor) for transportation of the API or intermediate knows and follows the appropriate transport and storage conditions.

A system should be in place by which the distribution of each batch of intermediate and/or API can be readily determined to permit its recall.

# XI. LABORATORY CONTROLS (11)

## A. GENERAL CONTROLS (11.1)

The independent quality unit(s) should have at its disposal adequate laboratory facilities.

There should be documented procedures describing sampling, testing, approval, or rejection of materials and recording and storage of laboratory data. Laboratory records should be maintained in accordance with Section 6.6.

All specifications, sampling plans, and test procedures should be scientifically sound and appropriate to ensure that raw materials, intermediates, APIs, and labels and packaging materials conform to established standards of quality and/or purity. Specifications and test procedures should be consistent with those included in the registration/filing. There can be specifications in addition to those in the registration/filing. Specifications, sampling plans, and test procedures, including changes to them, should be drafted by the appropriate organizational unit and reviewed and approved by the quality unit(s).

Appropriate specifications should be established for APIs in accordance with accepted standards and consistent with the manufacturing process. The specifications should include control of impurities (e.g., organic impurities, inorganic impurities, and residual solvents). If the API has a specification for microbiological purity, appropriate action limits for total microbial counts and objectionable organisms should be established and met. If the API has a specification for endotoxins, appropriate action limits should be established and met.

Laboratory controls should be followed and documented at the time of performance. Any departures from the above-described procedures should be documented and explained.

Any out-of-specification result obtained should be investigated and documented according to a procedure. This procedure should include analysis of the data, assessment of whether a significant problem exists, allocation of the tasks for corrective actions, and conclusions. Any resampling and/or retesting after OOS results should be performed according to a documented procedure.

Reagents and standard solutions should be prepared and labeled following written procedures. Use by dates should be applied, as appropriate, for analytical reagents or standard solutions.

Primary reference standards should be obtained, as appropriate, for the manufacture of APIs. The source of each primary reference standard should be documented. Records should be maintained of each primary reference standard's storage and use in accordance with the supplier's recommendations. Primary reference standards obtained from an officially recognized source are normally used without testing if stored under conditions consistent with the supplier's recommendations.

Where a primary reference standard is not available from an officially recognized source, an *in-house primary standard* should be established. Appropriate testing should be performed to establish fully the identity and purity of the primary reference standard. Appropriate documentation of this testing should be maintained.

Secondary reference standards should be appropriately prepared, identified, tested, approved, and stored. The suitability of each batch of secondary reference standard should be determined prior to first use by comparing against a primary reference standard. Each batch of secondary reference standard should be periodically requalified in accordance with a written protocol.

## B. TESTING OF INTERMEDIATES AND APIs (11.2)

For each batch of intermediate and API, appropriate laboratory tests should be conducted to determine conformance to specifications.

An impurity profile describing the identified and unidentified impurities present in a typical batch produced by a specific controlled production process should normally be established for each API. The impurity profile should include the identity or some qualitative analytical designation (e.g., retention time), the range of each impurity observed, and classification of each identified impurity (e.g., inorganic, organic, solvent). The impurity profile is normally dependent upon the production process and origin of the API. Impurity profiles are normally not necessary for APIs from herbal or animal tissue origin. Biotechnology considerations are covered in ICH guidance Q6B.

The impurity profile should be compared at appropriate intervals against the impurity profile in the regulatory submission or compared against historical data to detect changes to the API resulting from modifications in raw materials, equipment operating parameters, or the production process.

Appropriate microbiological tests should be conducted on each batch of intermediate and API where microbial quality is specified.

## C. Validation of Analytical Procedures

See Section 12. (11.3)

## D. Certificates of Analysis (11.4)

Authentic certificates of analysis should be issued for each batch of intermediate or API on request.

Information on the name of the intermediate or API including, where appropriate, its grade, the batch number, and the date of release should be provided on the certificate of analysis. For intermediates or APIs with an expiry date, the expiry date should be provided on the label and certificate of analysis. For intermediates or APIs with a retest date, the retest date should be indicated on the label and/or certificate of analysis.

The certificate should list each test performed in accordance with compendial or customer requirements, including the acceptance limits, and the numerical results obtained (if test results are numerical).

Certificates should be dated and signed by authorized personnel of the quality unit(s) and should show the name, address, and telephone number of the original manufacturer. Where the analysis has been carried out by a repacker or reprocessor, the certificate of analysis should show the name, address, and telephone number of the repacker/reprocessor and reference the name of the original manufacturer.

If new certificates are issued by or on behalf of repackers/reprocessors, agents or brokers, these certificates should show the name, address and telephone number of the laboratory that performed the analysis. They should also contain a reference to the name and address of the original manufacturer and to the original batch certificate, a copy of which should be attached.

## E. Stability Monitoring of APIs (11.5)

A documented, on-going testing program should be established to monitor the stability characteristics of APIs, and the results should be used to confirm appropriate storage conditions and retest or expiry dates.

The test procedures used in stability testing should be validated and be stability indicating.

Stability samples should be stored in containers that simulate the market container. For example, if the API is marketed in bags within fiber drums, stability samples can be packaged in bags of the same material and in small-scale drums of similar or identical material composition to the market drums.

Normally, the first three commercial production batches should be placed on the stability monitoring program to confirm the retest or expiry date. However, where data from previous studies show that the API is expected to remain stable for at least 2 years, fewer than three batches can be used.

Thereafter, at least one batch per year of API manufactured (unless none is produced that year) should be added to the stability monitoring program and tested at least annually to confirm the stability.

For APIs with short shelf-lives, testing should be done more frequently. For example, for those biotechnological/biologic and other APIs with shelf-lives of one year or less, stability samples should be obtained and should be tested monthly for the first 3 months, and at 3-month intervals after that. When data exist that confirm that the stability of the API is not compromised, elimination of specific test intervals (e.g., 9-month testing) can be considered.

Where appropriate, the stability storage conditions should be consistent with the ICH guidances on stability.

## F. Expiry and Retest Dating (11.6)

When an intermediate is intended to be transferred outside the control of the manufacturer's material management system and an expiry or retest date is assigned, supporting stability information should be available (e.g., published data, test results).

An API expiry or retest date should be based on an evaluation of data derived from stability studies. Common practice is to use a retest date, not an expiration date.

Preliminary API expiry or retest dates can be based on pilot scale batches if (1) the pilot batches employ a method of manufacture and procedure that simulates the final process to be used on a commercial manufacturing scale and (2) the quality of the API represents the material to be made on a commercial scale.

A representative sample should be taken for the purpose of performing a retest.

## G. Reserve/Retention Samples (11.7)

The packaging and holding of reserve samples is for the purpose of potential future evaluation of the quality of batches of API and not for future stability testing purposes.

Appropriately identified reserve samples of each API batch should be retained for 1 year after the expiry date of the batch assigned by the manufacturer, or for 3 years after distribution of the batch, whichever is longer. For APIs with retest dates, similar reserve samples should be retained for 3 years after the batch is completely distributed by the manufacturer.

The reserve sample should be stored in the same packaging system in which the API is stored or in one that is equivalent to or more protective than the marketed packaging system. Sufficient quantities should be retained to conduct at least two full compendial analyses or, when there is no pharmacopoeial monograph, two full specification analyses.

# XII. VALIDATION (12)

## A. Validation Policy (12.1)

The company's overall policy, intentions, and approach to validation, including the validation of production processes, cleaning procedures, analytical methods, in-process control test procedures, computerized systems, and persons responsible for design, review, approval, and documentation of each validation phase, should be documented.

The critical parameters/attributes should normally be identified during the development stage or from historical data, and the necessary ranges for the reproducible operation should be defined. This should include:

- Defining the API in terms of its critical product attributes
- Identifying process parameters that could affect the critical quality attributes of the API
- Determining the range for each critical process parameter expected to be used during routine manufacturing and process control

Validation should extend to those operations determined to be critical to the quality and purity of the API.

## B. VALIDATION DOCUMENTATION (12.2)

A written validation protocol should be established that specifies how validation of a particular process will be conducted. The protocol should be reviewed and approved by the quality unit(s) and other designated units.

The validation protocol should specify critical process steps and acceptance criteria as well as the type of validation to be conducted (e.g., retrospective, prospective, concurrent) and the number of process runs.

A validation report that cross-references the validation protocol should be prepared, summarizing the results obtained, commenting on any deviations observed, and drawing the appropriate conclusions, including recommending changes to correct deficiencies.

Any variations from the validation protocol should be documented with appropriate justification.

## C. QUALIFICATION (12.3)

Before initiating process validation activities, appropriate qualification of critical equipment and ancillary systems should be completed. Qualification is usually carried out by conducting the following activities, individually or combined:

- Design Qualification (DQ): documented verification that the proposed design of the facilities, equipment, or systems is suitable for the intended purpose
- Installation Qualification (IQ): documented verification that the equipment or systems, as installed or modified, comply with the approved design, the manufacturer's recommendations and/or user requirements
- Operational Qualification (OQ): documented verification that the equipment or systems, as installed or modified, perform as intended throughout the anticipated operating ranges
- Performance Qualification (PQ): documented verification that the equipment and ancillary systems, as connected together, can perform effectively and reproducibly based on the approved process method and specifications

## D. APPROACHES TO PROCESS VALIDATION (12.4)

Process Validation (PV) is the documented evidence that the process, operated within established parameters, can perform effectively and reproducibly to produce an intermediate or API meeting its predetermined specifications and quality attributes.

There are three approaches to validation. Prospective validation is the preferred approach, but there are situations where the other approaches can be used. These approaches and their applicability are discussed here.

Prospective validation should normally be performed for all API processes as defined in 12.1. Prospective validation of an API process should be completed before the commercial distribution of the final drug product manufactured from that API.

Concurrent validation can be conducted when data from replicate production runs are unavailable because only a limited number of API batches have been produced, API batches are produced infrequently, or API batches are produced by a validated process that has been modified. Prior to the completion of concurrent validation, batches can be released and used in final drug product for commercial distribution based on thorough monitoring and testing of the API batches.

An exception can be made for retrospective validation of well-established processes that have been used without significant changes to API quality due to changes in raw materials, equipment, systems, facilities, or the production process. This validation approach may be used where:

1. Critical quality attributes and critical process parameters have been identified
2. Appropriate in-process acceptance criteria and controls have been established
3. There have not been significant process/product failures attributable to causes other than operator error or equipment failures unrelated to equipment suitability
4. Impurity profiles have been established for the existing API

Batches selected for retrospective validation should be representative of all batches produced during the review period, including any batches that failed to meet specifications, and should be sufficient in number to demonstrate process consistency. Retained samples can be tested to obtain data to retrospectively validate the process.

## E. PROCESS VALIDATION PROGRAM (12.5)

The number of process runs for validation should depend on the complexity of the process or the magnitude of the process change being considered. For prospective and concurrent validation, three consecutive successful production batches should be used as a guide, but there may be situations where additional process runs are warranted to prove consistency of the process (e.g., complex API processes or API processes with prolonged completion times). For retrospective validation, generally data from 10 to 30 consecutive batches should be examined to assess process consistency, but fewer batches can be examined if justified.

Critical process parameters should be controlled and monitored during process validation studies. Process parameters unrelated to quality, such as variables controlled to minimize energy consumption or equipment use, need not be included in the process validation.

Process validation should confirm that the impurity profile for each API is within the limits specified. The impurity profile should be comparable to, or better than, historical data and, where applicable, the profile determined during process development or for batches used for pivotal clinical and toxicological studies.

## F. PERIODIC REVIEW OF VALIDATED SYSTEMS (12.6)

Systems and processes should be periodically evaluated to verify that they are still operating in a valid manner. Where no significant changes have been made to the system or process, and a quality review confirms that the system or process is consistently producing material meeting its specifications, there is normally no need for revalidation.

## G. CLEANING VALIDATION (12.7)

Cleaning procedures should normally be validated. In general, cleaning validation should be directed to situations or process steps where contamination or carryover of materials poses the greatest risk to API quality. For example, in early production it may be unnecessary to validate equipment cleaning procedures where residues are removed by subsequent purification steps.

Validation of cleaning procedures should reflect actual equipment usage patterns. If various APIs or intermediates are manufactured in the same equipment and the equipment is cleaned by the same process, a representative intermediate or API can be selected for cleaning validation. This selection should be based on the solubility and difficulty of cleaning and the calculation of residue limits based on potency, toxicity, and stability.

The cleaning validation protocol should describe the equipment to be cleaned, procedures, materials, acceptable cleaning levels, parameters to be monitored and controlled, and analytical methods. The protocol should also indicate the type of samples to be obtained and how they are collected and labeled.

Sampling should include swabbing, rinsing, or alternative methods (e.g., direct extraction), as appropriate, to detect both insoluble and soluble residues. The sampling methods used should be capable of quantitatively measuring levels of residues remaining on the equipment surfaces after cleaning. Swab sampling may be impractical when product contact surfaces are not easily accessible due to equipment design and/or process limitations (e.g., inner surfaces of hoses, transfer pipes, reactor tanks with small ports or handling toxic materials, and small intricate equipment such as micronizers and microfluidizers).

Validated analytical methods having sensitivity to detect residues or contaminants should be used. The detection limit for each analytical method should be sufficiently sensitive to detect the established acceptable level of the residue or contaminant. The method's attainable recovery level should be established. Residue limits should be practical, achievable, verifiable, and based on the most deleterious residue. Limits can be established based on the minimum known pharmacological, toxicological, or physiological activity of the API or its most deleterious component.

Equipment cleaning/sanitation studies should address microbiological and endotoxin contamination for those processes where there is a need to reduce total microbiological count or endotoxins in the API, or other processes where such contamination could be of concern (e.g., non-sterile APIs used to manufacture sterile products).

Cleaning procedures should be monitored at appropriate intervals after validation to ensure that these procedures are effective when used during routine production. Equipment cleanliness can be monitored by analytical testing and visual examination, where feasible. Visual inspection can allow detection of gross contamination concentrated in small areas that could otherwise go undetected by sampling and/or analysis.

## H. VALIDATION OF ANALYTICAL METHODS (12.8)

Analytical methods should be validated unless the method employed is included in the relevant pharmacopoeia or other recognized standard reference. The suitability of all testing methods used should nonetheless be verified under actual conditions of use and documented.

Methods should be validated to include consideration of characteristics included within the ICH guidances on validation of analytical methods. The degree of analytical validation performed should reflect the purpose of the analysis and the stage of the API production process.

Appropriate qualification of analytical equipment should be considered before initiating validation of analytical methods.

Complete records should be maintained of any modification of a validated analytical method. Such records should include the reason for the modification and appropriate data to verify that the modification produces results that are as accurate and reliable as the established method.

## XIII. CHANGE CONTROL (13)

A formal change control system should be established to evaluate all changes that could affect the production and control of the intermediate or API.

Written procedures should provide for the identification, documentation, appropriate review, and approval of changes in raw materials, specifications, analytical methods, facilities, support systems, equipment (including computer hardware), processing steps, labeling and packaging materials, and computer software.

Any proposals for GMP relevant changes should be drafted, reviewed, and approved by the appropriate organizational units and reviewed and approved by the quality unit(s).

The potential impact of the proposed change on the quality of the intermediate or API should be evaluated. A classification procedure may help in determining the level of testing, validation, and documentation needed to justify changes to a validated process. Changes can be classified (e.g., as minor or major) depending on the nature and extent of the changes, and the effects these changes may impart on the process. Scientific judgment should determine what additional testing and validation studies are appropriate to justify a change in a validated process.

When implementing approved changes, measures should be taken to ensure that all documents affected by the changes are revised.

After the change has been implemented, there should be an evaluation of the first batches produced or tested under the change.

The potential for critical changes to affect established retest or expiry dates should be evaluated. If necessary, samples of the intermediate or API produced by the modified process can be placed on an accelerated stability program and/or can be added to the stability monitoring program.

Current dosage form manufacturers should be notified of changes from established production and process control procedures that can affect the quality of the API.

## XIV. REJECTION AND RE-USE OF MATERIALS (14)

### A. REJECTION (14.1)

Intermediates and APIs failing to meet established specifications should be identified as such and quarantined. These intermediates or APIs can be reprocessed or reworked as described below. The final disposition of rejected materials should be recorded.

### B. REPROCESSING (14.2)

Introducing an intermediate or API, including one that does not conform to standards or specifications, back into the process and reprocessing by repeating a crystallization step or other appropriate chemical or physical manipulation steps (e.g., distillation, filtration, chromatography, milling) that are part of the established manufacturing process is generally considered acceptable. However, if such reprocessing is used for a majority of batches, such reprocessing should be included as part of the standard manufacturing process.

Continuation of a process step after an in-process control test has shown that the step is incomplete is considered to be part of the normal process. This is not considered to be reprocessing.

Introducing unreacted material back into a process and repeating a chemical reaction is considered to be reprocessing unless it is part of the established process. Such reprocessing should be preceded by careful evaluation to ensure that the quality of the intermediate or API is not adversely affected due to the potential formation of by-products and over-reacted materials.

## C. Reworking (14.3)

Before a decision is taken to rework batches that do not conform to established standards or specifications, an investigation into the reason for nonconformance should be performed.

Batches that have been reworked should be subjected to appropriate evaluation, testing, stability testing if warranted, and documentation to show that the reworked product is of equivalent quality to that produced by the original process. Concurrent validation is often the appropriate validation approach for rework procedures. This allows a protocol to define the rework procedure, how it will be carried out, and the expected results. If there is only one batch to be reworked, a report can be written and the batch released once it is found to be acceptable.

Procedures should provide for comparing the impurity profile of each reworked batch against batches manufactured by the established process. Where routine analytical methods are inadequate to characterize the reworked batch, additional methods should be used.

## D. Recovery of Materials and Solvents (14.4)

Recovery (e.g., from mother liquor or filtrates) of reactants, intermediates, or the API is considered acceptable, provided that approved procedures exist for the recovery and the recovered materials meet specifications suitable for their intended use.

Solvents can be recovered and reused in the same processes or in different processes, provided that the recovery procedures are controlled and monitored to ensure that solvents meet appropriate standards before reuse or commingling with other approved materials.

Fresh and recovered solvents and reagents can be combined if adequate testing has shown their suitability for all manufacturing processes in which they may be used.

The use of recovered solvents, mother liquors, and other recovered materials should be adequately documented.

## E. Returns (14.5)

Returned intermediates or APIs should be identified as such and quarantined.

If the conditions under which returned intermediates or APIs have been stored or shipped before or during their return or the condition of their containers casts doubt on their quality, the returned intermediates or APIs should be reprocessed, reworked, or destroyed, as appropriate.

Records of returned intermediates or APIs should be maintained. For each return, documentation should include:

- Name and address of the consignee
- Intermediate or API, batch number, and quantity returned
- Reason for return
- Use or disposal of the returned intermediate or API

## XV. COMPLAINTS AND RECALLS (15)

All quality-related complaints, whether received orally or in writing, should be recorded and investigated according to a written procedure.

Complaint records should include:

- Name and address of complainant
- Name (and, where appropriate, title) and phone number of person submitting the complaint
- Complaint nature (including name and batch number of the API)
- Date complaint is received
- Action initially taken (including dates and identity of person taking the action)
- Any follow-up action taken
- Response provided to the originator of complaint (including date response sent)
- Final decision on intermediate or API batch or lot

Records of complaints should be retained to evaluate trends, product-related frequencies, and severity with a view to taking additional, and if appropriate, immediate corrective action.

There should be a written procedure that defines the circumstances under which a recall of an intermediate or API should be considered.

The recall procedure should designate who should be involved in evaluating the information, how a recall should be initiated, who should be informed about the recall, and how the recalled material should be treated.

In the event of a serious or potentially life-threatening situation, local, national, and/or international authorities should be informed and their advice sought.

## XVI. CONTRACT MANUFACTURERS (INCLUDING LABORATORIES) (16)

All contract manufacturers (including laboratories) should comply with the GMP defined in this guidance. Special consideration should be given to the prevention of cross-contamination and to maintaining traceability.

Companies should evaluate any contractors (including laboratories) to ensure GMP compliance of the specific operations occurring at the contractor sites.

There should be a written and approved contract or formal agreement between a company and its contractors that defines in detail the GMP responsibilities, including the quality measures, of each party.

A contract should permit a company to audit its contractor's facilities for compliance with GMP.

Where subcontracting is allowed, a contractor should not pass to a third party any of the work entrusted to it under the contract without the company's prior evaluation and approval of the arrangements.

Manufacturing and laboratory records should be kept at the site where the activity occurs and be readily available.

Changes in the process, equipment, test methods, specifications, or other contractual requirements should not be made unless the contract giver is informed and approves the changes.

## XVII. AGENTS, BROKERS, TRADERS, DISTRIBUTORS, REPACKERS, AND RELABELLERS (17)

### A. APPLICABILITY (17.1)

This section applies to any party other than the original manufacturer who may trade and/or take possession, repack, relabel, manipulate, distribute, or store an API or intermediate.

All agents, brokers, traders, distributors, repackers, and relabelers should comply with GMP as defined in this guidance.

### B. TRACEABILITY OF DISTRIBUTED APIS AND INTERMEDIATES (17.2)

Agents, brokers, traders, distributors, repackers, or relabelers should maintain complete traceability of APIs and intermediates that they distribute. Documents that should be retained and available include:

- Identity of original manufacturer
- Address of original manufacturer
- Purchase orders
- Bills of lading (transportation documentation)
- Receipt documents
- Name or designation of API or intermediate
- Manufacturer's batch number
- Transportation and distribution records
- All authentic Certificates of Analysis, including those of the original manufacturer
- Retest or expiry date

### C. QUALITY MANAGEMENT (17.3)

Agents, brokers, traders, distributors, repackers, or relabelers should establish, document and implement an effective system of managing quality, as specified in Section 2.

### D. REPACKAGING, RELABELING, AND HOLDING OF APIS AND INTERMEDIATES (17.4)

Repackaging, relabeling, and holding APIs and intermediates should be performed under appropriate GMP controls, as stipulated in this guidance, to avoid mix-ups and loss of API or intermediate identity or purity.

Repackaging should be conducted under appropriate environmental conditions to avoid contamination and cross-contamination.

### E. STABILITY (17.5)

Stability studies to justify assigned expiration or retest dates should be conducted if the API or intermediate is repackaged in a different type of container than that used by the API or intermediate manufacturer.

## F. TRANSFER OF INFORMATION (17.6)

Agents, brokers, distributors, repackers, or relabelers should transfer all quality or regulatory information received from an API or intermediate manufacturer to the customer, and from the customer to the API or intermediate manufacturer.

The agent, broker, trader, distributor, repacker, or relabeler who supplies the API or intermediate to the customer should provide the name of the original API or intermediate manufacturer and the batch number(s) supplied.

The agent should also provide the identity of the original API or intermediate manufacturer to regulatory authorities upon request. The original manufacturer can respond to the regulatory authority directly or through its authorized agents, depending on the legal relationship between the authorized agents and the original API or intermediate manufacturer. (In this context *authorized* refers to authorized by the manufacturer.)

The specific guidance for certificate of analysis included in Section 11.4 should be met.

## G. HANDLING OF COMPLAINTS AND RECALLS (17.7)

Agents, brokers, traders, distributors, repackers, or relabelers should maintain records of complaints and recalls, as specified in Section 15, for all complaints and recalls that come to their attention.

If the situation warrants, the agents, brokers, traders, distributors, repackers, or relabelers should review the complaint with the original API or intermediate manufacturer to determine whether any further action, either with other customers who may have received this API or intermediate or with the regulatory authority, or both, should be initiated. The investigation into the cause for the complaint or recall should be conducted and documented by the appropriate party.

Where a complaint is referred to the original API or intermediate manufacturer, the record maintained by the agents, brokers, traders, distributors, repackers, or relabelers should include any response received from the original API or intermediate manufacturer (including date and information provided).

## H. HANDLING OF RETURNS (17.8)

Returns should be handled as specified in Section 14.5. The agents, brokers, traders, distributors, repackers, or relabelers should maintain documentation of returned APIs and intermediates.

# XVIII. SPECIFIC GUIDANCE FOR APIS MANUFACTURED BY CELL CULTURE/FERMENTATION (18)

## A. GENERAL (18.1)

Section 18 is intended to address specific controls for APIs or intermediates manufactured by cell culture or fermentation using natural or recombinant organisms and that have not been covered adequately in the previous sections. It is not intended to be a stand-alone section. In general, the GMP principles in the other sections of this document apply. Note that the principles of fermentation for *classical* processes for production of small molecules and for processes using recombinant and nonrecombinant organisms for production of proteins and/or polypeptides are the same, although the degree of control will differ. Where practical, this section will address these differences. In general, the degree of control for biotechnological processes used to produce proteins and polypeptides is greater than that for classical fermentation processes.

The term *biotechnological process* (biotech) refers to the use of cells or organisms that have been generated or modified by recombinant DNA, hybridoma, or other technology to produce APIs. The APIs produced by biotechnological processes normally consist of high molecular weight substances, such as proteins and polypeptides, for which specific guidance is given in this Section. Certain APIs of low molecular weight, such as antibiotics, amino acids, vitamins, and carbohydrates, can also be produced by recombinant DNA technology. The level of control for these types of APIs is similar to that employed for classical fermentation.

The term *classical fermentation* refers to processes that use microorganisms existing in nature and/or modified by conventional methods (e.g., irradiation or chemical mutagenesis) to produce APIs. APIs produced by *classical fermentation* are normally low molecular weight products such as antibiotics, amino acids, vitamins, and carbohydrates.

Production of APIs or intermediates from cell culture or fermentation involves biological processes such as cultivation of cells or extraction and purification of material from living organisms. Note that there may be additional process steps, such as physicochemical modification, that are part of the manufacturing process. The raw materials used (media, buffer components) may provide the potential for growth of microbiological contaminants. Depending on the source, method of preparation, and the intended use of the API or intermediate, control of bioburden, viral contamination, and/or endotoxins during manufacturing and monitoring of the process at appropriate stages may be necessary.

Appropriate controls should be established at all stages of manufacturing to ensure intermediate and/or API quality. While this guidance starts at the cell culture/fermentation step, prior steps (e.g., cell banking) should be performed under appropriate process controls. This guidance covers cell culture/fermentation from the point at which a vial of the cell bank is retrieved for use in manufacturing.

Appropriate equipment and environmental controls should be used to minimize the risk of contamination. The acceptance criteria for determining environmental quality and the frequency of monitoring should depend on the step in production and the production conditions (open, closed, or contained systems).

In general, process controls should take into account:

- Maintenance of the working cell bank (where appropriate)
- Proper inoculation and expansion of the culture
- Control of the critical operating parameters during fermentation/cell culture
- Monitoring of the process for cell growth, viability (for most cell culture processes) and productivity, where appropriate
- Harvest and purification procedures that remove cells, cellular debris and media components while protecting the intermediate or API from contamination (particularly of a microbiological nature) and from loss of quality
- Monitoring of bioburden and, where needed, endotoxin levels at appropriate stages of production
- Viral safety concerns as described in ICH guidance Q5A Quality of Biotechnological Products: Viral Safety Evaluation of Biotechnology Products Derived from Cell Lines of Human or Animal Origin

Where appropriate, the removal of media components, host cell proteins, other process-related impurities, product-related impurities and contaminants should be demonstrated.

## B. Cell Bank Maintenance and Record Keeping (18.2)

Access to cell banks should be limited to authorized personnel.

Cell banks should be maintained under storage conditions designed to maintain viability and prevent contamination.

Records of the use of the vials from the cell banks and storage conditions should be maintained.

Where appropriate, cell banks should be periodically monitored to determine suitability for use.

See ICH guidance Q5D *Quality of Biotechnological Products: Derivation and Characterization of Cell Substrates Used for Production of Biotechnological/Biological Products* for a more complete discussion of cell banking.

## C. Cell Culture/Fermentation (18.3)

Where cell substrates, media, buffers, and gases are to be added under aseptic conditions, closed or contained systems should be used where possible. If the inoculation of the initial vessel or subsequent transfers or additions (media, buffers) are performed in open vessels, there should be controls and procedures in place to minimize the risk of contamination.

Where the quality of the API can be affected by microbial contamination, manipulations using open vessels should be performed in a biosafety cabinet or similarly controlled environment.

Personnel should be appropriately gowned and take special precautions handling the cultures.

Critical operating parameters (for example temperature, pH, agitation rates, addition of gases, pressure) should be monitored to ensure consistency with the established process. Cell growth, viability (for most cell culture processes), and, where appropriate, productivity should also be monitored. Critical parameters will vary from one process to another, and for classical fermentation, certain parameters (cell viability, for example) may not need to be monitored.

Cell culture equipment should be cleaned and sterilized after use. As appropriate, fermentation equipment should be cleaned, sanitized, or sterilized.

Culture media should be sterilized before use, when necessary, to protect the quality of the API.

Appropriate procedures should be in place to detect contamination and determine the course of action to be taken. Procedures should be available to determine the impact of the contamination on the product and to decontaminate the equipment and return it to a condition to be used in subsequent batches. Foreign organisms observed during fermentation processes should be identified, as appropriate, and the effect of their presence on product quality should be assessed, if necessary. The results of such assessments should be taken into consideration in the disposition of the material produced.

Records of contamination events should be maintained.

Shared (multi-product) equipment may warrant additional testing after cleaning between product campaigns, as appropriate, to minimize the risk of cross-contamination.

## D. Harvesting, Isolation and Purification (18.4)

Harvesting steps, either to remove cells or cellular components or to collect cellular components after disruption should be performed in equipment and areas designed to minimize the risk of contamination.

Harvest and purification procedures that remove or inactivate the producing organism, cellular debris and media components (while minimizing degradation, contamination, and loss of quality) should be adequate to ensure that the intermediate or API is recovered with consistent quality.

All equipment should be properly cleaned and, as appropriate, sanitized after use. Multiple successive hatching without cleaning can be used if intermediate or API quality is not compromised.

If open systems are used, purification should be performed under environmental conditions appropriate for the preservation of product quality.

Additional controls, such as the use of dedicated chromatography resins or additional testing, may be appropriate if equipment is to be used for multiple products.

## E. Viral Removal/Inactivation Steps (18.5)

See ICH guidance Q5A Quality of Biotechnological Products: Viral Safety Evaluation of Biotechnology Products Derived from Cell Lines of Human or Animal Origin for more specific information.

Viral removal and viral inactivation steps are critical processing steps for some processes and should be performed within their validated parameters.

Appropriate precautions should be taken to prevent potential viral contamination from previral to postviral removal/inactivation steps. Therefore, open processing should be performed in areas that are separate from other processing activities and have separate air handling units.

The same equipment is not normally used for different purification steps. However, if the same equipment is to be used, the equipment should be appropriately cleaned and sanitized before reuse. Appropriate precautions should be taken to prevent potential virus carry-over (e.g., through equipment or environment) from previous steps.

# XIX. APIS FOR USE IN CLINICAL TRIALS (19)

## A. General (19.1)

Not all the controls in the previous sections of this guidance are appropriate for the manufacture of a new API for investigational use during its development. Section XIX (19) provides specific guidance unique to these circumstances.

The controls used in the manufacture of APIs for use in clinical trials should be consistent with the stage of development of the drug product incorporating the API. Process and test procedures should be flexible to provide for changes as knowledge of the process increases and clinical testing of a drug product progresses from pre-clinical stages through clinical stages. Once drug development reaches the stage where the API is produced for use in drug products intended for clinical trials, manufacturers should ensure that APIs are manufactured in suitable facilities using appropriate production and control procedures to ensure the quality of the API.

## B. Quality (19.2)

Appropriate GMP concepts should be applied in the production of APIs for use in clinical trials with a suitable mechanism for approval of each batch.

A quality unit(s) independent from production should be established for the approval or rejection of each batch of API for use in clinical trials.

Some of the testing functions commonly performed by the quality unit(s) can be performed within other organizational units.

Quality measures should include a system for testing of raw materials, packaging materials, intermediates, and APIs.

Process and quality problems should be evaluated.

Labeling for APIs intended for use in clinical trials should be appropriately controlled and should identify the material as being for investigational use.

## C. Equipment and Facilities (19.3)

During all phases of clinical development, including the use of small-scale facilities or laboratories to manufacture batches of APIs for use in clinical trials, procedures should be in place to ensure that equipment is calibrated, clean, and suitable for its intended use.

Procedures for the use of facilities should ensure that materials are handled in a manner that minimizes the risk of contamination and cross-contamination.

## D. Control of Raw Materials (19.4)

Raw materials used in production of APIs for use in clinical trials should be evaluated by testing, or received with a supplier's analysis and subjected to identity testing. When a material is considered hazardous, a supplier's analysis should suffice.

In some instances, the suitability of a raw material can be determined before use based on acceptability in small-scale reactions (i.e., use testing) rather than on analytical testing alone.

## E. Production (19.5)

The production of APIs for use in clinical trials should be documented in laboratory notebooks, batch records, or by other appropriate means. These documents should include information on the use of production materials, equipment, processing, and scientific observations.

Expected yields can be more variable and less defined than the expected yields used in commercial processes. Investigations into yield variations are not expected.

## F. Validation (19.6)

Process validation for the production of APIs for use in clinical trials is normally inappropriate, where a single API batch is produced or where process changes during API development make batch replication difficult or inexact. The combination of controls, calibration, and, where appropriate, equipment qualification ensures API quality during this development phase.

Process validation should be conducted in accordance with Section 12 when batches are produced for commercial use, even when such batches are produced on a pilot or small scale.

## G. Changes (19.7)

Changes are expected during development, as knowledge is gained and the production is scaled up. Every change in the production, specifications, or test procedures should be adequately recorded.

## H. Laboratory Controls (19.8)

While analytical methods performed to evaluate a batch of API for clinical trials may not yet be validated, they should be scientifically sound.

A system for retaining reserve samples of all batches should be in place. This system should ensure that a sufficient quantity of each reserve sample is retained for an appropriate length of time after approval, termination, or discontinuation of an application.

Expiry and retest dating as defined in Section 11.6 applies to existing APIs used in clinical trials. For new APIs, Section 11.6 does not normally apply in early stages of clinical trials.

## I. DOCUMENTATION (19.9)

A system should be in place to ensure that information gained during the development and the manufacture of APIs for use in clinical trials is documented and available.

The development and implementation of the analytical methods used to support the release of a batch of API for use in clinical trials should be appropriately documented.

A system for retaining production and control records and documents should be used. This system should ensure that records and documents are retained for an appropriate length of time after the approval, termination, or discontinuation of an application.

## GLOSSARY (20)

**Acceptance Criteria:** Numerical limits, ranges, or other suitable measures for acceptance of test results.

**Active Pharmaceutical Ingredient (API) (*or Drug Substance*):** Any substance or mixture of substances intended to be used in the manufacture of a drug (medicinal) product and that, when used in the production of a drug, becomes an active ingredient of the drug product. Such substances are intended to furnish pharmacological activity or other direct effect in the diagnosis, cure, mitigation, treatment, or prevention of disease or to affect the structure and function of the body.

**API Starting Material:** A raw material, intermediate, or an API that is used in the production of an API and that is incorporated as a significant structural fragment into the structure of the API. An API starting material can be an article of commerce, a material purchased from one or more suppliers under contract or commercial agreement, or produced in-house. API starting materials are normally of defined chemical properties and structure.

**Batch (or Lot):** A specific quantity of material produced in a process or series of processes so that it is expected to be homogeneous within specified limits. In the case of continuous production, a batch may correspond to a defined fraction of the production. The batch size can be defined either by a fixed quantity or by the amount produced in a fixed time interval.

**Batch Number (or Lot Number):** A unique combination of numbers, letters, and/or symbols that identifies a batch (or lot) and from which the production and distribution history can be determined.

**Bioburden:** The level and type (e.g., objectionable or not) of microorganisms that can be present in raw materials, API starting materials, intermediates or APIs. Bioburden should not be considered contamination unless the levels have been exceeded or defined objectionable organisms have been detected.

**Calibration:** The demonstration that a particular instrument or device produces results within specified limits by comparison with results produced by a reference or traceable standard over an appropriate range of measurements.

**Computer System:** A group of hardware components and associated software designed and assembled to perform a specific function or group of functions.

**Computerized System:** A process or operation integrated with a computer system.

**Contamination:** The undesired introduction of impurities of a chemical or microbiological nature, or of foreign matter, into or onto a raw material, intermediate, or API during production, sampling, packaging, or repackaging, storage or transport.

**Contract Manufacturer:** A manufacturer who performs some aspect of manufacturing on behalf of the original manufacturer.

**Critical:** Describes a process step, process condition, test requirement, or other relevant parameter or item that must be controlled within predetermined criteria to ensure that the API meets its specification.

**Cross-Contamination:** Contamination of a material or product with another material or product.

**Deviation:** Departure from an approved instruction or established standard.

**Drug (Medicinal) Product:** The dosage form in the final immediate packaging intended for marketing. (Reference Q1A)

**Drug Substance:** See Active Pharmaceutical Ingredient.

**Expiry Date (or Expiration Date):** The date placed on the container/labels of an API designating the time during which the API is expected to remain within established shelf life specifications if stored under defined conditions and after which it should not be used.

**Impurity:** Any component present in the intermediate or API that is not the desired entity.

**Impurity Profile:** A description of the identified and unidentified impurities present in an API.

**In-Process Control (or Process Control):** Checks performed during production to monitor and, if appropriate, to adjust the process and/or to ensure that the intermediate or API conforms to its specifications.

**Intermediate:** A material produced during steps of the processing of an API that undergoes further molecular change or purification before it becomes an API. Intermediates may or may not be isolated. (Note: this guidance only addresses those intermediates produced after the point that a company has defined as the point at which the production of the API begins.)

**Lot:** See Batch

**Lot Number:** See Batch Number

**Manufacture:** All operations of receipt of materials, production, packaging, repackaging, labeling, relabeling, quality control, release, storage, and distribution of APIs and related controls.

**Material:** A general term used to denote raw materials (starting materials, reagents, solvents), process aids, intermediates, APIs, and packaging and labeling materials.

**Mother Liquor:** The residual liquid that remains after the crystallization or isolation processes. A mother liquor may contain unreacted materials, intermediates, levels of the API, and/or impurities. It can be used for further processing.

**Packaging Material:** Any material intended to protect an intermediate or API during storage and transport.

**Procedure:** A documented description of the operations to be performed, the precautions to be taken, and measures to be applied directly or indirectly related to the manufacture of an intermediate or API.

**Process Aids:** Materials, excluding solvents, used as an aid in the manufacture of an intermediate or API that do not themselves participate in a chemical or biological reaction (e.g., filter aid, activated carbon).

**Process Control:** See In Process Control

**Production:** All operations involved in the preparation of an API from receipt of materials through processing and packaging of the API.

**Qualification:** Action of proving and documenting that equipment or ancillary systems are properly installed, work correctly, and actually lead to the expected results. Qualification is part of validation, but the individual qualification steps alone do not constitute process validation.

**Quality Assurance (QA):** The sum total of the organized arrangements made with the object of ensuring that all APIs are of the quality required for their intended use and that quality systems are maintained.

**Quality Control (QC):** Checking or testing that specifications are met.

**Quality Unit(s):** An organizational unit independent of production that fulfills both quality assurance and quality control responsibilities. This can be in the form of separate QA and QC units or a single individual or group, depending upon the size and structure of the organization.

**Quarantine:** The status of materials isolated physically or by other effective means pending a decision on their subsequent approval or rejection.

**Raw Material:** A general term used to denote starting materials, reagents, and solvents intended for use in the production of intermediates or APIs.

**Reference Standard, Primary:** A substance that has been shown by an extensive set of analytical tests to be authentic material that should be of high purity. This standard can be: (1) obtained from an officially recognized source, (2) prepared by independent synthesis, (3) obtained from existing production material of high purity, or (4) prepared by further purification of existing production material.

**Reference Standard, Secondary:** A substance of established quality and purity, as shown by comparison to a primary reference standard, used as a reference standard for routine laboratory analysis.

**Reprocessing:** Introducing an intermediate or API, including one that does not conform to standards or specifications, back into the process and repeating a crystallization step or other appropriate chemical or physical manipulation steps (e.g., distillation, filtration, chromatography, milling) that are part of the established manufacturing process. Continuation of a process step after an in-process control test has shown that the step is incomplete, is considered to be part of the normal process, and is not reprocessing.

**Retest Date:** The date when a material should be re-examined to ensure that it is still suitable for use.

**Reworking:** Subjecting an intermediate or API that does not conform to standards or specifications to one or more processing steps that are different from the established manufacturing process to obtain acceptable quality intermediate or API (e.g., recrystallizing with a different solvent).

**Signature (signed):** See definition for signed.

**Signed (signature):** The record of the individual who performed a particular action or review. This record can be initials, full handwritten signature, personal seal, or authenticated and secure electronic signature.

**Solvent:** An inorganic or organic liquid used as a vehicle for the preparation of solutions or suspensions in the manufacture of an intermediate or API.

**Specification:** A list of tests, references to analytical procedures, and appropriate acceptance criteria that are numerical limits, ranges, or other criteria for the test described. It establishes the set of criteria to which a material should conform to be considered acceptable for its intended use. *Conformance to specification* means that the material, when tested according to the listed analytical procedures, will meet the listed acceptance criteria.

**Validation:** A documented program that provides a high degree of assurance that a specific process, method, or system will consistently produce a result meeting predetermined acceptance criteria.

**Validation Protocol:** A written plan stating how validation will be conducted and defining acceptance criteria. For example, the protocol for a manufacturing process identifies processing equipment, critical process parameters and/or operating ranges, product characteristics, sampling, test data to be collected, number of validation runs, and acceptable test results.

**Yield, Expected:** The quantity of material or the percentage of theoretical yield anticipated at any appropriate phase of production based on previous laboratory, pilot scale, or manufacturing data.

**Yield, Theoretical:** The quantity that would be produced at any appropriate phase of production based upon the quantity of material to be used, in the absence of any loss or error in actual production.

# Part 1301

## Registration of Manufacturers, Distributors, and Dispensers of Controlled Substances

## CONTENTS

### SECTION 1301.01 SCOPE OF THIS PART 1301

Procedures governing the registration of manufacturers, distributors, dispensers, importers, and exporters of controlled substances pursuant to sections 301-304 and 1007-1008 of the Act (21 U.S.C. 821-824 and 957-958) are set forth generally by those sections and specifically by the sections of this part.

[62 13945, Mar. 24, 1997]

### SECTION 1301.02 DEFINITIONS

Any term used in this part shall have the definition set forth in section 102 of the Act (21 U.S.C. 802) or part 1300 of this chapter.

### SECTION 1301.03 INFORMATION; SPECIAL INSTRUCTIONS

Information regarding procedures under these rules and instructions supplementing these rules will be furnished upon request by writing to the Registration Unit, Drug Enforcement Administration, Department of Justice, Post Office Box 28083, Central Station, Washington, D.C. 20005.

### SECTION 1301.11 PERSONS REQUIRED TO REGISTER

(a) Every person who manufactures, distributes, dispenses, imports, or exports any controlled substance or who proposes to engage in the manufacture, distribution, dispensing, importation or exportation of any controlled substance shall obtain a registration unless exempted by law or pursuant to §1301.22-1301.26. Only persons actually engaged in such activities are required to obtain a registration; related or affiliated persons who are not engaged in such activities are not required to be registered. (For example, a stockholder or parent corporation of a corporation manufacturing controlled substances is not required to obtain a registration.)

(b) [Reserved]

## Section 1301.12 Separate Registrations for Separate Locations

(a) A separate registration is required for each principal place of business or professional practice at one general physical location where controlled substances are manufactured, distributed, imported, exported, or dispensed by a person.

(b) The following locations shall be deemed not to be places where controlled substances are manufactured, distributed, or dispensed:

(1) A warehouse where controlled substances are stored by or on behalf of a registered person, unless such substances are distributed directly from such warehouse to registered locations other than the registered location from which the substances were delivered or to persons not required to register by virtue of subsection 302(c)(2) or subsection 1007(b)(1)(B) of the Act (21 U.S.C. 822(c)(2) or 957(b)(1)(B));

(2) An office used by agents of a registrant where sales of controlled substances are solicited, made, or supervised but which neither contains such substances (other than substances for display purposes or lawful distribution as samples only) nor serves as a distribution point for filling sales orders; and

(3) An office used by a practitioner (who is registered at another location) where controlled substances are prescribed but neither administered nor otherwise dispensed as a regular part of the professional practice of the practitioner at such office, and where no supplies of controlled substances are maintained.

(4) A freight forwarding facility, as defined in §1300.01 of this part, provided that the distributing registrant operating the facility has submitted written notice of intent to operate the facility by registered mail, return receipt requested (or other suitable means of documented delivery) and such notice has been approved. The notice shall be submitted to the Special Agent in Charge of the Administration's offices in both the area in which the facility is located and each area in which the distributing registrant maintains a registered location that will transfer controlled substances through the facility. The notice shall detail the registered locations that will utilize the facility, the location of the facility, the hours of operation, the individual(s) responsible for the controlled substances, the security and recordkeeping procedures that will be employed, and whether controlled substances returns will be processed through the facility. The notice must also detail what state licensing requirements apply to the facility and the registrant's actions to comply with any such requirements. The Special Agent in Charge of the DEA Office in the area where the freight forwarding facility will be operated will provide written notice of approval or disapproval to the person within thirty days after confirmed receipt of the notice. Registrants that are currently operating freight forwarding facilities under a memorandum of understanding with the Administration must provide notice as required by this section no later than September 18, 2000 and receive written approval from the Special Agent in Charge of the DEA Office in the area in which the freight forwarding facility is operated in order to continue operation of the facility.

### Section 1301.13 Application for Registration; Time for Application; Expiration Date; Registration for Independent Activities; Application Forms, Fees, Contents and Signature; Coincident Activities

(a) Any person who is required to be registered and who is not so registered may apply for registration at any time. No person required to be registered shall engage in any activity for which registration is required until the application for registration is granted and a Certificate of Registration is issued by the Administrator to such person.

(b) Any person who is registered may apply to be reregistered not more than 60 days before the expiration date of his/her registration, except that a bulk manufacturer of Schedule I or II controlled substances or an importer of Schedule I or II controlled substances may apply to be reregistered no more than 120 days before the expiration date of their registration.

(c) At the time a manufacturer, distributor, researcher, analytical lab, importer, exporter or narcotic treatment program is first registered, that business activity shall be assigned to one of twelve groups, which shall correspond to the months of the year. The expiration date of the registrations of all registrants within any group will be the last date of the month designated for that group. In assigning any of the above business activities to a group, the Administration may select a group the expiration date of which is less than one year from the date such business activity was registered. If the business activity is assigned to a group which has an expiration date less than three months from the date of which the business activity is registered, the registration shall not expire until one year from that expiration date; in all other cases, the registration shall expire on the expiration date following the date on which the business activity is registered.

(d) At the time a retail pharmacy, hospital/clinic, practitioner or teaching institution is first registered, that business activity shall be assigned to one of twelve groups, which shall correspond to the months of the year. The expiration date of the registrations of all registrants within any group will be the last day of the month designated for that group. In assigning any of the above business activities to a group, the Administration may select a group the expiration date of which is not less than 28 months nor more than 39 months from the date such business activity was registered. After the initial registration period, the registration shall expire 36 months from the initial expiration date.

(e) Any person who is required to be registered and who is not so registered, shall make application for registration for one of the following groups of controlled substances activities, which are deemed to be independent of each other. Application for each registration shall be made on the indicated form, and shall be accompanied by the indicated fee. Fee payments shall be made in the form of a personal, certified, or cashier's check or money order made payable to the "Drug Enforcement Administration." The application fees are not refundable. Any person, when registered to engage in the activities described in each subparagraph in this paragraph, shall be authorized to engage in the coincident activities described without obtaining a registration to engage in such coincident activities, provided that, unless specifically exempted, he/she complies with all requirements and duties prescribed by law for persons registered to engage in such coincident activities. Any person who engages in more than one group of independent activities shall obtain a separate registration for each group of activities, except as provided in this paragraph under coincident activities. A single registration to engage in any group of independent activities listed below may include one or more controlled substances listed in the schedules authorized in that group of independent activities. A person registered to conduct research with controlled substances listed in Schedule I may conduct research with any substances listed in Schedule I for which he/she has filed and had approved a research protocol.

(1)

| Business Activity | Controlled Substances | DEA Application Forms | Application Fee (dollars) | Registration Period (years) | Coincident Activities Allowed |
|---|---|---|---|---|---|
| (i) Manufacturing | Schedules I through V | New — 225<br>Renewal — 22 | 875<br>875 | 1 | Schedules I through V: May distribute that substance or class for which registration was issued; may not distribute any substance or class for which not registered.<br>Schedules II through V: May conduct chemical analysis and preclinical research (including quality control analysis) with substances listed in those schedules for which authorization as a manufacturer was issued. |
| (ii) Distributing | Schedules I through V | New — 225<br>Renewal — 225a | 438<br>438 | 1 | |
| (iii) Dispensing or Instructing (Includes Practitioner Hospital/Clinic, Retail Pharmacy, Teaching Institution) | Schedules II through V | New — 224<br>Renewal — 224 | 210<br>210 | 3 | May conduct any instructional activities for which registration was granted, except that a mid-level practitioner may conduct such research only to the extent expressly authorized under state statute. A pharmacist may manufacture an aqueous or oleaginous solution or solid dosage form containing a narcotic controlled substance in Schedule II through V in a proportion not exceeding 20 percent of the complete solution, compound, or mixture. |
| (iv) Research | Schedule I | New — 225<br>Renewal — 225a | 70<br>70 | 1 | A researcher may manufacture or import the basic class of substance or substances for which registration was issued, provided that such manufacture or import is set forth in the protocol required in Section 1301.18 and to distribute such class to person registered or authorized to conduct research with such class of substance or registered or authorized to conduct chemical analysis with controlled substances. |
| (v) Research | Schedules II through V | New — 225<br>Renewal — 225a | 70<br>70 | 1 | May conduct chemical analysis with controlled substances in those schedules for which registration was issued; manufacture such substances if and to the extent that such manufacture is set forth in a statement filed with the application for registration or reregistration and provided that the manufacture is not for the purposes of dosage form development; import such substances for research purposes; distribute such substances to persons registered or authorized to conduct chemical analysis, instructional activities, or research with such substances, and to persons exempted from registration pursuant to Section 1301.24; and conduct instructional activities with controlled substances. |

| Business Activity | Controlled Substances | DEA Application Forms | Application Fee (dollars) | Registration Period (years) | Coincident Activities Allowed |
|---|---|---|---|---|---|
| (vi) Narcotic Treatment Program (including compounder) | Narcotic Drugs in Schedules II through V | New — 363<br>Renewal — 363a | 70<br>70 | 1 | |
| (vii) Importing | Schedules I through V | New — 225<br>Renewal — 225 | 438<br>438 | 1 | May distribute that substance or class for which registration was issued; may not distribute any substance or class for which not registered. |
| (viii) Exporting | Schedules I through V | New — 225<br>Renewal — 225a | 438<br>438 | 1 | |
| (ix) Chemical Analysis | Schedules I through V | New — 225<br>Renewal — 225a | 70<br>70 | 1 | May manufacture and import controlled substances for analytical or instructional activities; may distribute such substances to persons registered or authorized to conduct chemical analysis, instructional activities, or research with such substances and to persons exempted from registration pursuant to Section 1301.24; may export such substances to persons in other countries performing chemical analysis or enforcing laws relating to controlled substances or drugs in those countries; and may conduct instructional activities with controlled substances. |

(2) DEA Forms 224, 225, and 363 may be obtained at any area office of the Administration or by writing to the Registration Unit, Drug Enforcement Administration, Department of Justice, Post Office Box 28083, Central Station, Washington, D.C. 20005.

(3) DEA Forms 224a, 225a, and 363a will be mailed, as applicable, to each registered person approximately 60 days before the expiration date of his/her registration; if any registered person does not receive such forms within 45 days before the expiration date of his/her registration, he/she must promptly give notice of such fact and request such forms by writing to the Registration Unit of the Administration at the foregoing address.

(f) Each application for registration to handle any basic class of controlled substance listed in Schedule I (except to conduct chemical analysis with such classes), and each application for registration to manufacture a basic class of controlled substance listed in Schedule II shall include the Administration Controlled Substances Code Number, as set forth in part 1308 of this chapter, for each basic class to be covered by such registration.

(g) Each application for registration to import or export controlled substances shall include the Administration Controlled Substances Code Number, as set forth in part 1308 of this chapter, for each controlled substance whose importation or exportation is to be authorized by such registration. Registration as an importer or exporter shall not entitle a registrant to import or export any controlled substance not specified in such registration.

(h) Each application for registration to conduct research with any basic class of controlled substance listed in Schedule II shall include the Administration Controlled Substances Code Number, as set forth in part 1308 of this chapter, for each such basic class to be manufactured or imported as a coincident activity of that registration. A statement listing the quantity of each such basic class of controlled substance to be imported or manufactured during the registration period for which application is being made shall be included with each such application. For purposes of this paragraph only, manufacturing is defined as the production of a controlled substance by synthesis, extraction or by agricultural/horticultural means.

(i) Each application shall include all information called for in the form, unless the item is not applicable, in which case this fact shall be indicated.

(j) Each application, attachment, or other document filed as part of an application, shall be signed by the applicant, if an individual; by a partner of the applicant, if a partnership; or by an officer of the applicant, if a corporation, corporate division, association, trust or other entity. An applicant may authorize one or more individuals, who would not otherwise be authorized to do so, to sign applications for the applicant by filing with the Registration Unit of the Administration a power of attorney for each such individual. The power of attorney shall be signed by a person who is authorized to sign applications under this paragraph and shall contain the signature of the individual being authorized to sign applications. The power of attorney shall be valid until revoked by the applicant.

## Section 1301.14 Filing of Application; Acceptance for Filing; Defective Applications

(a) All applications for registration shall be submitted for filing to the Registration Unit, Drug Enforcement Administration, Department of Justice, Post Office Box 28083, Central Station, Washington, D.C. 20005. The appropriate registration fee and any required attachments must accompany the application.

(b) Any person required to obtain more than one registration may submit all applications in one package. Each application must be complete and should not refer to any accompanying application for required information.

(c) Applications submitted for filing are dated upon receipt. If found to be complete, the application will be accepted for filing. Applications failing to comply with the requirements of this part will not generally be accepted for filing. In the case of minor defects as to completeness, the Administrator may accept the application for filing with a request to the applicant for additional information. A defective application will be returned to the applicant within 10 days following its receipt with a statement of the reason for not accepting the application for filing. A defective application may be corrected and resubmitted for filing at any time; the Administrator shall accept for filing any application upon resubmission by the applicant, whether complete or not.

(d) Accepting an application for filing does not preclude any subsequent request for additional information pursuant to §1301.15 and has no bearing on whether the application will be granted.

## Section 1301.15 Additional Information

The Administrator may require an applicant to submit such documents or written statements of fact relevant to the application as he/she deems necessary to determine whether the application should be granted. The failure of the applicant to provide such documents or statements within a reasonable time after being requested to do so shall be deemed to be a waiver by the applicant of an opportunity to present such documents or facts for consideration by the Administrator in granting or denying the application.

## Section 1301.16 Amendments to and Withdrawal of Applications

(a) An application may be amended or withdrawn without permission of the Administrator at any time before the date on which the applicant receives an order to show cause pursuant to §1301.37. An application may be amended or withdrawn with permission of the Administrator at any time where good cause is shown by the applicant or where the amendment or withdrawal is in the public interest.

(b) After an application has been accepted for filing, the request by the applicant that it be returned or the failure of the applicant to respond to official correspondence regarding the application, when sent by registered or certified mail, return receipt requested, shall be deemed to be a withdrawal of the application.

## Section 1301.17 Special Procedures for Certain Applications

(a) If, at the time of application for registration of a new pharmacy, the pharmacy has been issued a license from the appropriate State licensing agency, the applicant may include with his/her application an affidavit as to the existence of the State license in the following form:

### Affidavit for New Pharmacy

I, _____ , the _____ (Title of officer, official, partner, or other position) of _____ (Corporation, partnership, or sole proprietor), doing business as _____ (Store name) at _____ (Number and Street), _____ (City) _____ (State) _____ (Zip code), hereby certify that said store was issued a pharmacy permit No. _____ by the _____ (Board of Pharmacy or Licensing Agency) of the State of _____ on _____ (Date).

This statement is submitted in order to obtain a Drug Enforcement Administration registration number. I understand that if any information is false, the Administration may immediately suspend the registration for this store and commence proceedings to revoke under 21 U.S.C. 824(a) because of the danger to public health and safety. I further understand that any false information contained in this affidavit may subject me personally and the above-named corporation/partnership/business to prosecution under 21 U.S.C. 843, the penalties for conviction of which include imprisonment for up to 4 years, a fine of not more than $30,000 or both.

_____

Signature (Person who signs Application for Registration)

State of _____

County of _____

Subscribed to and sworn before me this _____ day of _____ , 19____ .

_____

Notary Public

(b) Whenever the ownership of a pharmacy is being transferred from one person to another, if the transferee owns at least one other pharmacy licensed in the same State as the one the ownership of which is being transferred, the transferee may apply for registration prior to the date of transfer. The Administrator may register the applicant and authorize him to obtain controlled substances at the time of transfer. Such registration shall not authorize the transferee to dispense controlled substances until the pharmacy has been issued a valid State license. The transferee shall include with his/her application the following affidavit:

### Affidavit for Transfer of Pharmacy

I, _____ , the _____ (Title of officer, official, partner or other position) of _____ (Corporation, partnership, or sole proprietor), doing business as _____ (Store name) hereby certify:

(1) That said company was issued a pharmacy permit No. _____ by the _____ (Board of Pharmacy of Licensing Agency) of the State of _____ and a DEA Registration Number _____ for a pharmacy located at _____ (Number and Street), _____ (City) _____ (State) _____ (Zip code); and

(2) That said company is acquiring the pharmacy business of _____ (Name of Seller) doing business as _____ with DEA Registration Number _____ on or about _____ (Date of Transfer) and that said company has applied (or will apply on _____ (Date) for a pharmacy permit from the board of pharmacy (or licensing agency) of the State of _____ to do business as _____ (Store name) at _____ (Number and Street) _____ (City) _____ (State) _____ (Zip Code).

This statement is submitted in order to obtain a Drug Enforcement Administration registration number.

I understand that if a DEA registration number is issued, the pharmacy may acquire controlled substances but may not dispense them until a pharmacy permit or license is issued by the State board of pharmacy or licensing agency.

I understand that if any information is false, the Administration may immediately suspend the registration for this store and commence proceedings to revoke under 21 U.S.C. 824(a) because of the danger to public health and safety. I further understand that any false information contained in this affidavit may subject me personally to prosecution under 21 U.S.C. 843, the penalties for conviction of which include imprisonment for up to 4 years, a fine of not more than $30,000 or both.

_____

Signature (Person who signs Application for Registration)

State of _____

County of _____

Subscribed to and sworn before me this _____ day of _____ , 19_____ .

_____

Notary Public

(c) The Administrator shall follow the normal procedures for approving an application to verify the statements in the affidavit. If the statements prove to be false, the Administrator may revoke the registration on the basis of section 304(a)(1) of the Act (21 U.S.C. 824(a)(1)) and suspend the registration immediately by pending revocation on the basis of section 304(d) of the Act (21 U.S.C. 824(d)). At the same time, the Administrator may seize and place under seal all controlled substances possessed by the applicant under section 304(f) of the Act (21 U.S.C. 824(f)). Intentional misuse of the affidavit procedure may subject the applicant to prosecution for fraud under section 403(a)(4) of the Act (21 U.S.C. 843(a)(4)), and obtaining controlled substances through registration by fraudulent means may subject the applicant to prosecution under section 403(a)(3) of the Act (21 U.S.C. 843(a)(3)). The penalties for conviction of either offense include imprisonment for up to 4 years, a fine not exceeding $30,000 or both.

## Section 1301.18 Research Protocols

(a) A protocol to conduct research with controlled substances listed in Schedule I shall be in the following form and contain the following information where applicable:

    (1) Investigator:
        (i) Name, address, and DEA registration number; if any
        (ii) Institutional affiliation
        (iii) Qualifications, including a curriculum vitae and an appropriate bibliography (list of publications).

    (2) Research project:
        (i) Title of project
        (ii) Statement of the purpose
        (iii) Name of the controlled substances or substances involved and the amount of each needed.
        (iv) Description of the research to be conducted, including the number and species of research subjects, the dosage to be administered, the route and method of administration, and the duration of the project.
        (v) Location where the research will be conducted.
        (vi) Statement of the security provisions for storing the controlled substances (in accordance with §1301.75) and for dispensing the controlled substances in order to prevent diversion.
        (vii) If the investigator desires to manufacture or import any controlled substance listed in paragraph (a)(2)(iii) of this section, a statement of the quantity to be manufactured or imported and the sources of the chemicals to be used or the substance to be imported.

    (3) Authority:
        (i) Institutional approval
        (ii) Approval of a Human Research Committee for human studies
        (iii) Indication of an approved active Notice of Claimed Investigational Exemption for a New Drug (number)
        (iv) Indication of an approved funded grant (number), if any

(b) In the case of a clinical investigation with controlled substances listed in Schedule I, the applicant shall submit three copies of a Notice of Claimed Investigational Exemption for a New Drug (IND) together with a statement of the security provisions (as proscribed in paragraph (a)(2)(vi) of this section for a research protocol) to, and have such submission approved by, the Food and Drug Administration as required in 21 U.S.C. 355(i) and §130.3 of this title. Submission of this Notice and statement to the Food and Drug Administration shall be in lieu of a research protocol to the Administration as required in paragraph (a) of this section. The applicant, when applying for registration with the Administration, shall indicate that such notice has been submitted to the Food and Drug Administration by submitting to the Administration with his/her DEA Form 225 three copies of the following certificate:

I hereby certify that on _____ (Date), pursuant to 21 U.S.C. 355(i) and 21 CFR 130.3, I, _____ (Name and Address of IND Sponsor) submitted a Notice of Claimed Investigational Exemption for a New Drug (IND) to the Food and Drug Administration for:

_____

(Name of Investigational Drug).

_____

(Date)

_____

(Signature of Applicant).

(c) In the event that the registrant desires to increase the quantity of a controlled substance used for an approved research project, he/she shall submit a request to the Registration Unit, Drug Enforcement Administration, Post Office Box 28083, Central Station, Washington, DC 20005, by registered mail, return receipt requested. The request shall contain the following information: DEA registration number; name of the controlled substance or substances and the quantity of each authorized in the approved protocol; and the additional quantity of each desired. Upon return of the receipt, the registrant shall be authorized to purchase the additional quantity of the controlled substance or substances specified in the request. The Administration shall review the letter and forward it to the Food and Drug Administration together with the Administration comments. The Food and Drug Administration shall approve or deny the request as an amendment to the protocol and so notify the registrant. Approval of the letter by the Food and Drug Administration shall authorize the registrant to use the additional quantity of the controlled substance in the research project.

(d) In the event the registrant desires to conduct research beyond the variations provided in the registrant's approved protocol (excluding any increase in the quantity of the controlled substance requested for his/her research project as outlined in paragraph (c) of this section), he/she shall submit three copies of a supplemental protocol in accordance with paragraph (a) of this section describing the new research and omitting information in the supplemental protocol which has been stated in the original protocol. Supplemental protocols shall be processed and approved or denied in the same manner as original research protocols.

## EXCEPTIONS TO REGISTRATION AND FEES

### SECTION 1301.21 EXEMPTION FROM FEES

(a) The Administrator shall exempt from payment of an application fee for registration or reregistration:

(1) Any hospital or other institution which is operated by an agency of the United States (including the U.S. Army, Navy, Marine Corps, Air Force, and Coast Guard), of any State, or any political subdivision or agency thereof.

(2) Any individual practitioner who is required to obtain an individual registration in order to carry out his or her duties as an official of an agency of the United States (including the U.S. Army, Navy, Marine Corps, Air Force, and Coast Guard), of any State, or any political subdivision or agency thereof.

(b) In order to claim exemption from payment of a registration or reregistration application fee, the registrant shall have completed the certification on the appropriate application form, wherein the registrant's superior (if the registrant is an individual) or officer (if the registrant is an agency) certifies to the status and address of the registrant and to the authority of the registrant to acquire, possess, or handle controlled substances.

(c) Exemption from payment of a registration or reregistration application fee does not relieve the registrant of any other requirements or duties prescribed by law.

## Section 1301.22 Exemption of Agents and Employees; Affiliated Practitioners

(a) The requirement of registration is waived for any agent or employee of a person who is registered to engage in any group of independent activities, if such agent or employee is acting in the usual course of his/her business or employment.

(b) An individual practitioner who is an agent or employee of another practitioner (other than a mid-level practitioner) registered to dispense controlled substances may, when acting in the normal course of business or employment, administer or dispense (other than by issuance of prescription) controlled substances if and to the extent that such individual practitioner is authorized or permitted to do so by the jurisdiction in which he or she practices, under the registration of the employer or principal practitioner in lieu of being registered him/herself.

(c) An individual practitioner who is an agent or employee of a hospital or other institution may, when acting in the normal course of business or employment, administer, dispense, or prescribe controlled substances under the registration of the hospital or other institution which is registered in lieu of being registered him/herself, provided that:

(1) Such dispensing, administering or prescribing is done in the usual course of his/her professional practice;

(2) Such individual practitioner is authorized or permitted to do so by the jurisdiction in which he/she is practicing;

(3) The hospital or other institution by whom he/she is employed has verified that the individual practitioner is so permitted to dispense, administer, or prescribe drugs within the jurisdiction;

(4) Such individual practitioner is acting only within the scope of his/her employment in the hospital or institution;

(5) The hospital or other institution authorizes the individual practitioner to administer, dispense or prescribe under the hospital registration and designates a specific internal code number for each individual practitioner so authorized. The code number shall consist of numbers, letters, or a combination thereof and shall be a suffix to the institution's DEA registration number, preceded by a hyphen (e.g., APO123456-10 or APO123456-A12); and

(6) A current list of internal codes and the corresponding individual practitioners is kept by the hospital or other institution and is made available at all times to other registrants and law enforcement agencies upon request for the purpose of verifying the authority of the prescribing individual practitioner.

## Section 1301.23 Exemption of Certain Military and Other Personnel

(a) The requirement of registration is waived for any official of the U.S. Army, Navy, Marine Corps, Air Force, Coast Guard, Public Health Service, or Bureau of Prisons who is authorized to prescribe, dispense, or administer, but not to procure or purchase, controlled substances in the course of his/her official duties. Such officials shall follow procedures set forth in part 1306 of this chapter regarding prescriptions, but shall state the branch of service or agency (e.g., "U.S. Army" or "Public Health Service") and the service identification number of the issuing official in lieu of the registration number required on prescription forms. The service identification number for a Public Health Service employee is his/her Social Security identification number.

(b) The requirement of registration is waived for any official or agency of the U.S. Army, Navy, Marine Corps, Air Force, Coast Guard, or Public Health Service who or which is authorized to import or export controlled substances in the course of his/her official duties.

(c) If any official exempted by this section also engages as a private individual in any activity or group of activities for which registration is required, such official shall obtain a registration for such private activities.

## SECTION 1301.24 EXEMPTION OF LAW ENFORCEMENT OFFICIALS

(a) The requirement of registration is waived for the following persons in the circumstances described in this section:

(1) Any officer or employee of the Administration, any officer of the U.S. Customs Service, any officer or employee of the United States Food and Drug Administration, and any other Federal officer who is lawfully engaged in the enforcement of any Federal law relating to controlled substances, drugs or customs, and is duly authorized to possess or to import or export controlled substances in the course of his/her official duties; and

(2) Any officer or employee of any State, or any political subdivision or agency thereof, who is engaged in the enforcement of any State or local law relating to controlled substances and is duly authorized to possess controlled substances in the course of his/her official duties.

(b) Any official exempted by this section may, when acting in the course of his/her official duties, procure any controlled substance in the course of an inspection, in accordance with §1316.03(d) of this chapter, or in the course of any criminal investigation involving the person from whom the substance was procured, and may possess any controlled substance and distribute any such substance to any other official who is also exempted by this section and acting in the course of his/her official duties.

(c) In order to enable law enforcement agency laboratories, including laboratories of the Administration, to obtain and transfer controlled substances for use as standards in chemical analysis, such laboratories shall obtain annually a registration to conduct chemical analysis. Such laboratories shall be exempted from payment of a fee for registration. Laboratory personnel, when acting in the scope of their official duties, are deemed to be officials exempted by this section and within the activity described in section 515(d) of the Act (21 U.S.C. 885(d)). For purposes of this paragraph, laboratory activities shall not include field or other preliminary chemical tests by officials exempted by this section.

(d) In addition to the activities authorized under a registration to conduct chemical analysis pursuant to §1301.13(e)(1)(ix), laboratories of the Administration shall be authorized to manufacture or import controlled substances for any lawful purpose, to distribute or export such substances to any person, and to import and export such substances in emergencies without regard to the requirements of part 1312 of this chapter if a report concerning the importation or exportation is made to the Drug Operations Section of the Administration within 30 days of such importation or exportation.

## Section 1301.25 Registration Regarding Ocean Vessels, Aircraft, and Other Entities

(a)  If acquired by and dispensed under the general supervision of a medical officer described in paragraph (b) of this section, or the master or first officer of the vessel under the circumstances described in paragraph (d) of this section, controlled substances may be held for stocking, be maintained in, and dispensed from medicine chests, first aid packets, or dispensaries:

   (1)  On board any vessel engaged in international trade or in trade between ports of the United States and any merchant vessel belonging to the U.S. Government;

   (2)  On board any aircraft operated by an air carrier under a certificate of permit issued pursuant to the Federal Aviation Act of 1958 (49 U.S.C. 1301); and

   (3)  In any other entity of fixed or transient location approved by the Administrator as appropriate for application of this section (e.g., emergency kits at field sites of an industrial firm).

(b)  A medical officer shall be:

   (1)  Licensed in a state as a physician;

   (2)  Employed by the owner or operator of the vessel, aircraft or other entity; and

   (3)  Registered under the Act at either of the following locations:

      (i)  The principal office of the owner or operator of the vessel, aircraft or other entity or

      (ii) At any other location provided that the name, address, registration number and expiration date as they appear on his/her Certificate of Registration (DEA Form 223) for this location are maintained for inspection at said principal office in a readily retrievable manner.

(c)  A registered medical officer may serve as medical officer for more than one vessel, aircraft, or other entity under a single registration, unless he/she serves as medical officer for more than one owner or operator, in which case he/she shall either maintain a separate registration at the location of the principal office of each such owner or operator or utilize one or more registrations pursuant to paragraph (b)(3)(ii) of this section.

(d)  If no medical officer is employed by the owner or operator of a vessel, or in the event such medical officer is not accessible and the acquisition of controlled substances is required, the master or first officer of the vessel, who shall not be registered under the Act, may purchase controlled substances from a registered manufacturer or distributor, or from an authorized pharmacy as described in paragraph (f) of this section, by following the procedure outlined below:

   (1)  The master or first officer of the vessel must personally appear at the vendor's place of business, present proper identification (e.g., Seaman's photographic identification card) and a written requisition for the controlled substances.

   (2)  The written requisition must be on the vessel's official stationery or purchase order form and must include the name and address of the vendor, the name of the controlled substance, description of the controlled substance (dosage form, strength and number or volume per container) number of containers ordered, the name of the vessel, the vessel's official number and country of registry, the owner or operator of the vessel, the port at which the vessel is located, signature of the vessel's officer who is ordering the controlled substances and the date of the requisition.

(3) The vendor may, after verifying the identification of the vessel's officer requisitioning the controlled substances, deliver the control substances to that officer. The transaction shall be documented, in triplicate, on a record of sale in a format similar to that outlined in paragraph (d)(4) of this section. The vessel's requisition shall be attached to copy 1 of the record of sale and filed with the controlled substances records of the vendor, copy 2 of the record of sale shall be furnished to the officer of the vessel and retained aboard the vessel, copy 3 of the record of sale shall be forwarded to the nearest DEA Division Office within 15 days after the end of the month in which the sale is made.

(4) The vendor's record of sale should be similar to, and must include all the information contained in, the below listed format.

### SALE OF CONTROLLED SUBSTANCES TO VESSELS

(Name of registrant) _____

(Address of registrant) _____

(DEA registration number) _____

| Line No. | Number of packages distributed | Size of packages | Name of product | Packages distributed | Date |
|----------|-------------------------------|------------------|-----------------|----------------------|------|
| 1 | | | | | |
| 2 | | | | | |
| 3 | | | | | |

Footnote: Line numbers may be continued according to needs of the vendor.

Number of lines completed _____

Name of vessel _____

Vessel's official number _____

Vessel's country of registry _____

Owner or operator of the vessel _____

Name and title of vessel's officer who presented the requisition _____

Signature of vessel's officer who presented the requisition _____

(e) Any medical officer described in paragraph (b) of this section shall, in addition to complying with all requirements and duties prescribed for registrants generally, prepare an annual report as of the date on which his/her registration expires, which shall give in detail an accounting for each vessel, aircraft, or other entity, and a summary accounting for all vessels, aircraft, or other entities under his/her supervision for all controlled substances purchased, dispensed or disposed of during the year. The medical officer shall maintain this report with other records required to be kept under the Act and, upon request, deliver a copy of the report to the Administration. The medical officer need not be present when controlled substances are dispensed, if the person who actually dispensed the controlled substances is responsible to the medical officer to justify his/her actions.

(f) Any registered pharmacy that wishes to distribute controlled substances pursuant to this section shall be authorized to do so, provided:

    (1) The registered pharmacy notifies the nearest Division Office of the Administration of its intention to so distribute controlled substances prior to the initiation of such activity. This notification shall be by registered mail and shall contain the name, address, and registration number of the pharmacy as well as the date upon which such activity will commence; and

    (2) Such activity is authorized by state law; and

    (3) The total number of dosage units of all controlled substances distributed by the pharmacy during any calendar year in which the pharmacy is registered to dispense does not exceed the limitations imposed upon such distribution by §1307.11(a)(4) and (b) of this chapter.

(g) Owners or operators of vessels, aircraft, or other entities described in this section shall not be deemed to possess or dispense any controlled substance acquired, stored and dispensed in accordance with this section. Additionally, owners or operators of vessels, aircraft, or other entities described in this section or in Article 32 of the Single Convention on Narcotic Drugs, 1961, or in Article 14 of the Convention on Psychotropic Substances, 1971, shall not be deemed to import or export any controlled substances purchased and stored in accordance with that section or applicable article.

(h) The Master of a vessel shall prepare a report for each calendar year which shall give in detail an accounting for all controlled substances purchased, dispensed, or disposed of during the year. The Master shall file this report with the medical officer employed by the owner or operator of his/her vessel, if any, or, if not, he/she shall maintain this report with other records required to be kept under the Act and, upon request, deliver a copy of the report to the Administration.

(i) Controlled substances acquired and possessed in accordance with this section shall not be distributed to persons not under the general supervision of the medical officer employed by the owner or operator of the vessel, aircraft, or other entity, except in accordance with Sec. 1307.21 of this chapter.

## SECTION 1301.26 EXEMPTIONS FROM IMPORT OR EXPORT REQUIREMENTS FOR PERSONAL MEDICAL USE

Any individual who has in his/her possession a controlled substance listed in schedules II, III, IV, or V, which he/she has lawfully obtained for his/her personal medical use, or for administration to an animal accompanying him/her, may enter or depart the United States with such substance notwithstanding sections 1002-1005 of the Act (21 U.S.C. 952-955), providing the following conditions are met:

(a) The controlled substance is in the original container in which it was dispensed to the individual; and

(b) The individual makes a declaration to an appropriate official of the U.S. Customs Service stating:

    (1) That the controlled substance is possessed for his/her personal use, or for an animal accompanying him/her; and

    (2) The trade or chemical name and the symbol designating the schedule of the controlled substance if it appears on the container label, or, if such name does not appear on the

label, the name and address of the pharmacy or practitioner who dispensed the substance and the prescription number, if any; and

(c) The importation of the controlled substance for personal medical use is authorized or permitted under other Federal laws and state law.

## ACTION ON APPLICATION FOR REGISTRATION: REVOCATION OR SUSPENSION OF REGISTRATION

### Section 1301.31 Administrative Review Generally

The Administrator may inspect, or cause to be inspected, the establishment of an applicant or registrant, pursuant to subpart A of part 1316 of this chapter. The Administrator shall review the application for registration and other information gathered by the Administrator regarding an applicant in order to determine whether the applicable standards of section 303 (21 U.S.C. 823) or section 1008 (21 U.S.C. 958) of the Act have been met by the applicant.

### Section 1301.32 Action on Applications for Research in Schedule I Substances

(a) In the case of an application for registration to conduct research with controlled substances listed in Schedule I, the Administrator shall process the application and protocol and forward a copy of each to the Secretary of Health and Human Services (Secretary) within 7 days after receipt. The Secretary shall determine the qualifications and competency of the applicant, as well as the merits of the protocol (and shall notify the Administrator of his/her determination) within 21 days after receipt of the application and complete protocol, except that in the case of a clinical investigation, the Secretary shall have 30 days to make such determination and notify the Administrator. The Secretary, in determining the merits of the protocol, shall consult with the Administrator as to effective procedures to safeguard adequately against diversion of such controlled substances from legitimate medical or scientific use.

(b) An applicant whose protocol is defective shall be notified by the Secretary within 21 days after receipt of such protocol from the Administrator (or in the case of a clinical investigation within 30 days), and he/she shall be requested to correct the existing defects before consideration shall be given to his/her submission.

(c) If the Secretary determines the applicant qualified and competent and the research protocol meritorious, he/she shall notify the Administrator in writing of such determination. The Administrator shall issue a certificate of registration within 10 days after receipt of this notice, unless he/she determines that the certificate of registration should be denied on a ground specified in section 304(a) of the Act (21 U.S.C. 824(a)). In the case of a supplemental protocol, a replacement certificate of registration shall be issued by the Administrator.

(d) If the Secretary determines that the protocol is not meritorious and/or the applicant is not qualified or competent, he/she shall notify the Administrator in writing setting forth the reasons for such determination. If the Administrator determines that grounds exist for the denial of the application, he/she shall within 10 days issue an order to show cause pursuant to Sec. 1301.37 and, if requested by the applicant, hold a hearing on the application pursuant to Sec. 1301.41. If the grounds for denial of the application include a determination by the

Secretary, the Secretary or his duly authorized agent shall furnish testimony and documents pertaining to his determination at such hearing.

(e) Supplemental protocols will be processed in the same manner as original research protocols. If the processing of an application or research protocol is delayed beyond the time limits imposed by this section, the applicant shall be so notified in writing.

## SECTION 1301.33 APPLICATION FOR BULK MANUFACTURE OF SCHEDULE I AND II SUBSTANCES

(a) In the case of an application for registration or reregistration to manufacture in bulk a basic class of controlled substance listed in Schedule I or II, the Administrator shall, upon the filing of such application, publish in the Federal Register a notice naming the applicant and stating that such applicant has applied to be registered as a bulk manufacturer of a basic class of narcotic or nonnarcotic controlled substance, which class shall be identified. A copy of said notice shall be mailed simultaneously to each person registered as a bulk manufacturer of that basic class and to any other applicant therefor. Any such person may, within 60 days from the date of publication of the notice in the Federal Register, file with the Administrator written comments on or objections to the issuance of the proposed registration.

(b) In order to provide adequate competition, the Administrator shall not be required to limit the number of manufacturers in any basic class to a number less than that consistent with maintenance of effective controls against diversion solely because a smaller number is capable of producing an adequate and uninterrupted supply.

(c) This section shall not apply to the manufacture of basic classes of controlled substances listed in Schedules I or II as an incident to research or chemical analysis as authorized in §1301.13(e)(1).

## SECTION 1301.34 APPLICATION FOR IMPORTATION OF SCHEDULE I AND II SUBSTANCES

(a) In the case of an application for registration or reregistration to import a controlled substance listed in Schedule I or II, under the authority of section 1002(a)(2)(B) of the Act (21 U.S.C. 952(a)(2)(B)), the Administrator shall, upon the filing of such application, publish in the Federal Register a notice naming the applicant and stating that such applicant has applied to be registered as an importer of a Schedule I or II controlled substance, which substance shall be identified. A copy of said notice shall be mailed simultaneously to each person registered as a bulk manufacturer of that controlled substance and to any other applicant therefor. Any such person may, within 30 days from the date of publication of the notice in the Federal Register, file written comments on or objections to the issuance of the proposed registration, and may, at the same time, file a written request for a hearing on the application pursuant to §1301.43. If a hearing is requested, the Administrator shall hold a hearing on the application in accordance with §1301.41. Notice of the hearing shall be published in the Federal Register, and shall be mailed simultaneously to the applicant and to all persons to whom notice of the application was mailed. Any such person may participate in the hearing by filing a notice of appearance in accordance with §1301.43 of this chapter. Notice of the hearing shall contain a summary of all comments and objections filed regarding the application and shall state the time and place for the hearing, which shall not be less than 30 days after the date of publication of such notice in the Federal Register. A hearing pursuant to this section may be consolidated with a hearing held pursuant to §1301.35 or §1301.36 of this part.

(b) The Administrator shall register an applicant to import a controlled substance listed in Schedule I or II if he/she determines that such registration is consistent with the public interest and with U.S. obligations under international treaties, conventions, or protocols in effect on May 1, 1971. In determining the public interest, the following factors shall be considered:

(1) Maintenance of effective controls against diversion of particular controlled substances and any controlled substance in Schedule I or II compounded therefrom into other than legitimate medical, scientific research, or industrial channels, by limiting the importation and bulk manufacture of such controlled substances to a number of establishments which can produce an adequate and uninterrupted supply of these substances under adequately competitive conditions for legitimate medical, scientific, research, and industrial purposes;

(2) Compliance with applicable State and local law;

(3) Promotion of technical advances in the art of manufacturing these substances and the development of new substances;

(4) Prior conviction record of applicant under Federal and State laws relating to the manufacture, distribution, or dispensing of such substances;

(5) Past experience in the manufacture of controlled substances, and the existence in the establishment of effective control against diversion;

(6) That the applicant will be permitted to import only:

(i) Such amounts of crude opium, poppy straw, concentrate of poppy straw, and coca leaves as the Administrator finds to be necessary to provide for medical, scientific, or other legitimate purposes; or

(ii) Such amounts of any controlled substances listed in Schedule I or II as the Administrator shall find to be necessary to provide for the medical, scientific, or other legitimate needs of the United States during an emergency in which domestic supplies of such substances are found by the Administrator to be inadequate; or

(iii) Such amounts of any controlled substance listed in Schedule I or II as the Administrator shall find to be necessary to provide for the medical, scientific, or other legitimate needs of the United States in any case in which the Administrator finds that competition among domestic manufacturers of the controlled substance is inadequate and will not be rendered adequate by the registration of additional manufacturers under section 303 of the Act (21 U.S.C. 823); or

(iv) Such limited quantities of any controlled substance listed in Schedule I or II as the Administrator shall find to be necessary for scientific, analytical or research uses; and

(7) Such other factors as may be relevant to and consistent with the public health and safety.

(c) In determining whether the applicant can and will maintain effective controls against diversion within the meaning of paragraph (b) of this section, the Administrator shall consider among other factors:

(1) Compliance with the security requirements set forth in §1301.71-1301.76; and

(2) Employment of security procedures to guard against in-transit losses within and without the jurisdiction of the United States.

(d) In determining whether competition among the domestic manufacturers of a controlled substance is adequate within the meaning of paragraphs (b)(1) and (b)(6)(iii) of this section, as well as section 1002(a)(2)(B) of the Act (21 U.S.C. 952(a)(2)(B)), the Administrator shall consider:

(1) The extent of price rigidity in the light of changes in:
   (i) raw materials and other costs and
   (ii) conditions of supply and demand;
(2) The extent of service and quality competition among the domestic manufacturers for shares of the domestic market including:
   (i) Shifts in market shares and
   (ii) Shifts in individual customers among domestic manufacturers;
(3) The existence of substantial differentials between domestic prices and the higher of prices generally prevailing in foreign markets or the prices at which the applicant for registration to import is committed to undertake to provide such products in the domestic market in conformity with the Act. In determining the existence of substantial differentials hereunder, appropriate consideration should be given to any additional costs imposed on domestic manufacturers by the requirements of the Act and such other cost-related and other factors as the Administrator may deem relevant. In no event shall an importer's offering prices in the United States be considered if they are lower than those prevailing in the foreign market or markets from which the importer is obtaining his/her supply;
(4) The existence of competitive restraints imposed upon domestic manufacturers by governmental regulations; and
(5) Such other factors as may be relevant to the determinations required under this paragraph.
(e) In considering the scope of the domestic market, consideration shall be given to substitute products which are reasonably interchangeable in terms of price, quality and use.
(f) The fact that the number of existing manufacturers is small shall not demonstrate, in and of itself, that adequate competition among them does not exist.

## Section 1301.35 Certificate of Registration; Denial of Registration

(a) The Administrator shall issue a Certificate of Registration (DEA Form 223) to an applicant if the issuance of registration or reregistration is required under the applicable provisions of sections 303 or 1008 of the Act (21 U.S.C. 823 and 958). In the event that the issuance of registration or reregistration is not required, the Administrator shall deny the application. Before denying any application, the Administrator shall issue an order to show cause pursuant to §1301.37 and, if requested by the applicant, shall hold a hearing on the application pursuant to §1301.41.
(b) If in response to a show cause order a hearing is requested by an applicant for registration or reregistration to manufacture in bulk a basic class of controlled substance listed in Schedule I or II, notice that a hearing has been requested shall be published in the Federal Register and shall be mailed simultaneously to the applicant and to all persons to whom notice of the application was mailed. Any person entitled to file comments or objections to the issuance of the proposed registration pursuant to §1301.33(a) may participate in the hearing by filing notice of appearance in accordance with §1301.43. Such persons shall have 30 days to file a notice of appearance after the date of publication of the notice of a request for a hearing in the Federal Register.
(c) The Certificate of Registration (DEA Form 223) shall contain the name, address, and registration number of the registrant, the activity authorized by the registration, the schedules and/or Administration Controlled Substances Code Number (as set forth in part 1308 of this

chapter) of the controlled substances which the registrant is authorized to handle, the amount of fee paid (or exemption), and the expiration date of the registration. The registrant shall maintain the certificate of registration at the registered location in a readily retrievable manner and shall permit inspection of the certificate by any official, agent or employee of the Administration or of any Federal, State, or local agency engaged in enforcement of laws relating to controlled substances.

### SECTION 1301.36 SUSPENSION OR REVOCATION OF REGISTRATION; SUSPENSION OF REGISTRATION PENDING FINAL ORDER; EXTENSION OF REGISTRATION PENDING FINAL ORDER

(a) For any registration issued under section 303 of the Act (21 U.S.C. 823), the Administrator may:
  (1) Suspend the registration pursuant to section 304(a) of the Act (21 U.S.C. 824(a)) for any period of time.
  (2) Revoke the registration pursuant to section 304(a) of the Act (21 U.S.C. 824(a)).
(b) For any registration issued under section 1008 of the Act (21 U.S.C. 958), the Administrator may:
  (1) Suspend the registration pursuant to section 1008(d) of the Act (21 U.S.C. 958(d)) for any period of time.
  (2) Revoke the registration pursuant to section 1008(d) of the Act (21 U.S.C. 958(d)) if he/she determines that such registration is inconsistent with the public interest as defined in section 1008 or with the United States obligations under international treaties, conventions, or protocols in effect on October 12, 1984.
(c) The Administrator may limit the revocation or suspension of a registration to the particular controlled substance, or substances, with respect to which grounds for revocation or suspension exist.
(d) Before revoking or suspending any registration, the Administrator shall issue an order to show cause pursuant to §1301.37 and, if requested by the registrant, shall hold a hearing pursuant to §1301.41.
(e) The Administrator may suspend any registration simultaneously with or at any time subsequent to the service upon the registrant of an order to show cause why such registration should not be revoked or suspended, in any case where he/she finds that there is an imminent danger to the public health or safety. If the Administrator so suspends, he/she shall serve with the order to show cause pursuant to §1301.37 an order of immediate suspension which shall contain a statement of his findings regarding the danger to public health or safety.
(f) Upon service of the order of the Administrator suspending or revoking registration, the registrant shall immediately deliver his/her Certificate of Registration, any order forms, and any import or export permits in his/her possession to the nearest office of the Administration. The suspension or revocation of a registration shall suspend or revoke any individual manufacturing or procurement quota fixed for the registrant pursuant to part 1303 of this chapter and any import or export permits issued to the registrant pursuant to part 1312 of this chapter. Also, upon service of the order of the Administrator revoking or suspending registration, the registrant shall, as instructed by the Administrator:
  (1) Deliver all controlled substances in his/her possession to the nearest office of the Administration or to authorized agents of the Administration; or
  (2) Place all controlled substances in his/her possession under seal as described in sections 304(f) or 1008(d)(6) of the Act (21 U.S.C. 824(f) or 958(d)(6)).

(g) In the event that revocation or suspension is limited to a particular controlled substance or substances, the registrant shall be given a new Certificate of Registration for all substances not affected by such revocation or suspension; no fee shall be required to be paid for the new Certificate of Registration. The registrant shall deliver the old Certificate of Registration and, if appropriate, any order forms in his/her possession to the nearest office of the Administration. The suspension or revocation of a registration, when limited to a particular basic class or classes of controlled substances, shall suspend or revoke any individual manufacturing or procurement quota fixed for the registrant for such class or classes pursuant to part 1303 of this chapter and any import or export permits issued to the registrant for such class or classes pursuant to part 1312 of this chapter. Also, upon service of the order of the Administrator revoking or suspending registration, the registrant shall, as instructed by the Administrator:

   (1) Deliver to the nearest office of the Administration or to authorized agents of the Administration all of the particular controlled substance or substances affected by the revocation or suspension which are in his/her possession; or

   (2) Place all of such substances under seal as described in sections 304(f) or 958(d)(6) of the Act (21 U.S.C. 824(f) or 958(d)(6)).

(h) Any suspension shall continue in effect until the conclusion of all proceedings upon the revocation or suspension, including any judicial review thereof, unless sooner withdrawn by the Administrator or dissolved by a court of competent jurisdiction. Any registrant whose registration is suspended under paragraph (e) of this section may request a hearing on the revocation or suspension of his/her registration at a time earlier than specified in the order to show cause pursuant to §1301.37. This request shall be granted by the Administrator, who shall fix a date for such hearing as early as reasonably possible.

(i) In the event that an applicant for reregistration (who is doing business under a registration previously granted and not revoked or suspended) has applied for reregistration at least 45 days before the date on which the existing registration is due to expire, and the Administrator has issued no order on the application on the date on which the existing registration is due to expire, the existing registration of the applicant shall automatically be extended and continue in effect until the date on which the Administrator so issues his/her order. The Administrator may extend any other existing registration under the circumstances contemplated in this section even though the registrant failed to apply for reregistration at least 45 days before expiration of the existing registration, with or without request by the registrant, if the Administrator finds that such extension is not inconsistent with the public health and safety.

## SECTION 1301.37 ORDER TO SHOW CAUSE

(a) If, upon examination of the application for registration from any applicant and other information gathered by the Administration regarding the applicant, the Administrator is unable to make the determinations required by the applicable provisions of section 303 and/or section 1008 of the Act (21 U.S.C. 823 and 958) to register the applicant, the Administrator shall serve upon the applicant an order to show cause why the registration should not be denied.

(b) If, upon information gathered by the Administration regarding any registrant, the Administrator determines that the registration of such registrant is subject to suspension or revocation pursuant to section 304 or section 1008 of the Act (21 U.S.C. 824 and 958), the Administrator shall serve upon the registrant an order to show cause why the registration should not be revoked or suspended.

(c) The order to show cause shall call upon the applicant or registrant to appear before the Administrator at a time and place stated in the order, which shall not be less than 30 days after the date of receipt of the order. The order to show cause shall also contain a statement of the legal basis for such hearing and for the denial, revocation, or suspension of registration and a summary of the matters of fact and law asserted.

(d) Upon receipt of an order to show cause, the applicant or registrant must, if he/she desires a hearing, file a request for a hearing pursuant to §1301.43. If a hearing is requested, the Administrator shall hold a hearing at the time and place stated in the order, pursuant to §1301.41.

(e) When authorized by the Administrator, any agent of the Administration may serve the order to show cause.

## HEARINGS

### Section 1301.41 Hearings Generally

(a) In any case where the Administrator shall hold a hearing on any registration or application therefor, the procedures for such hearing shall be governed generally by the adjudication procedures set forth in the Administrative Procedure Act (5 U.S.C. 551-559) and specifically by sections 303, 304, and 1008 of the Act (21 U.S.C. 823-824 and 958), by §1301.42-1301.46 of this part, and by the procedures for administrative hearings under the Act set forth in §1316.41-1316.67 of this chapter.

(b) Any hearing under this part shall be independent of, and not in lieu of, criminal prosecutions or other proceedings under the Act or any other law of the United States.

### Section 1301.42 Purpose of Hearing

If requested by a person entitled to a hearing, the Administrator shall hold a hearing for the purpose of receiving factual evidence regarding the issues involved in the denial, revocation, or suspension of any registration, and the granting of any application for registration to import or to manufacture in bulk a basic class of controlled substance listed in Schedule I or II. Extensive argument should not be offered into evidence but rather presented in opening or closing statements of counsel or in memoranda or proposed findings of fact and conclusions of law.

### Section 1301.43 Request for Hearing or Appearance; Waiver

(a) Any person entitled to a hearing pursuant to §1301.32 or §1301.34-1301.36 and desiring a hearing shall, within 30 days after the date of receipt of the order to show cause (or the date of publication of notice of the application for registration in the Federal Register in the case of §1301.34), file with the Administrator a written request for a hearing in the form prescribed in §1316.47 of this chapter.

(b) Any person entitled to participate in a hearing pursuant to §1301.34 or §1301.35(b) and desiring to do so shall, within 30 days of the date of publication of notice of the request for a hearing in the Federal Register, file with the Administrator a written notice of intent to participate in such hearing in the form prescribed in §1316.48 of this chapter. Any person filing a request for a hearing need not also file a notice of appearance.

(c) Any person entitled to a hearing or to participate in a hearing pursuant to §1301.32 or §1301.34-1301.36 may, within the period permitted for filing a request for a hearing or a notice of appearance, file with the Administrator a waiver of an opportunity for a hearing or to participate in a hearing, together with a written statement regarding such person's position on the matters of fact and law involved in such hearing. Such statement, if admissible, shall be made a part of the record and shall be considered in light of the lack of opportunity for cross-examination in determining the weight to be attached to matters of fact asserted therein.

(d) If any person entitled to a hearing or to participate in a hearing pursuant to §1301.32 or §1301.34-1301.36 fails to file a request for a hearing or a notice of appearance, or if such person so files and fails to appear at the hearing, such person shall be deemed to have waived the opportunity for a hearing or to participate in the hearing, unless such person shows good cause for such failure.

(e) If all persons entitled to a hearing or to participate in a hearing waive or are deemed to waive their opportunity for the hearing or to participate in the hearing, the Administrator may cancel the hearing, if scheduled, and issue his/her final order pursuant to §1301.46 without a hearing.

## Section 1301.44 Burden of Proof

(a) At any hearing on an application to manufacture any controlled substance listed in Schedule I or II, the applicant shall have the burden of proving that the requirements for such registration pursuant to section 303(a) of the Act (21 U.S.C. 823(a)) are satisfied. Any other person participating in the hearing pursuant to §1301.35(b) shall have the burden of proving any propositions of fact or law asserted by such person in the hearing.

(b) At any hearing on the granting or denial of an applicant to be registered to conduct a narcotic treatment program or as a compounder, the applicant shall have the burden of proving that the requirements for each registration pursuant to section 303(g) of the Act (21 U.S.C. 823(g)) are satisfied.

(c) At any hearing on the granting or denial of an application to be registered to import or export any controlled substance listed in Schedule I or II, the applicant shall have the burden of proving that the requirements for such registration pursuant to sections 1008(a) and (d) of the Act (21 U.S.C. 958 (a) and (d)) are satisfied. Any other person participating in the hearing pursuant to §1301.34 shall have the burden of proving any propositions of fact or law asserted by him/her in the hearings.

(d) At any other hearing for the denial of a registration, the Administration shall have the burden of proving that the requirements for such registration pursuant to section 303 or section 1008(c) and (d) of the Act (21 U.S.C. 823 or 958(c) and (d)) are not satisfied.

(e) At any hearing for the revocation or suspension of a registration, the Administration shall have the burden of proving that the requirements for such revocation or suspension pursuant to section 304(a) or section 1008(d) of the Act (21 U.S.C. 824(a) or 958(d)) are satisfied.

## Section 1301.45 Time and Place of Hearing

The hearing will commence at the place and time designated in the order to show cause or notice of hearing published in the Federal Register (unless expedited pursuant to §1301.36(h)) but thereafter it may be moved to a different place and may be continued from day to day or recessed to a later day without notice other than announcement thereof by the presiding officer at the hearing.

## Section 1301.46 Final Order

As soon as practicable after the presiding officer has certified the record to the Administrator, the Administrator shall issue his/her order on the granting, denial, revocation, or suspension of registration. In the event that an application for registration to import or to manufacture in bulk a basic class of any controlled substance listed in Schedule I or II is granted, or any application for registration is denied, or any registration is revoked or suspended, the order shall include the findings of fact and conclusions of law upon which the order is based. The order shall specify the date on which it shall take effect. The Administrator shall serve one copy of his/her order upon each party in the hearing.

Modification, Transfer and Termination of Registration

## Section 1301.51 Modification in Registration

Any registrant may apply to modify his/her registration to authorize the handling of additional controlled substances or to change his/her name or address, by submitting a letter of request to the Registration Unit, Drug Enforcement Administration, Department of Justice, Post Office Box 28083, Central Station, Washington, D.C. 20005. The letter shall contain the registrant's name, address, and registration number as printed on the certificate of registration, and the substances and/or schedules to be added to his/her registration or the new name or address and shall be signed in accordance with §1301.13(j). If the registrant is seeking to handle additional controlled substances listed in Schedule I for the purpose of research or instructional activities, he/she shall attach three copies of a research protocol describing each research project involving the additional substances, or two copies of a statement describing the nature, extent, and duration of such instructional activities, as appropriate. No fee shall be required to be paid for the modification. The request for modification shall be handled in the same manner as an application for registration. If the modification in registration is approved, the Administrator shall issue a new certificate of registration (DEA Form 223) to the registrant, who shall maintain it with the old certificate of registration until expiration.

## Section 1301.52 Termination of Registration; Transfer of Registration; Distribution upon Discontinuance of Business

(a) Except as provided in paragraph (b) of this section, the registration of any person shall terminate if and when such person dies, ceases legal existence, or discontinues business or professional practice. Any registrant who ceases legal existence or discontinues business or professional practice shall notify the Administrator promptly of such fact.

(b) No registration or any authority conferred thereby shall be assigned or otherwise transferred except upon such conditions as the Administration may specifically designate and then only pursuant to written consent. Any person seeking authority to transfer a registration shall submit a written request, providing full details regarding the proposed transfer of registration, to the Deputy Assistant Administrator, Office of Diversion Control, Drug Enforcement Administration, Department of Justice, Washington, D.C. 20537.

(c) Any registrant desiring to discontinue business activities altogether or with respect to controlled substances (without transferring such business activities to another person) shall return for cancellation his/her certificate of registration, and any unexecuted order forms in his/her possession, to the Registration Unit, Drug Enforcement Administration, Department of Justice, Post Office Box 28083, Central Station, Washington, D.C. 20005. Any controlled substances in his/her possession may be disposed of in accordance with §1307.21 of this chapter.

(d) Any registrant desiring to discontinue business activities altogether or with respect to controlled substance (by transferring such business activities to another person) shall submit in person or by registered or certified mail, return receipt requested, to the Special Agent in Charge in his/her area, at least 14 days in advance of the date of the proposed transfer (unless the Special Agent in Charge waives this time limitation in individual instances), the following information:

   (1) The name, address, registration number, and authorized business activity of the registrant discontinuing the business (registrant-transferor);

   (2) The name, address, registration number, and authorized business activity of the person acquiring the business (registrant-transferee);

   (3) Whether the business activities will be continued at the location registered by the person discontinuing business, or moved to another location (if the latter, the address of the new location should be listed);

   (4) Whether the registrant-transferor has a quota to manufacture or procure any controlled substance listed in Schedule I or II (if so, the basic class or class of the substance should be indicated); and

   (5) The date on which the transfer of controlled substances will occur.

(e) Unless the registrant-transferor is informed by the Special Agent in Charge, before the date on which the transfer was stated to occur, that the transfer may not occur, the registrant-transferor may distribute (without being registered to distribute) controlled substances in his/her possession to the registrant-transferee in accordance with the following:

   (1) On the date of transfer of the controlled substances, a complete inventory of all controlled substances being transferred shall be taken in accordance with §1304.11 of this chapter. This inventory shall serve as the final inventory of the registrant-transferor and the initial inventory of the registrant-transferee, and a copy of the inventory shall be included in the records of each person. It shall not be necessary to file a copy of the inventory with the Administration unless requested by the Special Agent in Charge. Transfers of any substances listed in Schedule I or II shall require the use of order forms in accordance with part 1305 of this chapter.

   (2) On the date of transfer of the controlled substances, all records required to be kept by the registrant-transferor with reference to the controlled substances being transferred, under part 1304 of this chapter, shall be transferred to the registrant-transferee. Responsibility for the accuracy of records prior to the date of transfer remains with the transferor, but responsibility for custody and maintenance shall be upon the transferee.

   (3) In the case of registrants required to make reports pursuant to part 1304 of this chapter, a report marked "Final" will be prepared and submitted by the registrant-transferor showing the disposition of all the controlled substances for which a report is required; no additional report will be required from him, if no further transactions involving controlled substances are consummated by him. The initial report of the registrant-transferee shall account for transactions beginning with the day next succeeding the date of discontinuance or transfer of business by the transferor-registrant and the substances transferred to him shall be reported as receipts in his/her initial report.

# SECURITY REQUIREMENTS

## SECTION 1301.71 SECURITY REQUIREMENTS GENERALLY

(a) All applicants and registrants shall provide effective controls and procedures to guard against theft and diversion of controlled substances. In order to determine whether a registrant has provided effective controls against diversion, the Administrator shall use the security requirements set forth in §1301.72-1301.76 as standards for the physical security controls and operating procedures necessary to prevent diversion. Materials and construction which will provide a structural equivalent to the physical security controls set forth in §1301.72, 1301.73 and 1301.75 may be used in lieu of the materials and construction described in those sections.

(b) Substantial compliance with the standards set forth in §1301.72-1301.76 may be deemed sufficient by the Administrator after evaluation of the overall security system and needs of the applicant or registrant. In evaluating the overall security system of a registrant or applicant, the Administrator may consider any of the following factors as he may deem relevant to the need for strict compliance with security requirements:

   (1) The type of activity conducted (e.g., processing of bulk chemicals, preparing dosage forms, packaging, labeling, cooperative buying, etc.);

   (2) The type and form of controlled substances handled (e.g., bulk liquids or dosage units, usable powders or nonusable powders);

   (3) The quantity of controlled substances handled;

   (4) The location of the premises and the relationship such location bears on security needs;

   (5) The type of building construction comprising the facility and the general characteristics of the building or buildings;

   (6) The type of vault, safe, and secure enclosures or other storage system (e.g., automatic storage and retrieval system) used;

   (7) The type of closures on vaults, safes, and secure enclosures;

   (8) The adequacy of key control systems and/or combination lock control systems;

   (9) The adequacy of electric detection and alarm systems, if any including use of supervised transmittal lines and standby power sources;

   (10) The extent of unsupervised public access to the facility, including the presence and characteristics of perimeter fencing, if any;

   (11) The adequacy of supervision over employees having access to manufacturing and storage areas;

   (12) The procedures for handling business guests, visitors, maintenance personnel, and non-employee service personnel;

   (13) The availability of local police protection or of the registrant's or applicant's security personnel, and;

   (14) The adequacy of the registrant's or applicant's system for monitoring the receipt, manufacture, distribution, and disposition of controlled substances in its operations.

(c) When physical security controls become inadequate as a result of a controlled substance being transferred to a different schedule, or as a result of a noncontrolled substance being listed on any schedule, or as a result of a significant increase in the quantity of controlled substances in the possession of the registrant during normal business operations, the physical security controls shall be expanded and extended accordingly. A registrant may adjust physical security controls within the requirements set forth in §1301.72-1301.76 when the need for such controls decreases as a result of a controlled substance being transferred to a different schedule, or a result of a controlled substance being removed from control, or as a result of a significant decrease in the quantity of controlled substances in the possession of the registrant during normal business operations.

(d) Any registrant or applicant desiring to determine whether a proposed security system substantially complies with, or is the structural equivalent of, the requirements set forth in §1301.72-1301.76 may submit any plans, blueprints, sketches or other materials regarding the proposed security system either to the Special Agent in Charge in the region in which the system will be used, or to the Diversion Operations Section, Drug Enforcement Administration, Department of Justice, Washington, D.C. 20537.

(e) Physical security controls of locations registered under the Harrison Narcotic Act or the Narcotics Manufacturing Act of 1960 on April 30, 1971, shall be deemed to comply substantially with the standards set forth in §1301.72, 1301.73 and 1301.75. Any new facilities or work or storage areas constructed or utilized for controlled substances, which facilities or work or storage areas have not been previously approved by the Administration, shall not necessarily be deemed to comply substantially with the standards set forth in §1301.72, 1301.73 and 1301.75, notwithstanding that such facilities or work or storage areas have physical security controls similar to those previously approved by the Administration.

## Section 1301.72 Physical Security Controls for Non-Practitioners; Narcotic Treatment Programs and Compounders for Narcotic Treatment Programs; Storage Areas

(a) Schedules I and II. Raw material, bulk materials awaiting further processing, and finished products which are controlled substances listed in Schedule I or II (except GHB that is manufactured or distributed in accordance with an exemption under section 505(i) of the FFDCA which shall be subject to the requirements of paragraph (b) of this section) shall be stored in one of the following secured areas:

(1) Where small quantities permit, a safe or steel cabinet;

   (i) Which safe or steel cabinet shall have the following specifications or the equivalent: 30 man-minutes against surreptitious entry, 10 man-minutes against forced entry, 20 man-hours against lock manipulation, and 20 man-hours against radiological techniques;

   (ii) Which safe or steel cabinet, if it weighs less than 750 pounds, is bolted or cemented to the floor or wall in such a way that it cannot be readily removed; and

   (iii) Which safe or steel cabinet, if necessary, depending upon the quantities and type of controlled substances stored, is equipped with an alarm system which, upon attempted unauthorized entry, shall transmit a signal directly to a central protection company or a local or State police agency which has a legal duty to respond, or a 24-hour control station operated by the registrant, or such other protection as the Administrator may approve.

(2) A vault constructed before, or under construction on, September 1, 1971, which is of substantial construction with a steel door, combination or key lock, and an alarm system; or

(3) A vault constructed after September 1, 1971:

   (i) The walls, floors, and ceilings of which vault are constructed of at least 8 inches of reinforced concrete or other substantial masonry, reinforced vertically and horizontally with 1/2-inch steel rods tied 6 inches on center, or the structural equivalent to such reinforced walls, floors, and ceilings;

(ii) The door and frame unit of which vault shall conform to the following specifications or the equivalent: 30 man-minutes against surreptitious entry, 10 man-minutes against forced entry, 20 man-hours against lock manipulation, and 20 man-hours against radiological techniques;

(iii) Which vault, if operations require it to remain open for frequent access, is equipped with a "day-gate" which is self-closing and self-locking, or the equivalent, for use during the hours of operation in which the vault door is open;

(iv) The walls or perimeter of which vault are equipped with an alarm, which upon unauthorized entry shall transmit a signal directly to a central station protection company, or a local or State police agency which has a legal duty to respond, or a 24-hour control station operated by the registrant, or such other protection as the Administrator may approve, and, if necessary, holdup buttons at strategic points of entry to the perimeter area of the vault;

(v) The door of which vault is equipped with contact switches; and

(vi) Which vault has one of the following: Complete electrical lacing of the walls, floor and ceilings; sensitive ultrasonic equipment within the vault; a sensitive sound accumulator system; or such other device designed to detect illegal entry as may be approved by the Administration.

(b) Schedules III, IV and V. Raw material, bulk materials awaiting further processing, and finished products which are controlled substances listed in Schedules III, IV, and V, and GHB when it is manufactured or distributed in accordance with an exemption under section 505(i) of the FFDCA, shall be stored in the following secure storage areas:

(1) A safe or steel cabinet as described in paragraph (a)(1) of this section;

(2) A vault as described in paragraph (a)(2) or (3) of this section equipped with an alarm system as described in paragraph (b)(4)(v) of this section;

(3) A building used for storage of Schedules III through V controlled substances with perimeter security which limits access during working hours and provides security after working hours and meets the following specifications:

(i) Has an electronic alarm system as described in paragraph (b)(4)(v) of this section,

(ii) Is equipped with self-closing, self-locking doors constructed of substantial material commensurate with the type of building construction, provided, however, a door which is kept closed and locked at all times when not in use and when in use is kept under direct observation of a responsible employee or agent of the registrant is permitted in lieu of a self-closing, self-locking door. Doors may be sliding or hinged. Regarding hinged doors, where hinges are mounted on the outside, such hinges shall be sealed, welded or otherwise constructed to inhibit removal. Locking devices for such doors shall be either of the multiple-position combination or key lock type and:

(a) In the case of key locks, shall require key control which limits

(b) In the case of combination locks, the combination shall be limited to a minimum number of employees and can be changed upon termination of employment of an employee having knowledge of the combination;

(4) A cage, located within a building on the premises, meeting the following specifications:

(i) Having walls constructed of not less than No. 10 gauge steel fabric mounted on steel posts, which posts are:

(a) At least one inch in diameter;

(b) Set in concrete or installed with lag bolts that are pinned or brazed; and

(c) Which are placed no more than ten feet apart with horizontal one and one-half inch reinforcements every sixty inches;

(ii) Having a mesh construction with openings of not more than two and one-half inches across the square,

(iii) Having a ceiling constructed of the same material, or in the alternative, a cage shall be erected which reaches and is securely attached to the structural ceiling of the building. A lighter gauge mesh may be used for the ceilings of large enclosed areas if walls are at least 14 feet in height,

(iv) Is equipped with a door constructed of No. 10 gauge steel fabric on a metal door frame in a metal door flange, and in all other respects conforms to all the requirements of 21 CFR 1301.72(b)(3)(ii), and

(v) Is equipped with an alarm system which upon unauthorized entry shall transmit a signal directly to a central station protection agency or a local or state police agency, each having a legal duty to respond, or to a 24-hour control station operated by the registrant, or to such other source of protection as the Administrator may approve;

(5) An enclosure of masonry or other material, approved in writing by the Administrator as providing security comparable to a cage;

(6) A building or enclosure within a building which has been inspected and approved by DEA or its predecessor agency, BND, and continues to provide adequate security against the diversion of Schedule III through V controlled substances, of which fact written acknowledgment has been made by the Special Agent in Charge of DEA for situated;

(7) Such other secure storage areas as may be approved by the Administrator after considering the factors listed in §1301.71(b), (1) through (14);

(8) (i) Schedule III through V controlled substances may be stored with Schedules I and II controlled substances under security measures provided by 21 CFR 1301.72(a);

(ii) Non-controlled drugs, substances and other materials may be stored with Schedule III through V controlled substances in any of the secure storage areas required by 21 CFR 1301.72(b), provided that permission for such storage of non-controlled items is obtained in advance, in writing, from the Special Agent in Charge of DEA for the area in which such storage area is situated. Any such permission tendered must be upon the Special Agent in Charge's written determination that such non-segregated storage does not diminish security effectiveness for Schedules III through V controlled substances.

(c) Multiple storage areas. Where several types or classes of controlled substances are handled separately by the registrant or applicant for different purposes (e.g., returned goods, or goods in process), the controlled substances may be stored separately, provided that each storage area complies with the requirements set forth in this section.

(d) Accessibility to storage areas. The controlled substances storage areas shall be accessible only to an absolute minimum number of specifically authorized employees. When it is necessary for employee maintenance personnel, nonemployee maintenance personnel, business guests, or visitors to be present in or pass through controlled substances storage areas, the registrant shall provide for adequate observation of the area by an employee specifically authorized in writing.

## Section 1301.73 Physical Security Controls for Non-Practitioners; Compounders for Narcotic Treatment Programs; Manufacturing and Compounding Areas

All manufacturing activities (including processing, packaging and labeling) involving controlled substances listed in any schedule and all activities of compounders shall be conducted in accordance with the following:

(a) All in-process substances shall be returned to the controlled substances storage area at the termination of the process. If the process is not terminated at the end of a workday (except where a continuous process or other normal manufacturing operation should not be interrupted), the processing area or tanks, vessels, bins or bulk containers containing such substances shall be securely locked, with adequate security for the area or building. If such security requires an alarm, such alarm, upon unauthorized entry, shall transmit a signal directly to a central station protection company, or local or state police agency which has a legal duty to respond, or a 24-hour control station operated by the registrant.

(b) Manufacturing activities with controlled substances shall be conducted in an area or areas of clearly defined limited access which is under surveillance by an employee or employees designated in writing as responsible for the area. "Limited access" may be provided, in the absence of physical dividers such as walls or partitions, by traffic control lines or restricted space designation. The employee designated as responsible for the area may be engaged in the particular manufacturing operation being conducted, provided, that he is able to provide continuous surveillance of the area in order that unauthorized persons may not enter or leave the area without his knowledge.

(c) During the production of controlled substances, the manufacturing areas shall be accessible to only those employees required for efficient operation. When it is necessary for employee maintenance personnel, nonemployee maintenance personnel, business guests, or visitors to be present in or pass through manufacturing areas during production of controlled substances, the registrant shall provide for adequate observation of the area by an employee specifically authorized in writing.

## Section 1301.74 Other Security Controls for Non-Practitioners; Narcotic Treatment Programs and Compounders for Narcotic Treatment Programs

(a) Before distributing a controlled substance to any person who the registrant does not know to be registered to possess the controlled substance, the registrant shall make a good faith inquiry either with the Administration or with the appropriate State controlled substances registration agency, if any, to determine that the person is registered to possess the controlled substance.

(b) The registrant shall design and operate a system to disclose to the registrant suspicious orders of controlled substances. The registrant shall inform the Field Division Office of the Administration in his area of suspicious orders when discovered by the registrant. Suspicious orders include orders of unusual size, orders deviating substantially from a normal pattern, and orders of unusual frequency.

(c) The registrant shall notify the Field Division Office of the Administration in his area of any theft or significant loss of any controlled substances upon discovery of such theft or loss. The supplier shall be responsible for reporting in-transit losses of controlled substances by the common or contract carrier selected pursuant to §1301.74(e), upon discovery of such theft or loss. The registrant shall also complete DEA Form 106 regarding such theft or loss. Thefts must be reported whether or not the controlled substances are subsequently recovered and/or the responsible parties are identified and action taken against them.

(d) The registrant shall not distribute any controlled substance listed in Schedules II through V as a complimentary sample to any potential or current customer (1) without the prior written request of the customer, (2) to be used only for satisfying the legitimate medical needs of patients of the customer, and (3) only in reasonable quantities. Such request must contain the name, address, and registration number of the customer and the name and quantity of the specific controlled substance desired. The request shall be preserved by the registrant with other records of distribution of controlled substances. In addition, the requirements of part 1305 of the chapter shall be complied with for any distribution of a controlled substance listed in Schedule II. For purposes of this paragraph, the term "customer" includes a person to whom a complimentary sample of a substance is given in order to encourage the pre- scribing or recommending of the substance by the person.

(e) When shipping controlled substances, a registrant is responsible for selecting common or contract carriers which provide adequate security to guard against in-transit losses. When storing controlled substances in a public warehouse, a registrant is responsible for selecting a warehouseman which will provide adequate security to guard against storage losses; wherever possible, the registrant shall store controlled substances in a public warehouse which complies with the requirements set forth in §1301.72. In addition, the registrant shall employ precautions (e.g., assuring that shipping containers do not indicate that contents are controlled substances) to guard against storage or in-transit losses.

(f) When distributing controlled substances through agents (e.g., detailmen), a registrant is responsible for providing and requiring adequate security to guard against theft and diversion while the substances are being stored or handled by the agent or agents.

(g) Before the initial distribution of carfentanil etorphine hydrochloride and/or diprenorphine to any person, the registrant must verify that the person is authorized to handle the sub- stances(s) by contacting the Drug Enforcement Administration.

(h) The acceptance of delivery of narcotic substances by a narcotic treatment program shall be made only by a licensed practitioner employed at the facility or other authorized individuals designated in writing. At the time of delivery, the licensed practitioner or other authorized individual designated in writing (excluding persons currently or previously dependent on narcotic drugs), shall sign for the narcotics and place his specific title (if any) on any invoice. Copies of these signed invoices shall be kept by the distributor.

(i) Narcotics dispensed or administered at a narcotic treatment program will be dispensed or administered directly to the patient by either (1) the licensed practitioner, (2) a registered nurse under the direction of the licensed practitioner, (3) a licensed practical nurse under the direction of the licensed practitioner, or (4) a pharmacist under the direction of the licensed practitioner.

(j) Persons enrolled in a narcotic treatment program will be required to wait in an area physically separated from the narcotic storage and dispensing area. This requirement will be enforced by the program physician and employees.

(k) All narcotic treatment programs must comply with standards established by the Secretary of Health and Human Services (after consultation with the Administration) respecting the quantities of narcotic drugs which may be provided to persons enrolled in a narcotic treatment program for unsupervised use.

(l) DEA may exercise discretion regarding the degree of security required in narcotic treatment programs based on such factors as the location of a program, the number of patients enrolled in a program and the number of physicians, staff members and security guards. Similarly, such factors will be taken into consideration when evaluating existing security or requiring new security at a narcotic treatment program.

## SECTION 1301.75 PHYSICAL SECURITY CONTROLS FOR PRACTITIONERS

(a) Controlled substances listed in Schedule I shall be stored in a securely locked, substantially constructed cabinet.

(b) Controlled substances listed in Schedules II, III, IV, and V shall be stored in a securely locked, substantially constructed cabinet. However, pharmacies and institutional practitioners may disperse such substances throughout the stock of noncontrolled substances in such a manner as to obstruct the theft or diversion of the controlled substances.

(c) This section shall also apply to nonpractitioners authorized to conduct research or chemical analysis under another registration.

(d) Carfentanil etorphine hydrochloride and diprenorphine shall be stored in a safe or steel cabinet equivalent to a U.S. Government Class V security container.

## SECTION 1301.76 OTHER SECURITY CONTROLS FOR PRACTITIONERS

(a) The registrant shall not employ, as an agent or employee who has access to controlled substances, any person who has been convicted of a felony offense relating to controlled substances or who, at any time, had an application for registration with the DEA denied, had a DEA registration revoked or has surrendered a DEA registration for cause. For purposes of this subsection, the term "for cause" means a surrender in lieu of, or as a consequence of, any federal or state administrative, civil or criminal action resulting from an investigation of the individual's handling of controlled substances.

(b) The registrant shall notify the Field Division Office of the Administration in his area of the theft or significant loss of any controlled substances upon discovery of such loss or theft. The registrant shall also complete DEA (or BND) Form 106 regarding such loss or theft.

(c) Whenever the registrant distributes a controlled substance (without being registered as a distributor, as permitted in §1301.13(e)(1) and/or §1307.11-1307.12) he/she shall comply with the requirements imposed on nonpractitioners in §1301.74 (a), (b), and (e).

## SECTION 1301.77 SECURITY CONTROLS FOR FREIGHT FORWARDING FACILITIES

(a) All Schedule II-V controlled substances that will be temporarily stored at the freight forwarding facility must be either:
   (1) stored in a segregated area under constant observation by designated responsible individual(s); or
   (2) stored in a secured area that meets the requirements of §1301.72(b) of this Part. For purposes of this requirement, a facility that may be locked down (i.e., secured against physical entry in a manner consistent with requirements of §1301.72(b)(3)(ii) of this part) and has a monitored alarm system or is subject to continuous monitoring by security personnel will be deemed to meet the requirements of §1301.72(b)(3) of this Part.

(b) Access to controlled substances must be kept to an absolute minimum number of specifically authorized individuals. Non-authorized individuals may not be present in or pass through controlled substances storage areas without adequate observation provided by an individual authorized in writing by the registrant.

(c) Controlled substances being transferred through a freight forwarding facility must be packed in sealed, unmarked shipping containers.

## EMPLOYEE SCREENING — NON-PRACTITIONERS

### SECTION 1301.90 EMPLOYEE SCREENING PROCEDURES

It is the position of DEA that the obtaining of certain information by non-practitioners is vital to fairly assess the likelihood of an employee committing a drug security breach. The need to know this information is a matter of business necessity, essential to overall controlled substances security. In this regard, it is believed that conviction of crimes and unauthorized use of controlled substances are activities that are proper subjects for inquiry. It is, therefore, assumed that the following questions will become a part of an employer's comprehensive employee screening program:

Question. Within the past five years, have you been convicted of a felony, or within the past two years, of any misdemeanor or are you presently formally charged with committing a criminal offense? (Do not include any traffic violations, juvenile offenses or military convictions, except by general court-martial.) If the answer is yes, furnish details of conviction, offense, location, date and sentence.

Question. In the past three years, have you ever knowingly used any narcotics, amphetamines or barbiturates, other than those prescribed to you by a physician? If the answer is yes, furnish details.

Advice. An authorization, in writing, that allows inquiries to be made of courts and law enforcement agencies for possible pending charges or convictions must be executed by a person who is allowed to work in an area where access to controlled substances clearly exists. A person must be advised that any false information or omission of information will jeopardize his or her position with respect to employment. The application for employment should inform a person that information furnished or recovered as a result of any inquiry will not necessarily preclude employment, but will be considered as part of an overall evaluation of the person's qualifications. The maintaining of fair employment practices, the protection of the person's right of privacy, and the assurance that the results of such inquiries will be treated by the employer in confidence will be explained to the employee.

### SECTION 1301.91 EMPLOYEE RESPONSIBILITY TO REPORT DRUG DIVERSION

Reports of drug diversion by fellow employees is not only a necessary part of an overall employee security program but also serves the public interest at large. It is, therefore, the position of DEA that an employee who has knowledge of drug diversion from his employer by a fellow employee has an obligation to report such information to a responsible security official of the employer. The employer shall treat such information as confidential and shall take all reasonable steps to protect the confidentiality of the information and the identity of the employee furnishing information. A failure to report information of drug diversion will be considered in determining the feasibility of continuing to allow an employee to work in a drug security area. The employer shall inform all employees concerning this policy.

## SECTION 1301.92 ILLICIT ACTIVITIES BY EMPLOYEES

It is the position of DEA that employees who possess, sell, use or divert controlled substances will subject themselves not only to State or Federal prosecution for any illicit activity, but shall also immediately become the subject of independent action regarding their continued employment. The employer will assess the seriousness of the employee's violation, the position of responsibility held by the employee, past record of employment, etc., in determining whether to suspend, transfer, terminate or take other action against the employee.

## SECTION 1301.93 SOURCES OF INFORMATION FOR EMPLOYEE CHECKS

DEA recommends that inquiries concerning employees' criminal records be made as follows: Local inquiries. Inquiries should be made by name, date and place of birth, and other identifying information, to local courts and law enforcement agencies for records of pending charges and convictions. Local practice may require such inquiries to be made in person, rather than by mail, and a copy of an authorization from the employee may be required by certain law enforcement agencies.

DEA inquiries. Inquiries supplying identifying information should also be furnished to DEA Field Division Offices along with written consent from the concerned individual for a check of DEA files for records of convictions. The Regional check will result in a national check being made by the Field Division Office.

# Part 1302
## Labeling and Packaging Requirements for Controlled Substances

## CONTENTS

### SECTION 1302.01 SCOPE OF PART 1302

Requirements governing the labeling and packaging of controlled substances pursuant to sections 1305 and 1008(d) of the Act (21 U.S.C. 825 and 958(d)) are set forth generally by those sections and specifically by the sections of this part.

### SECTION 1302.02 DEFINITIONS

Any term contained in this part shall have the definition set forth in section 102 of the Act (21 U.S.C. 802) or part 1300 of this chapter.

### SECTION 1302.03 SYMBOL REQUIRED; EXCEPTIONS

(a) Each commercial container of a controlled substance (except for a controlled substance excepted by the Administrator pursuant to §1308.31 of this chapter) shall have printed on the label the symbol designating the schedule in which such controlled substance is listed. Each such commercial container, if it otherwise has no label, must bear a label complying with the requirement of this part.

(b) Each manufacturer shall print upon the labeling of each controlled substance distributed by him the symbol designating the schedule in which such controlled substance is listed.

(c) The following symbols shall designate the schedule corresponding thereto:

Schedule

| Schedule I | CI or C-I |
| Schedule II | CII or C-II |
| Schedule III | CIII or C-III |
| Schedule IV | CIV or C-IV |
| Schedule V | CV or C-V |

The word "schedule" need not be used. No distinction need be made between narcotic and nonnarcotic substances.

(d) The symbol is not required on a carton or wrapper in which a commercial container is held if the symbol is easily legible through such carton or wrapper.

(e) The symbol is not required on a commercial container too small or otherwise unable to accommodate a label, if the symbol is printed on the box or package from which the commercial container is removed upon dispensing to an ultimate user.

(f) The symbol is not required on a commercial container containing, or on the labeling of, a controlled substance being utilized in clinical research involving blind and double blind studies.

## SECTION 1302.04 LOCATION AND SIZE OF SYMBOL ON LABEL AND LABELING

The symbol shall be prominently located on the label or the labeling of the commercial container and/or the panel of the commercial container normally displayed to dispensers of any controlled substance. The symbol on labels shall be clear and large enough to afford easy identification of the schedule of the controlled substance upon inspection without removal from the dispenser's shelf. The symbol on all other labeling shall be clear and large enough to afford prompt identification of the controlled substance upon inspection of the labeling.

## SECTION 1302.05 EFFECTIVE DATES OF LABELING REQUIREMENTS

All labels on commercial containers of, and all labeling of, a controlled substance which either is transferred to another schedule or is added to any schedule shall comply with the requirements of §1302.03, on or before the effective date established in the final order for the transfer or addition.

## SECTION 1302.06 SEALING OF CONTROLLED SUBSTANCES

On each bottle, multiple dose vial, or other commercial container of any controlled substance, there shall be securely affixed to the stopper, cap, lid, covering, or wrapper or such container a seal to disclose upon inspection any tampering or opening of the container.

## SECTION 1302.07 LABELING AND PACKAGING REQUIREMENTS FOR IMPORTED AND EXPORTED SUBSTANCES

(a) The symbol requirements of §1302.03-1302.05 apply to every commercial container containing, and to all labeling of, controlled substances imported into the jurisdiction of and/or the customs territory of the United States.

(b) The symbol requirements of §1302.03-1302.05 do not apply to any commercial containers containing, or any labeling of, a controlled substance intended for export from the jurisdiction of the United States.

(c) The sealing requirements of §1302.06 apply to every bottle, multiple dose vial, or other commercial container of any controlled substance listed in schedule I or II, or any narcotic controlled substance listed in schedule III or IV, imported into, exported from, or intended for export from, the jurisdiction of and/or the customs territory of the United States.

# Part 1303
## Quotas

## CONTENTS

### SECTION 1303.01 SCOPE OF PART 1303

Procedures governing the establishment of production and manufacturing quotas on basic classes of controlled substances listed in Schedules I and II pursuant to section 306 of the Act (21 U.S.C. 826) are governed generally by that section and specifically by the sections of this part.

### SECTION 1303.02 DEFINITIONS

Any term contained in this part shall have the definition set forth in section 102 of the Act (21 U.S.C. 802) or part 1300 of this chapter.

# AGGREGATE PRODUCTION AND PROCUREMENT QUOTAS

## SECTION 1303.11 AGGREGATE PRODUCTION QUOTAS

(a) The Administrator shall determine the total quantity of each basic class of controlled substance listed in Schedule I or II necessary to be manufactured during the following calendar year to provide for the estimated medical, scientific, research and industrial needs of the United States, for lawful export requirements, and for the establishment and maintenance of reserve stocks.

(b) In making his determinations, the Administrator shall consider the following factors:

(1) Total net disposal of the class by all manufacturers during the current and 2 preceding years;

(2) Trends in the national rate of net disposal of the class;

(3) Total actual (or estimated) inventories of the class and of all substances manufactured from the class, and trends in inventory accumulation;

(4) Projected demand for such class as indicated by procurement quotas requested pursuant to §1303.12; and

(5) Other factors affecting medical, scientific, research, and industrial needs in the United States and lawful export requirements, as the Administrator finds relevant, including changes in the currently accepted medical use in treatment with the class or the substances which are manufactured from it, the economic and physical availability of raw materials for use in manufacturing and for inventory purposes, yield and stability problems, potential disruptions to production (including possible labor strikes), and recent unforeseen emergencies such as floods and fires.

(c) The Administrator shall, on or before May 1 of each year, publish in the Federal Register, general notice of an aggregate production quota for any basic class determined by him under this section. A copy of said notice shall be mailed simultaneously to each person registered as a bulk manufacturer of the basic class. The Administrator shall permit any interested person to file written comments on or objections to the proposal and shall designate in the notice the time during which such filings may be made. The Administrator may, but shall not be required to, hold a public hearing on one or more issues raised by the comments and objections filed with him. In the event the Administrator decides to hold such a hearing, he shall publish notice of the hearing in the Federal Register, which notice shall summarize the issues to be heard and shall set the time for the hearing which shall not be less than 30 days after the date of publication of the notice. After consideration of any comments or objections, or after a hearing if one is ordered by the Administrator, the Administrator shall issue and publish in the Federal Register his final order determining the aggregate production quota for the basic class of controlled substance. The order shall include the findings of fact and conclusions of law upon which the order is based. The order shall specify the date on which it shall take effect. A copy of said order shall be mailed simultaneously to each person registered as a bulk manufacturer of the basic class.

## SECTION 1303.12 PROCUREMENT QUOTAS

(a) In order to determine the estimated needs for, and to insure an adequate and uninterrupted supply of, basic classes of controlled substances listed in Schedules I and II (except raw opium being imported by the registrant pursuant to an import permit) the Administrator shall issue procurement quotas authorizing persons to procure and use quantities of each basic class of such substances for the purpose of manufacturing such class into dosage forms or into other substances.

(b) Any person who is registered to manufacture controlled substances listed in any schedule and who desires to use during the next calendar year any basic class of controlled substances listed in Schedule I or II (except raw opium being imported by the registrant pursuant to an import permit) for purposes of manufacturing, shall apply on DEA Form 250 for a procurement quota for such basic class. A separate application must be made for each basic class desired to be procured or used. The applicant shall state whether he intends to manufacture the basic class himself or purchase it from another manufacturer. The applicant shall state separately each purpose for which the basic class is desired, the quantity desired for that purpose during the next calendar year, and the quantities used and estimated to be used, if any, for that purpose during the current and preceding 2 calendar years. If the purpose is to manufacture the basic class into dosage form, the applicant shall state the official name, common or usual name, chemical name, or brand name of that form. If the purpose is to manufacture another substance, the applicant shall state the official name, common or usual name, chemical name, or brand name of the substance, and, if a controlled substance listed in any schedule, the schedule number and Administration Controlled Substances Code Number, as set forth in part 1308 of this chapter, of the substance. If the purpose is to manufacture another basic class of controlled substance listed in Schedule I or II, the applicant shall also state the quantity of the other basic class which the applicant has applied to manufacture pursuant to §1303.22 and the quantity of the first basic class necessary to manufacture a specified unit of the second basic class. DEA Form 250 shall be filed on or before April 1 of the year preceding the calendar year for which the procurement quota is being applied. Copies of DEA Form 250 may be obtained from, and shall be filed with, the Drug and Chemical Evaluation Section, Drug Enforcement Administration, Department of Justice, Washington, D.C. 20537.

(c) The Administrator shall, on or before July 1 of the year preceding the calendar year during which the quota shall be effective, issue to each qualified applicant a procurement quota authorizing him to procure and use:

(1) All quantities of such class necessary to manufacture all quantities of other basic classes of controlled substances listed in Schedules I and II which the applicant is authorized to manufacture pursuant to §1303.23; and

(2) Such other quantities of such class as the applicant has applied to procure and use and are consistent with his past use, his estimated needs, and the total quantity of such class that will be produced.

(d) Any person to whom a procurement quota has been issued may at any time request an adjustment in the quota by applying to the Administrator with a statement showing the need for the adjustment. Such application shall be filed with the Drug and Chemical Evaluation Section, Drug Enforcement Administration, Department of Justice, Washington, D.C. 20537. The Administrator shall increase or decrease the procurement quota of such person if and to the extent that he finds, after considering the factors enumerated in paragraph (c) of this section and any occurrences since the issuance of the procurement quota, that the need justifies an adjustment.

(e) The following persons need not obtain a procurement quota:

(1) Any person who is registered to manufacture a basic class of controlled substance listed in Schedule I or II and who uses all of the quantity he manufactures in the manufacture of a substance not controlled under the Act;

(2) Any person who is registered or authorized to conduct chemical analysis with controlled substances (for controlled substances to be used in such analysis only); and

(3) Any person who is registered to conduct research with a basic class of controlled substance listed in Schedule I or II and who is authorized to manufacture a quantity of such class pursuant to §1301.13 of this chapter.

(f) Any person to whom a procurement quota has been issued, authorizing that person to procure and use a quantity of a basic class of controlled substances listed in Schedules I or II during the current calendar year, shall, at or before the time of giving an order to another manufacturer requiring the distribution of a quantity of such basic class, certify in writing to such other manufacturer that the quantity of such basic class ordered does not exceed the person's unused and available procurement quota of such basic class for the current calendar year. The written certification shall be executed by the same individual who signed the DEA Form 222 transmitting the order. Manufacturers shall not fill an order from persons required to apply for a procurement quota under paragraph (b) of this section unless the order is accompanied by a certification as required under this section. The certification required by this section shall contain the following: The date of the certification; the name and address of the bulk manufacturer to whom the certification is directed; a reference to the number of the DEA Form 222 to which the certification applies; the name of the person giving the order to which the certification applies; the name of the basic class specified in the DEA Form 222 to which the certification applies; the appropriate schedule within which is listed the basic class specified in the DEA Form 222 to which the certification applies; a statement that the quantity (expressed in grams) of the basic class specified in the DEA Form 222 to which the certification applies does not exceed the unused and available procurement quota of such basic class, issued to the person giving the order, for the current calendar year; and the signature of the individual who signed the DEA Form 222 to which the certification applies.

## SECTION 1303.13 ADJUSTMENTS OF AGGREGATE PRODUCTION QUOTAS

(a) The Administrator may at any time increase or reduce the aggregate production quota for a basic class of controlled substance listed in Schedule I or II which he has previously fixed pursuant to §1303.11.

(b) In determining to adjust the aggregate production quota, the Administrator shall consider the following factors:

(1) Changes in the demand for that class, changes in the national rate of net disposal of the class, and changes in the rate of net disposal of the class by registrants holding individual manufacturing quotas for that class;

(2) Whether any increased demand for that class, the national and/or individual rates of net disposal of that class are temporary, short term, or long term;

(3) Whether any increased demand for that class can be met through existing inventories, increased individual manufacturing quotas, or increased importation, without increasing the aggregate production quota, taking into account production delays and the probability that other individual manufacturing quotas may be suspended pursuant to §1303.24(b);

(4) Whether any decreased demand for that class will result in excessive inventory accumulation by all persons registered to handle that class (including manufacturers, distributors, practitioners, importers, and exporters), notwithstanding the possibility that individual manufacturing quotas may be suspended pursuant to §1303.24(b) or abandoned pursuant to §1303.27;

(5) Other factors affecting medical, scientific, research, and industrial needs in the United States and lawful export requirements, as the Administrator finds relevant, including changes in the currently accepted medical use in treatment with the class or the substances which are manufactured from it, the economic and physical availability of raw materials for use in manufacturing and for inventory purposes, yield and stability problems, potential disruptions to production (including possible labor strikes), and recent unforeseen emergencies such as floods and fires.

(c) The Administrator in the event he determines to increase or reduce the aggregate production quota for a basic class of controlled substance, shall publish in the Federal Register general notice of an adjustment in the aggregate production quota for that class determined by him under this section. A copy of said notice shall be mailed simultaneously to each person registered as a bulk manufacturer of the basic class. The Administrator shall permit any interested person to file written comments on or objections to the proposal and shall designate in the notice the time during which such filings may be made. The Administrator may, but shall not be required to, hold a public hearing on one or more issues raised by the comments and objections filed with him. In the event the Administrator decides to hold such a hearing, he shall publish notice of the hearing in the Federal Register, which notice shall summarize the issues to be heard and shall set the time for the hearing, which shall not be less than 10 days after the date of publication of the notice. After consideration of any comments or objections, or after a hearing if one is ordered by the Administrator, the Administrator shall issue and publish in the Federal Register his final order determining the aggregate production for the basic class of controlled substance. The order shall include the findings of fact and conclusions of law upon which the order is based. The order shall specify the date on which it shall take effect. A copy of said order shall be mailed simultaneously to each person registered as a bulk manufacturer of the basic class.

## INDIVIDUAL MANUFACTURING QUOTAS

### SECTION 1303.21 INDIVIDUAL MANUFACTURING QUOTAS

(a) The Administrator shall, on or before July 1 of each year, fix for and issue to each person who is registered to manufacture a basic class of controlled substance listed in Schedule I or II, and who applies for a manufacturing quota, an individual manufacturing quota authorizing that person to manufacture during the next calendar year a quantity of that basic class. Any manufacturing quota fixed and issued by the Administrator shall be subject to his authority to reduce or limit it at a later date pursuant to §1303.26 and to his authority to revoke or suspend it at any time pursuant to §1301.36 of this chapter.

(b) No individual manufacturing quota shall be required for registrants listed in §1303.12(e).

## Section 1303.22 Procedure for Applying for Individual Manufacturing Quotas

Any person who is registered to manufacture any basic class of controlled substance listed in Schedule I or II and who desires to manufacture a quantity of such class shall apply on DEA Form 189 for a manufacturing quota for such quantity of such class. Copies of DEA Form 189 may be obtained from, and shall be filed (on or before May 1 of the year preceding the calendar year for which the manufacturing quota is being applied) with, the Drug and Chemical Evaluation Section, Drug Enforcement Administration, Department of Justice, Washington, D.C. 20537. A separate application must be made for each basic class desired to be manufactured. The applicant shall state:

(a) The name and Administration Controlled Substances Code Number, as set forth in part 1308 of this chapter, of the basic class.
(b) For the basic class in each of the current and preceding 2 calendar years,
    (1) The authorized individual manufacturing quota, if any;
    (2) The actual or estimated quantity manufactured;
    (3) The actual or estimated net disposal;
    (4) The actual or estimated inventory allowance pursuant to §1303.24; and
    (5) The actual or estimated inventory as of December 31;
(c) For the basic class in the next calendar year,
    (1) The desired individual manufacturing quota; and
    (2) Any additional factors which the applicant finds relevant to the fixing of his individual manufacturing quota, including the trend of (and recent changes in) his and the national rates of net disposal, his production cycle and current inventory position, the economic and physical availability of raw materials for use in manufacturing and for inventory purposes, yield and stability problems, potential disruptions to production (including possible labor strikes) and recent unforeseen emergencies such as floods and fires.

## Section 1303.23 Procedure for Fixing Individual Manufacturing Quotas

(a) In fixing individual manufacturing quotas for a basic class of controlled substance listed in Schedule I or II, the Administrator shall allocate to each applicant who is currently manufacturing such class a quota equal to 100 percent of the estimated net disposal of that applicant for the next calendar year, adjusted —
    (1) By the amount necessary to increase or reduce the estimated inventory of the applicant on December 31 of the current year to his estimated inventory allowance for the next calendar year, pursuant to §1303.24, and
    (2) By any other factors which the Administrator deems relevant to the fixing of the individual manufacturing quota of the applicant, including the trend of (and recent changes in) his and the national rates of net disposal, his production cycle and current inventory position, the economic and physical availability of raw materials for use in manufacturing and for inventory purposes, yield and stability problems, potential disruptions to production (including possible labor strikes), and recent unforeseen emergencies such as floods and fires.
(b) In fixing individual manufacturing quotas for a basic class of controlled substance listed in Schedule I or II, the Administrator shall allocate to each applicant who is not currently manufacturing such class a quota equal to 100 percent of the reasonably estimated net

disposal of that applicant for the next calendar year, as determined by the Administrator, adjusted —

(1) By the amount necessary to provide the applicant his estimated inventory allowance for the next calendar year, pursuant to Sec. 1303.24, and

(2) By any other factors which the Administrator deems relevant to the fixing of the individual manufacturing quota of the applicant, including the trend of (and recent changes in) the national rate of net disposal, his production cycle and current inventory position, the economic and physical availability of raw materials for use in manufacturing and for inventory purposes, yield and stability problems, potential disruptions to production (including possible labor strikes), and recent unforeseen emergencies such as floods and fires.

(c) The Administrator shall, on or before March 1 of each year, adjust the individual manufacturing quota allocated for that year to each applicant in paragraph (a) of this section by the amount necessary to increase or reduce the actual inventory of the applicant to December 31 of the preceding year to his estimated inventory allowance for the current calendar year, pursuant to §1303.24.

## SECTION 1303.24 INVENTORY ALLOWANCE

(a) For the purpose of determining individual manufacturing quotas pursuant to §1303.23, each registered manufacturer shall be allowed as a part of such quota an amount sufficient to maintain an inventory equal to,

(1) For current manufacturers, 50 percent of his average estimated net disposal for the current calendar year and the last preceding calendar year; or

(2) For new manufacturers, 50 percent of his reasonably estimated net disposal for the next calendar year as determined by the Administrator.

(b) During each calendar year each registered manufacturer shall be allowed to maintain an inventory of a basic class not exceeding 65 percent of his estimated net disposal of that class for that year, as determined at the time his quota for that year was determined. At any time the inventory of a basic class held by a manufacturer exceeds 65 percent of his estimated net disposal, his quota for that class is automatically suspended and shall remain suspended until his inventory is less than 60 percent of his estimated net disposal. The Administrator may, upon application and for good cause shown, permit a manufacturer whose quota is, or is likely to be, suspended pursuant to this paragraph to continue manufacturing and to accumulate an inventory in excess of 65 percent of his estimated net disposal, upon such conditions and within such limitations as the Administrator may find necessary or desirable.

(c) If, during a calendar year, a registrant has manufactured the entire quantity of a basic class allocated to him under an individual manufacturing quota, and his inventory of that class is less than 40 percent of his estimated net disposal of that class for that year, the Administrator may, upon application pursuant to Sec. 1303.25, increase the quota of such registrant sufficiently to allow restoration of the inventory to 50 percent of the estimated net disposal for that year.

## SECTION 1303.25 INCREASE IN INDIVIDUAL MANUFACTURING QUOTAS

(a) Any registrant who holds an individual manufacturing quota for a basic class of controlled substance listed in Schedule I or II may file with the Administrator an application on

Administration Form 189 for an increase in such quota in order for him to meet his estimated net disposal, inventory and other requirements during the remainder of such calendar year.

(b) The Administrator, in passing upon a registrant's application for an increase in his individual manufacturing quota, shall take into consideration any occurrences since the filing of such registrant's initial quota application that may require an increased manufacturing rate by such registrant during the balance of the calendar year. In passing upon such application the Administrator may also take into consideration the amount, if any, by which his determination of the total quantity for the basic class of controlled substance to be manufactured under §1303.11 exceeds the aggregate of all the individual manufacturing quotas for the basic class of controlled substance, and the equitable distribution of such excess among other registrants.

## SECTION 1303.26 REDUCTION IN INDIVIDUAL MANUFACTURING QUOTAS

The Administrator may at any time reduce an individual manufacturing quota for a basic class of controlled substance listed in Schedule I or II which he has previously fixed in order to prevent the aggregate of the individual manufacturing quotas and import permits outstanding or to be granted from exceeding the aggregate production quota which has been established for that class pursuant to §1303.11, as adjusted pursuant to §1303.13. If a quota assigned to a new manufacturer pursuant to §1303.23(b), or if a quota assigned to any manufacturer is increased pursuant to §1303.24(c), or if an import permit issued to an importer pursuant to part 1312 of this chapter, causes the total quantity of a basic class to be manufactured and imported during the year to exceed the aggregate production quota which has been established for that class pursuant to §1303.11, as adjusted pursuant to §1303.13, the Administrator may proportionately reduce the individual manufacturing quotas and import permits of all other registrants to keep the aggregate production quota within the limits originally established, or, alternatively, the Administrator may reduce the individual manufacturing quota of any registrant whose quota is suspended pursuant to §1303.24(b) or §1301.36 of this chapter, or is abandoned pursuant to §1303.27.

## SECTION 1303.27 ABANDONMENT OF QUOTA

Any manufacturer assigned an individual manufacturing quota for any basic class pursuant to §1303.23 may at any time abandon his right to manufacture all or any part of such quota by filing with the Drug and Chemical Evaluation Section a written notice of such abandonment, stating the name and Administration Controlled Substances Code Number, as set forth in part 1308 of this chapter, of the substance and the amount which he has chosen not to manufacture. The Administrator may, in his discretion, allocate such amount among the other manufacturers in proportion to their respective quotas.

## HEARINGS

## SECTION 1303.31 HEARINGS GENERALLY

(a) In any case where the Administrator shall hold a hearing regarding the determination of an aggregate production quota pursuant to §1303.11(c), or regarding the adjustment of an aggregate production quota pursuant to §1303.13(c), the procedures for such hearing shall be governed generally by the rule making procedures set forth in the Administrative Procedure Act (5 U.S.C. 551-559) and specifically by section 306 of the Act (21 U.S.C. 826), by

§1303.32-1303.37, and by the procedures for administrative hearings under the Act set forth in §1316.41-1316.67 of this chapter.

(b) In any case where the Administrator shall hold a hearing regarding the issuance, adjustment, suspension, or denial of a procurement quota pursuant to §1303.12, or the issuance, adjustment, suspension, or denial of an individual manufacturing quota pursuant to §1303.21-1303.27, the procedures for such hearing shall be governed generally by the adjudication procedures set forth in the Administrative Procedures Act (5 U.S.C. 551-559) and specifically by section 306 of the Act (21 U.S.C. 826), by §1303.32-1303.37, and by the procedures for administrative hearings under the Act set forth in §1316.41-1316.67 of this chapter.

## SECTION 1303.32 PURPOSE OF HEARING

(a) The Administrator may, in his sole discretion, hold a hearing for the purpose of receiving factual evidence regarding any one or more issues (to be specified by him) involved in the determination or adjustment of any aggregate production quota.

(b) If requested by a person applying for or holding a procurement quota or an individual manufacturing quota, the Administrator shall hold a hearing for the purpose of receiving factual evidence regarding the issues involved in the issuance, adjustment, suspension, or denial of such quota to such person, but the Administrator need not hold a hearing on the suspension of a quota pursuant to §1301.36 of this chapter separate from a hearing on the suspension of registration pursuant to those sections.

(c) Extensive argument should not be offered into evidence but rather presented in opening or closing statements of counsel or in memoranda or proposed findings of fact and conclusions of law.

## SECTION 1303.33 WAIVER OR MODIFICATION OF RULES

The Administrator or the presiding officer (with respect to matters pending before him) may modify or waive any rule in this part by notice in advance of the hearing, if he determines that no party in the hearing will be unduly prejudiced and the ends of justice will thereby be served. Such notice of modification or waiver shall be made a part of the record of the hearing.

## SECTION 1303.34 REQUEST FOR HEARING OR APPEARANCE; WAIVER

(a) Any applicant or registrant who desires a hearing on the issuance, adjustment, suspension, or denial of his procurement and/or individual manufacturing quota shall, within 30 days after the date of receipt of the issuance, adjustment, suspension, or denial of such quota, file with the Administrator a written request for a hearing in the form prescribed in §1316.47 of this chapter. Any interested person who desires a hearing on the determination of an aggregate production quota shall, within the time prescribed in §1303.11(c), file with the Administrator a written request for a hearing in the form prescribed in §1316.47 of this chapter, including in the request a statement of the grounds for a hearing.

(b) Any interested person who desires to participate in a hearing on the determination or adjustment of an aggregate production quota, which hearing is ordered by the Administrator pursuant to §1303.11(c) or §1303.13(c) may do so by filing with the Administrator, within 30 days of the date of publication of notice of the hearing in the Federal Register, a written notice of his intention to participate in such hearing in the form prescribed in §1316.48 of this chapter.

(c) Any person entitled to a hearing or to participate in a hearing pursuant to paragraph (b) of this section, may, within the period permitted for filing a request for a hearing of notice of appearance, file with the Administrator a waiver of an opportunity for a hearing or to participate in a hearing, together with a written statement regarding his position on the matters of fact and law involved in such hearing. Such statement, if admissible, shall be made a part of the record and shall be considered in light of the lack of opportunity for cross-examination in determining the weight to be attached to matters of fact asserted therein.

(d) If any person entitled to a hearing or to participate in a hearing pursuant to paragraph (b) of this section, fails to file a request for a hearing or notice of appearance, or if he so files and fails to appear at the hearing, he shall be deemed to have waived his opportunity for the hearing or to participate in the hearing, unless he shows good cause for such failure.

(e) If all persons entitled to a hearing or to participate in a hearing waive or are deemed to waive their opportunity for the hearing or to participate in the hearing, the Administrator may cancel the hearing, if scheduled, and issue his final order pursuant to §1303.37 without a hearing.

## Section 1303.35 Burden of Proof

(a) At any hearing regarding the determination or adjustment of an aggregate production quota, each interested person participating in the hearing shall have the burden of proving any propositions of fact or law asserted by him in the hearing.

(b) At any hearing regarding the issuance, adjustment, suspension, or denial of a procurement or individual manufacturing quota, the Administration shall have the burden of proving that the requirements of this part for such issuance, adjustment, suspension, or denial are satisfied.

## Section 1303.36 Time and Place of Hearing

(a) If any applicant or registrant requests a hearing on the issuance, adjustment, suspension, or denial of his procurement and/or individual manufacturing quota pursuant to §1303.34, the Administrator shall hold such hearing. Notice of the hearing shall be given to the applicant or registrant of the time and place at least 30 days prior to the hearing, unless the applicant or registrant waives such notice and requests the hearing be held at an earlier time, in which case the Administrator shall fix a date for such hearing as early as reasonably possible.

(b) The hearing will commence at the place and time designated in the notice given pursuant to paragraph (a) of this section or in the notice of hearing published in the Federal Register pursuant to §1303.11(c) or §1303.13 (c), but thereafter it may be moved to a different place and may be continued from day to day or recessed to a later day without notice other than announcement thereof by the presiding officer at the hearing.

## Section 1303.37 Final Order

As soon as practicable after the presiding officer has certified the Record to the Administrator, the Administrator shall issue his order on the determination or adjustment of the aggregate production quota or on the issuance, adjustment, suspension, or denial of the procurement quota or individual manufacturing quota, as case may be. The order shall include the findings of fact and conclusions of law upon which the order is based. The order shall specify the date on which it shall take effect. The Administrator shall serve one copy of his order upon each party in the hearing.

# Part 1304
## Records and Reports of Registrants

## CONTENTS

### SECTION 1304.01 SCOPE OF PART 1304

Inventory and other records and reports required under section 307 or section 1008(d) of the Act (21 U.S.C. 827 and 958(d)) shall be in accordance with, and contain the information required by, those sections and by the sections of this part.

### SECTION 1304.02 DEFINITIONS

Any term contained in this part shall have the definition set forth in section 102 of the Act (21 U.S.C. 802) or part 1300 of this chapter.

### SECTION 1304.03 PERSONS REQUIRED TO KEEP RECORDS AND FILE REPORTS

(a) Each registrant shall maintain the records and inventories and shall file the reports required by this part, except as exempted by this section. Any registrant who is authorized to conduct other activities without being registered to conduct those activities, either pursuant to

§1301.22(b) of this chapter or pursuant to §1307.11-1307.15 of this chapter, shall maintain the records and inventories and shall file the reports required by this part for persons registered to conduct such activities. This latter requirement should not be construed as requiring stocks of controlled substances being used in various activities under one registration to be stored separately, nor that separate records are required for each activity. The intent of the Administration is to permit the registrant to keep one set of records which are adapted by the registrant to account for controlled substances used in any activity. Also, the Administration does not wish to acquire separate stocks of the same substance to be purchased and stored for separate activities. Otherwise, there is no advantage gained by permitting several activities under one registration. Thus, when a researcher manufactures a controlled item, he must keep a record of the quantity manufactured; when he distributes a quantity of the item, he must use and keep invoices or order forms to document the transfer; when he imports a substance, he keeps as part of his records the documentation required of an importer; and when substances are used in chemical analysis, he need not keep a record of this because such a record would not be required of him under a registration to do chemical analysis. All of these records may be maintained in one consolidated record system. Similarly, the researcher may store all of his controlled items in one place, and every two years take inventory of all items on hand, regardless of whether the substances were manufactured by him, imported by him, or purchased domestically by him, of whether the substances will be administered to subjects, distributed to other researchers, or destroyed during chemical analysis.

(b) A registered individual practitioner is required to keep records, as described in §1304.04, of controlled substances in Schedules II, III, IV, and V which are dispensed, other than by prescribing or administering in the lawful course of professional practice.

(c) A registered individual practitioner is not required to keep records of controlled substances in Schedules II, III, IV, and V which are prescribed in the lawful course of professional practice, unless such substances are prescribed in the course of maintenance or detoxification treatment of an individual.

(d) A registered individual practitioner is not required to keep records of controlled substances listed in Schedules II, III, IV and V which are administered in the lawful course of professional practice unless the practitioner regularly engages in the dispensing or administering of controlled substances and charges patients, either separately or together with charges for other professional services, for substances so dispensed or administered. Records are required to be kept for controlled substances administered in the course of maintenance or detoxification treatment of an individual.

(e) Each registered mid-level practitioner shall maintain in a readily retrievable manner those documents required by the state in which he/she practices which describe the conditions and extent of his/her authorization to dispense controlled substances and shall make such documents available for inspection and copying by authorized employees of the Administration. Examples of such documentation include protocols, practice guidelines or practice agreements.

(f) Registered persons using any controlled substances while conducting preclinical research, in teaching at a registered establishment which maintains records with respect to such substances or conducting research in conformity with an exemption granted under section 505(i) or 512(j) of the Federal Food, Drug, and Cosmetic Act (21 U.S.C. 355(i) or 360b(j)) at a registered establishment which maintains records in accordance with either of those sections, are not required to keep records if he/she notifies the Administration of the name, address, and registration number of the establishment maintaining such records. This notification shall be given at the time the person applies for registration or reregistration and shall be made in the form of an attachment to the application, which shall be filed with the application.

(g) A distributing registrant who utilizes a freight forwarding facility shall maintain records to reflect transfer of controlled substances through the facility. These records must contain the date, time of transfer, number of cartons, crates, drums or other packages in which commercial containers of controlled substances are shipped and authorized signatures for each transfer. A distributing registrant may, as part of the initial request to operate a freight forwarding facility, request permission to store records at a central location. Approval of the request to maintain central records would be implicit in the approval of the request to operate the facility. Otherwise, a request to maintain records at a central location must be submitted in accordance with §1304.04 of this part. These records must be maintained for a period of two years.

## Section 1304.04 Maintenance of Records and Inventories

(a) Every inventory and other records required to be kept under this part shall be kept by the registrant and be available, for at least 2 years from the date of such inventory or records, for inspection and copying by authorized employees of the Administration, except that financial and shipping records (such as invoices and packing slips but not executed order forms subject to §1305.13 of this chapter) may be kept at a central location, rather than at the registered location, if the registrant has notified the Administration of his intention to keep central records. Written notification must be submitted by registered or certified mail, return receipt requested, in triplicate, to the Special Agent in Charge of the Administration in the area in which the registrant is located. Unless the registrant is informed by the Special Agent in Charge that permission to keep central records is denied, the registrant may maintain central records commencing 14 days after receipt of his notification by the Special Agent in Charge. All notifications must include:
   (1) The nature of the records to be kept centrally.
   (2) The exact location where the records will be kept.
   (3) The name, address, DEA registration number and type of DEA registration of the registrant whose records are being maintained centrally.
   (4) Whether central records will be maintained in a manual, or computer readable form.
(b) All registrants that are authorized to maintain a central recordkeeping system shall be subject to the following conditions:
   (1) The records to be maintained at the central record location shall not include executed order forms, prescriptions and/or inventories which shall be maintained at each registered location.
   (2) If the records are kept on microfilm, computer media or in any form requiring special equipment to render the records easily readable, the registrant shall provide access to such equipment with the records. If any code system is used (other than pricing information), a key to the code shall be provided to make the records understandable.
   (3) The registrant agrees to deliver all or any part of such records to the registered location within two business days upon receipt of a written request from the Administration for such records, and if the Administration chooses to do so in lieu of requiring delivery of such records to the registered location, to allow authorized employees of the Administration

to inspect such records at the central location upon request by such employees without a warrant of any kind.

(4) In the event that a registrant fails to comply with these conditions, the Special Agent in Charge may cancel such central recordkeeping authorization, and all other central record-keeping authorizations held by the registrant without a hearing or other procedures. In the event of a cancellation of central recordkeeping authorizations under this paragraph the registrant shall, within the time specified by the Special Agent in Charge, comply with the requirements of this section that all records be kept at the registered location.

(c) Registrants need not notify the Special Agent in Charge or obtain central recordkeeping approval in order to maintain records on an in-house computer system.

(d) ARCOS participants who desire authorization to report from other than their registered locations must obtain a separate central reporting identifier. Request for central reporting identifiers will be submitted to: ARCOS Unit, P.O. Box 28293, Central Station, Washington, D.C. 20005.

(e) All central recordkeeping permits previously issued by the Administration expired September 30, 1980.

(f) Each registered manufacturer, distributor, importer, exporter, narcotic treatment program and compounder for narcotic treatment program shall maintain inventories and records of controlled substances as follows:

(1) Inventories and records of controlled substances listed in Schedules I and II shall be maintained separately from all of the records of the registrant; and

(2) Inventories and records of controlled substances listed in Schedules III, IV, and V shall be maintained either separately from all other records of the registrant or in such form that the information required is readily retrievable from the ordinary business records of the registrant.

(g) Each registered individual practitioner required to keep records and institutional practitioner shall maintain inventories and records of controlled substances in the manner prescribed in paragraph (f) of this section.

(h) Each registered pharmacy shall maintain the inventories and records of controlled substances as follows:

(1) Inventories and records of all controlled substances listed in Schedules I and II shall be maintained separately from all other records of the pharmacy, and prescriptions for such substances shall be maintained in a separate prescription file; and

(2) Inventories and records of controlled substances listed in Schedules III, IV, and V shall be maintained either separately from all other records of the pharmacy or in such form that the information required is readily retrievable from ordinary business records of the pharmacy, and prescriptions for such substances shall be maintained either in a separate prescription file for controlled substances listed in Schedules III, IV, and V only or in such form that they are readily retrievable from the other prescription records of the pharmacy. Prescriptions will be deemed readily retrievable if, at the time they are initially filed, the face of the prescription is stamped in red ink in the lower right corner with the letter "C" no less than 1 inch high and filed either in the prescription file for controlled substances listed in Schedules I and II or in the usual consecutively numbered prescription file for non-controlled substances. However, if a pharmacy employs an ADP system or other electronic recordkeeping system for prescriptions which permits identification by prescription number and retrieval of original documents by prescriber's name, patient's name, drug dispensed, and date filled, then the requirement to mark the hard copy prescription with a red "C" is waived.

# INVENTORY REQUIREMENTS

## SECTION 1304.11 INVENTORY REQUIREMENTS

(a) General requirements. Each inventory shall contain a complete and accurate record of all controlled substances on hand on the date the inventory is taken, and shall be maintained in written, typewritten, or printed form at the registered location. An inventory taken by use of an oral recording device must be promptly transcribed. Controlled substances shall be deemed to be "on hand" if they are in the possession of or under the control of the registrant, including substances returned by a customer, ordered by a customer but not yet invoiced, stored in a warehouse on behalf of the registrant, and substances in the possession of employees of the registrant and intended for distribution as complimentary samples. A separate inventory shall be made for each registered location and each independent activity registered, except as provided in paragraph (e)(4) of this section. In the event controlled substances in the possession or under the control of the registrant are stored at a location for which he/she is not registered, the substances shall be included in the inventory of the registered location to which they are subject to control or to which the person possessing the substance is responsible. The inventory may be taken either as of opening of business or as of the close of business on the inventory date and it shall be indicated on the inventory.

(b) Initial inventory date. Every person required to keep records shall take an inventory of all stocks of controlled substances on hand on the date he/she first engages in the manufacture, distribution, or dispensing of controlled substances, in accordance with paragraph (e) of this section as applicable. In the event a person commences business with no controlled substances on hand, he/she shall record this fact as the initial inventory.

(c) Biennial inventory date. After the initial inventory is taken, the registrant shall take a new inventory of all stocks of controlled substances on hand at least every two years. The biennial inventory may be taken on any date which is within two years of the previous biennial inventory date.

(d) Inventory date for newly controlled substances. On the effective date of a rule by the Administrator pursuant to §1308.45, 1308.46, or 1308.47 of this chapter adding a substance to any schedule of controlled substances, which substance was, immediately prior to that date, not listed on any such schedule, every registrant required to keep records who possesses that substance shall take an inventory of all stocks of the substance on hand. Thereafter, such substance shall be included in each inventory made by the registrant pursuant to paragraph (c) of this section.

(e) Inventories of manufacturers, distributors, dispensers, researchers, importers, exporters and chemical analysts. Each person registered or authorized (by §1301.13 or §1307.11-1307.13 of this chapter) to manufacture, distribute, dispense, import, export, conduct research or chemical analysis with controlled substances and required to keep records pursuant to §1304.03 shall include in the inventory the information listed below.

  (1) Inventories of manufacturers. Each person registered or authorized to manufacture controlled substances shall include the following information in the inventory:

  (i) For each controlled substance in bulk form to be used in (or capable of use in) the manufacture of the same or other controlled or non-controlled substances in finished form, the inventory shall include:

  (A) The name of the substance and

        (B)  The total quantity of the substance to the nearest metric unit weight consistent with unit size.

   (ii)  For each controlled substance in the process of manufacture on the inventory date, the inventory shall include:

        (A)  The name of the substance;

        (B)  The quantity of the substance in each batch and/or stage of manufacture, identified by the batch number or other appropriate identifying number; and

        (C)  The physical form which the substance is to take upon completion of the manufacturing process (e.g., granulations, tablets, capsules, or solutions), identified by the batch number or other appropriate identifying number, and if possible the finished form of the substance (e.g., 10-milligram tablet or 10-milligram concentration per fluid ounce or milliliter) and the number or volume thereof.

  (iii)  For each controlled substance in finished form the inventory shall include:

        (A)  The name of the substance;

        (B)  Each finished form of the substance (e.g., 10-milligram tablet or 10-milligram concentration per fluid ounce or milliliter);

        (C)  The number of units or volume of each finished form in each commercial container (e.g., 100-tablet bottle or 3-milliliter vial); and

        (D)  The number of commercial containers of each such finished form (e.g. four 100-tablet bottles or six 3-milliliter vials).

  (iv)  For each controlled substance not included in paragraphs (e)(1) (i), (ii) or (iii) of this section (e.g., damaged, defective or impure substances awaiting disposal, substances held for quality control purposes, or substances maintained for extemporaneous compoundings) the inventories shall include:

        (A)  The name of the substance;

        (B)  The total quantity of the substance to the nearest metric unit weight or the total number of units of finished form; and

        (C)  The reason for the substance being maintained by the registrant and whether such substance is capable of use in the manufacture of any controlled substance in finished form.

 (2)  Inventories of distributors. Each person registered or authorized to distribute controlled substances shall include in the inventory the same information required of manufacturers pursuant to paragraphs (e)(1) (iii) and (iv) of this section.

 (3)  Inventories of dispensers and researchers. Each person registered or authorized to dispense or conduct research with controlled substances shall include in the inventory the same information required of manufacturers pursuant to paragraphs (e)(1) (iii) and (iv) of this section. In determining the number of units of each finished form of a controlled substance in a commercial container which has been opened, the dispenser shall do as follows:

   (i)  If the substance is listed in Schedule I or II, make an exact count or measure of the contents, or

  (ii)  If the substance is listed in Schedule III, IV or V, make an estimated count or measure of the contents, unless the container holds more than 1,000 tablets or capsules in which case he/she must make an exact count of the contents.

 (4)  Inventories of importers and exporters. Each person registered or authorized to import or export controlled substances shall include in the inventory the same information

required of manufacturers pursuant to paragraphs (e)(1) (iii) and (iv) of this section. Each such person who is also registered as a manufacturer or as a distributor shall include in his/her inventory as an importer or exporter only those stocks of controlled substances that are actually separated from his stocks as a manufacturer or as a distributor (e.g., in transit or in storage for shipment).

(5) Inventories of chemical analysts. Each person registered or authorized to conduct chemical analysis with controlled substances shall include in his inventory the same information required of manufacturers pursuant to paragraphs (e)(1) (iii) and (iv) of this section as to substances which have been manufactured, imported, or received by such person. If less than 1 kilogram of any controlled substance (other than a hallucinogenic controlled substance listed in Schedule I), or less than 20 grams of a hallucinogenic substance listed in Schedule I (other than lysergic acid diethylamide), or less than 0.5 gram of lysergic acid diethylamide, is on hand at the time of inventory, that substance need not be included in the inventory. Laboratories of the Administration may possess up to 150 grams of any hallucinogenic substance in Schedule I without regard to a need for an inventory of those substances. No inventory is required of known or suspected controlled substances received as evidentiary materials for analysis.

## CONTINUING RECORDS

### Section 1304.21 General Requirements for Continuing Records

(a) Every registrant required to keep records pursuant to Section 1304.03 shall maintain on a current basis a complete and accurate record of each such substance manufactured, imported, received, sold, delivered, exported, or otherwise disposed of by him/her, except that no registrant shall be required to maintain a perpetual inventory.

(b) Separate records shall be maintained by a registrant for each registered location except as provided in §1304.04 (a). In the event controlled substances are in the possession or under the control of a registrant at a location for which he is not registered, the substances shall be included in the records of the registered location to which they are subject to control or to which the person possessing the substance is responsible.

(c) Separate records shall be maintained by a registrant for each independent activity for which he/she is registered, except as provided in §1304.22(d).

(d) In recording dates of receipt, importation, distribution, exportation, or other transfers, the date on which the controlled substances are actually received, imported, distributed, exported, or otherwise transferred shall be used as the date of receipt or distribution of any documents of transfer (e.g., invoices or packing slips).

### Section 1304.22 Records for Manufacturers, Distributors, Dispensers, Researchers, Importers and Exporters

Each person registered or authorized (by §1301.13(e) or §1307.11-1307.13 of this chapter) to manufacture, distribute, dispense, import, export or conduct research with controlled substances shall maintain records with the information listed below.

(a) Records for manufacturers. Each person registered or authorized to manufacture controlled substances shall maintain records with the following information:

(1) For each controlled substance in bulk form to be used in, or capable of use in, or being used in, the manufacture of the same or other controlled or noncontrolled substances in finished form,

(i) The name of the substance;

(ii) The quantity manufactured in bulk form by the registrant, including the date, quantity and batch or other identifying number of each batch manufactured;

(iii) The quantity received from other persons, including the date and quantity of each receipt and the name, address, and registration number of the other person from whom the substance was received;

(iv) The quantity imported directly by the registrant (under a registration as an importer) for use in manufacture by him/her, including the date, quantity, and import permit or declaration number for each importation;

(v) The quantity used to manufacture the same substance in finished form, including:

(A) The date and batch or other identifying number of each manufacture;

(B) The quantity used in the manufacture;

(C) The finished form (e.g., 10-milligram tablets or 10-milligram concentration per fluid ounce or milliliter);

(D) The number of units of finished form manufactured;

(E) The quantity used in quality control;

(F) The quantity lost during manufacturing and the causes therefore, if known;

(G) The total quantity of the substance contained in the finished form;

(H) The theoretical and actual yields; and

(I) Such other information as is necessary to account for all controlled substances used in the manufacturing process;

(vi) The quantity used to manufacture other controlled and noncontrolled substances, including the name of each substance manufactured and the information required in paragraph (a)(1)(v) of this section;

(vii) The quantity distributed in bulk form to other persons, including the date and quantity of each distribution and the name, address, and registration number of each person to whom a distribution was made;

(viii) The quantity exported directly by the registrant (under a registration as an exporter), including the date, quantity, and export permit or declaration number of each exportation;

(ix) The quantity distributed or disposed of in any other manner by the registrant (e.g., by distribution of complimentary samples or by destruction), including the date and manner of distribution or disposal, the name, address, and registration number of the person to whom distributed, and the quantity distributed or disposed; and

(x) The originals of all written certifications of available procurement quotas submitted by other persons (as required by §1303.12(f) of this chapter) relating to each order requiring the distribution of a basic class of controlled substance listed in Schedule I or II.

(2) For each controlled substance in finished form,

(i) The name of the substance;

(ii) Each finished form (e.g., 10-milligram tablet or 10-milligram concentration per fluid ounce or milliliter) and the number of units or volume of finished form in each commercial container (e.g., 100-tablet bottle or 3-milliliter vial);

(iii) The number of containers of each such commercial finished form manufactured from bulk form by the registrant, including the information required pursuant to paragraph (a)(1)(v) of this section;

(iv) The number of units of finished forms and/or commercial containers acquired from other persons, including the date of and number of units and/or commercial containers in each acquisition to inventory and the name, address, and registration number of the person from whom the units were acquired;

(v) The number of units of finished forms and/or commercial containers imported directly by the person (under a registration or authorization to import), including the date of, the number of units and/or commercial containers in, and the import permit or declaration number for, each importation;

(vi) The number of units and/or commercial containers manufactured by the registrant from units in finished form received from others or imported, including:

(A) The date and batch or other identifying number of each manufacture;

(B) The operation performed (e.g., repackaging or relabeling);

(C) The number of units of finished form used in the manufacture, the number manufactured and the number lost during manufacture, with the causes for such losses, if known; and

(D) Such other information as is necessary to account for all controlled substances used in the manufacturing process;

(vii) The number of commercial containers distributed to other persons, including the date of and number of containers in each reduction from inventory, and the name, address, and registration number of the person to whom the containers were distributed;

(viii) The number of commercial containers exported directly by the registrant (under a registration as an exporter), including the date, number of containers and export permit or declaration number for each exportation; and

(ix) The number of units of finished forms and/or commercial containers distributed or disposed of in any other manner by the registrant (e.g., by distribution of complimentary samples or by destruction), including the date and manner of distribution or disposal, the name, address, and registration number of the person to whom distributed, and the quantity in finished form distributed or disposed.

(b) Records for distributors. Each person registered or authorized to distribute controlled substances shall maintain records with the same information required of manufacturers pursuant to paragraphs (a)(2) (i), (ii), (iv), (v), (vii), (viii) and (ix) of this section.

(c) Records for dispensers and researchers. Each person registered or authorized to dispense or conduct research with controlled substances shall maintain records with the same information required of manufacturers pursuant to paragraph (a)(2) (i), (ii), (iv), (vii), and (ix) of this section. In addition, records shall be maintained of the number of units or volume of such finished form dispensed, including the name and address of the person to whom it was dispensed, the date of dispensing, the number of units or volume dispensed, and the written or typewritten name or initials of the individual who dispensed or administered the substance on behalf of the dispenser.

(d) Records for importers and exporters. Each person registered or authorized to import or export controlled substances shall maintain records with the same information required of manufacturers pursuant to paragraphs (a)(2) (i), (iv), (v) and (vii) of this section. In addition, the quantity disposed of in any other manner by the registrant (except quantities used in manufacturing by an importer under a registration as a manufacturer), which quantities are to be recorded pursuant to paragraphs (a)(1) (iv) and (v) of this section; and the quantity (or number of units or volume in finished form) exported, including the date, quantity (or number of units or volume), and the export permit or declaration number for each exportation, but excluding all quantities (and number of units and volumes) manufactured by an exporter under a registration as a manufacturer, which quantities (and numbers of units and volumes) are to be recorded pursuant to paragraphs (a)(1)(xiii) or (a)(2)(xiii) of this section.

## SECTION 1304.23 RECORDS FOR CHEMICAL ANALYSTS

(a) Each person registered or authorized (by §1301.22(b) of this chapter) to conduct chemical analysis with controlled substances shall maintain records with the following information (to the extent known and reasonably ascertainable by him) for each controlled substance:
  (1) The name of the substance;
  (2) The form or forms in which the substance is received, imported, or manufactured by the registrant (e.g., powder, granulation, tablet, capsule, or solution) and the concentration of the substance in such form (e.g., C.P., U.S.P., N.F., 10-milligram tablet or 10-milligram concentration per milliliter);
  (3) The total number of the forms received, imported or manufactured (e.g., 100 tablets, thirty 1-milliliter vials, or 10 grams of powder), including the date and quantity of each receipt, importation, or manufacture and the name, address, and registration number, if any, of the person from whom the substance was received;
  (4) The quantity distributed, exported, or destroyed in any manner by the registrant (except quantities used in chemical analysis or other laboratory work), including the date and manner of distribution, exportation, or destruction, and the name, address, and registration number, if any, of each person to whom the substance was distributed or exported.
(b) Records of controlled substances used in chemical analysis or other laboratory work are not required.
(c) Records relating to known or suspected controlled substances received as evidentiary material for analysis are not required under paragraph (a) of this section.

## SECTION 1304.24 RECORDS FOR MAINTENANCE TREATMENT PROGRAMS AND DETOXIFICATION TREATMENT PROGRAMS

(a) Each person registered or authorized (by §1301.22 of this chapter) to maintain and/or detoxify controlled substance users in a narcotic treatment program shall maintain records with the following information for each narcotic controlled substance:
  (1) Name of substance;
  (2) Strength of substance;
  (3) Dosage form;
  (4) Date dispensed;

(5) Adequate identification of patient (consumer);

(6) Amount consumed;

(7) Amount and dosage form taken home by patient; and

(8) Dispenser's initials.

(b) The records required by paragraph (a) of this section will be maintained in a dispensing log at the narcotic treatment program site and will be maintained in compliance with §1304.22 without reference to §1304.03.

(c) All sites which compound a bulk narcotic solution from bulk narcotic powder to liquid for on-site use must keep a separate batch record of the compounding.

(d) Records of identity, diagnosis, prognosis, or treatment of any patients which are maintained in connection with the performance of a narcotic treatment program shall be confidential, except that such records may be disclosed for purposes and under the circumstances authorized by part 310 and 42 CFR part 2.

## Section 1304.25 Records for Treatment Programs which Compound Narcotics for Treatment Programs and Other Locations

Each person registered or authorized by §1301.22 of this chapter to compound narcotic drugs for off-site use in a narcotic treatment program shall maintain records which include the following information for each narcotic drug:

(a) For each narcotic controlled substance in bulk form to be used in, or capable of use in, or being used in, the compounding of the same or other noncontrolled substances in finished form:

(1) The name of the substance;

(2) The quantity compounded in bulk form by the registrant, including the date, quantity and batch or other identifying number of each batch compounded;

(3) The quantity received from other persons, including the date and quantity of each receipt and the name, address and registration number of the other person from whom the substance was received;

(4) The quantity imported directly by the registrant (under a registration as an importer) for use in compounding by him, including the date, quantity and import permit or declaration number of each importation;

(5) The quantity used to compound the same substance in finished form, including:

(i) The date and batch or other identifying number of each compounding;

(ii) The quantity used in the compound;

(iii) The finished form (e.g., 10-milligram tablets or 10-milligram concentration per fluid ounce or milliliter;

(iv) The number of units of finished form compounded;

(v) The quantity used in quality control;

(vi) The quantity lost during compounding and the causes therefore, if known;

(vii) The total quantity of the substance contained in the finished form;

(viii) The theoretical and actual yields; and

(ix) Such other information as is necessary to account for all controlled substances used in the compounding process;

(6) The quantity used to manufacture other controlled and non-controlled substances; including the name of each substance manufactured and the information required in paragraph (a)(5) of this section;

(7) The quantity distributed in bulk form to other programs, including the date and quantity of each distribution and the name, address and registration number of each program to whom a distribution was made;

(8) The quantity exported directly by the registrant (under a registration as an exporter), including the date, quantity, and export permit or declaration number of each exploration; and

(9) The quantity disposed of by destruction, including the reason, date and manner of destruction. All other destruction of narcotic controlled substances will comply with §1307.22.

(b) For each narcotic controlled substance in finished form:

    (1) The name of the substance;

    (2) Each finished form (e.g., 10-milligram tablet or 10 milligram concentration per fluid ounce or milliliter) and the number of units or volume or finished form in each commercial container (e.g., 100-tablet bottle or 3-milliliter vial);

    (3) The number of containers of each such commercial finished form compounded from bulk form by the registrant, including the information required pursuant to paragraph (a)(5) of this section;

    (4) The number of units of finished forms and/or commercial containers received from other persons, including the date of and number of units and/or commercial containers in each receipt and the name, address and registration number of the person from whom the units were received;

    (5) The number of units of finished forms and/or commercial containers imported directly by the person (under a registration or authorization to import), including the date of, the number of units and/or commercial containers in, and the import permit or declaration number for, each importation;

    (6) The number of units and/or commercial containers compounded by the registrant from units in finished form received from others or imported, including:

        (i) The date and batch or other identifying number of each compounding;

        (ii) The operation performed (e.g., repackaging or relabeling);

        (iii) The number of units of finished form used in the compound, the number compounded and the number lost during compounding, with the causes for such losses, if known; and

        (iv) Such other information as is necessary to account for all controlled substances used in the compounding process;

    (7) The number of containers distributed to other programs, including the date, the number of containers in each distribution, and the name, address and registration number of the program to whom the containers were distributed;

    (8) The number of commercial containers exported directly by the registrant (under a registration as an exporter), including the date, number of containers and export permit or declaration number for each exportation; and

    (9) The number of units of finished forms and/or commercial containers destroyed in any manner by the registrant, including the reason, the date and manner of destruction. All other destruction of narcotic controlled substances will comply with §1307.22.

# REPORTS

## SECTION 1304.31 REPORTS FROM MANUFACTURERS IMPORTING NARCOTIC RAW MATERIAL

(a) Every manufacturer which imports or manufactures from narcotic raw material (opium, poppy straw, and concentrate of poppy straw) shall submit information which accounts for the importation and for all manufacturing operations performed between importation and the production in bulk or finished marketable products, standardized in accordance with the U.S. Pharmacopeia, National Formulary or other recognized medical standards. Reports shall be signed by the authorized official and submitted quarterly on company letterhead to the Drug Enforcement Administration, Drug and Chemical Evaluation Section, Washington, D.C. 20537, on or before the 15th day of the month immediately following the period for which it is submitted.

(b) The following information shall be submitted for each type of narcotic raw material (quantities are expressed as grams of anhydrous morphine alkaloid):
   (1) Beginning inventory;
   (2) Gains on reweighing;
   (3) Imports;
   (4) Other receipts;
   (5) Quantity put into process;
   (6) Losses on reweighing;
   (7) Other dispositions and
   (8) Ending inventory.

(c) The following information shall be submitted for each narcotic raw material derivative including morphine, codeine, thebaine, oxycodone, hydrocodone, medicinal opium, manufacturing opium, crude alkaloids and other derivatives (quantities are expressed as grams of anhydrous base or anhydrous morphine alkaloid for manufacturing opium and medicinal opium):
   (1) Beginning inventory;
   (2) Gains on reweighing;
   (3) Quantity extracted from narcotic raw material;
   (4) Quantity produced/manufactured/synthesized;
   (5) Quantity sold;
   (6) Quantity returned to conversion processes for reworking;
   (7) Quantity used for conversion;
   (8) Quantity placed in process;
   (9) Other dispositions;
   (10) Losses on reweighing and
   (11) Ending inventory.

(d) The following information shall be submitted for importation of each narcotic raw material:
   (1) Import permit number;
   (2) Date shipment arrived at the United States port of entry;
   (3) Actual quantity shipped;
   (4) Assay (percent) of morphine, codeine and thebaine and
   (5) Quantity shipped, expressed as anhydrous morphine alkaloid.

(e) Upon importation of crude opium, samples will be selected and assays made by the importing manufacturer in the manner and according to the method specified in the U.S. Pharmacopoeia. Where final assay data is not determined at the time of rendering report, the report shall be made on the basis of the best data available, subject to adjustment, and the necessary adjusting entries shall be made on the next report.

(f) Where factory procedure is such that partial withdrawals of opium are made from individual containers, there shall be attached to each container a stock record card on which shall be kept a complete record of all withdrawals therefrom.

(g) All in-process inventories should be expressed in terms of end-products and not precursors. Once precursor material has been changed or placed into process for the manufacture of a specified end-product, it must no longer be accounted for as precursor stocks available for conversion or use, but rather as end-product in-process inventories.

## SECTION 1304.32 REPORTS OF MANUFACTURERS IMPORTING COCA LEAVES

(a) Every manufacturer importing or manufacturing from raw coca leaves shall submit information accounting for the importation and for all manufacturing operations performed between the importation and the manufacture of bulk or finished products standardized in accordance with U.S. Pharmacopoeia, National Formulary, or other recognized standards. The reports shall be submitted quarterly on company letterhead to the Drug Enforcement Administration, Drug and Chemical Evaluation Section, Washington, D.C. 20537, on or before the 15th day of the month immediately following the period for which it is submitted.

(b) The following information shall be submitted for raw coca leaf, ecgonine, ecgonine for conversion or further manufacture, benzoylecgonine, manufacturing coca extracts (list for tinctures and extracts; and others separately), other crude alkaloids and other derivatives (quantities should be reported as grams of actual quantity involved and the cocaine alkaloid content or equivalency):

  (1) Beginning inventory;

  (2) Imports;

  (3) Gains on reweighing;

  (4) Quantity purchased;

  (5) Quantity produced;

  (6) Other receipts;

  (7) Quantity returned to processes for reworking;

  (8) Material used in purification for sale;

  (9) Material used for manufacture or production;

  (10) Losses on reweighing;

  (11) Material used for conversion;

  (12) Other dispositions and

  (13) Ending inventory.

(c) The following information shall be submitted for importation of coca leaves:

  (1) Import permit number;

  (2) Date the shipment arrived at the United States port of entry;

  (3) Actual quantity shipped;

  (4) Assay (percent) of cocaine alkaloid and

  (5) Total cocaine alkaloid content.

(d) Upon importation of coca leaves, samples will be selected and assays made by the importing manufacturer in accordance with recognized chemical procedures. These assays shall form the basis of accounting for such coca leaves, which shall be accounted for in terms of their cocaine alkaloid content or equivalency or their total anhydrous coca alkaloid content. Where final assay data is not determined at the time of submission, the report shall be made on the basis of the best data available, subject to adjustment, and the necessary adjusting entries shall be made on the next report.

(e) Where factory procedure is such that partial withdrawals of medicinal coca leaves are made from individual containers, there shall be attached to the container a stock record card on which shall be kept a complete record of withdrawals therefrom.

(f) All in-process inventories should be expressed in terms of end-products and not precursors. Once precursor material has been changed or placed into process for the manufacture of a specified end-product, it must no longer be accounted for as precursor stocks available for conversion or use, but rather as end-product in-process inventories.

## SECTION 1304.33 REPORTS TO ARCOS

(a) Reports generally. All reports required by this section shall be filed with the ARCOS Unit, P.O. Box 28293, Central Station, Washington, D.C. 20005 on DEA Form 333, or on media which contains the data required by DEA Form 333 and which is acceptable to the ARCOS Unit.

(b) Frequency of reports. Acquisition/Distribution transaction reports shall be filed every quarter not later than the 15th day of the month succeeding the quarter for which it is submitted; except that a registrant may be given permission to file more frequently (but not more frequently than monthly), depending on the number of transactions being reported each time by that registrant. Inventories shall provide data on the stocks of each reported controlled substance on hand as of the close of business on December 31 of each year, indicating whether the substance is in storage or in process of manufacturing. These reports shall be filed not later than January 15 of the following year. Manufacturing transaction reports shall be filed annually for each calendar year not later than January 15 of the following year, except that a registrant may be given permission to file more frequently (but not more frequently than quarterly).

(c) Persons reporting. For controlled substances in Schedules I, II or narcotic controlled substances in Schedule III, each person who is registered to manufacture in bulk or dosage form, or to package, repackage, label or relabel, and each person who is registered to distribute shall report acquisition/distribution transactions. In addition to reporting acquisition/distribution transactions, each person who is registered to manufacture controlled substances in bulk or dosage form shall report manufacturing transactions on controlled substances in Schedules I and II, each narcotic controlled substance listed in Schedules III, IV, and V, and on each psychotropic controlled substance listed in Schedules III and IV as identified in paragraph (d) of this section.

(d) Substances covered:

    (1) Manufacturing and acquisition/distribution transaction reports shall include data on each controlled substance listed in Schedules I and II and on each narcotic controlled substance listed in Schedule III (but not on any material, compound, mixture or preparation containing a quantity of a substance having a stimulant effect on the central nervous system, which material, compound, mixture or preparation is listed in Schedule III or on any narcotic controlled substance listed in Schedule V). Additionally, reports on manufacturing transactions shall include the following psychotropic controlled substances listed in Schedules III and IV:

      (i) Schedule III

        (A) Benzphetamine;

        (B) Cyclobarbital;

        (C) Methyprylon; and

        (D) Phendimetrazine.

     (ii) Schedule IV

        (A) Barbital;

        (B) Diethylpropion (Amfepramone);

        (C) Ethchlorvynol;

        (D) Ethinamate;

        (E) Lefetamine (SPA);

        (F) Mazindol;

        (G) Meprobamate;

        (H) Methylphenobarbital;

        (I) Phenobarbital;

        (J) Phentermine; and

        (K) Pipradrol.

(2) Data shall be presented in such a manner as to identify the particular form, strength, and trade name, if any, of the product containing the controlled substance for which the report is being made. For this purpose, persons filing reports shall utilize the National Drug Code Number assigned to the product under the National Drug Code System of the Food and Drug Administration.

(e) Transactions reported. Acquisition/distribution transaction reports shall provide data on each acquisition to inventory (identifying whether it is, e.g., by purchase or transfer, return from a customer, or supply by the Federal Government) and each reduction from inventory (identifying whether it is, e.g., by sale or transfer, theft, destruction or seizure by Government agencies). Manufacturing reports shall provide data on material manufactured, manufacture from other material, use in manufacturing other material and use in producing dosage forms.

(f) Exceptions. A registered institutional practitioner who repackages or relabels exclusively for distribution or who distributes exclusively to (for dispensing by) agents, employees, or affiliated institutional practitioners of the registrant may be exempted from filing reports under this section by applying to the ARCOS Unit of the Administration.

# Part 1305
## Order Forms

## CONTENTS

### SECTION 1305.01 SCOPE OF PART 1305

Procedures governing the issuance, use, and preservation of order forms pursuant to section 1308 of the Act (21 U.S.C. 828) are set forth generally by that section and specifically by the sections of this part.

### SECTION 1305.02 DEFINITIONS

Any term contained in this part shall have the definition set forth in section 102 of the Act (21 U.S.C. 802) or part 1300 of this chapter.

### SECTION 1305.03 DISTRIBUTIONS REQUIRING ORDER FORMS

An order form (DEA Form 222) is required for each distribution of a Schedule I or II controlled substance except to persons exempted from registration under part 1301 of this chapter; which are exported from the United States in conformity with the Act; or for delivery to a registered analytical laboratory, or its agent approved by DEA.

## SECTION 1305.04 PERSONS ENTITLED TO OBTAIN AND EXECUTE ORDER FORMS

(a) Order forms may be obtained only by persons who are registered under section 303 of the Act (21 U.S.C. 823) to handle controlled substances listed in Schedules I and II, and by persons who are registered under section 1008 of the Act (21 U.S.C. 958) to export such substances. Persons not registered to handle controlled substances listed in Schedule I or II and persons registered only to import controlled substances listed in any schedule are not entitled to obtain order forms.

(b) An order form may be executed only on behalf of the registrant named thereon and only if his/her registration as to the substances being purchased has not expired or been revoked or suspended.

## SECTION 1305.05 PROCEDURE FOR OBTAINING ORDER FORMS

(a) Order Forms are issued in mailing envelopes containing either seven or fourteen forms, each form containing an original duplicate and triplicate copy (respectively, Copy 1, Copy 2, and Copy 3). A limit, which is based on the business activity of the registrant, will be imposed on the number of order forms which will be furnished on any requisition unless additional forms are specifically requested and a reasonable need for such additional forms is shown.

(b) Any person applying for a registration which would entitle him/her to obtain order forms may requisition such forms by so indicating on the application form; order forms will be supplied upon the registration of the applicant. Any person holding a registration entitling him/her to obtain order forms may requisition such forms for the first time by contacting any Division Office or the Registration Unit of the Administration. Any person already holding order forms may requisition additional forms on DEA Form 222a which is mailed to a registrant approximately 30 days after each shipment of order forms to that registrant or by contacting any Division Office or the Registration Unit of the Administration. All requisition forms (DEA Form 222a) shall be submitted to the Registration Unit, Drug Enforcement Administration, Department of Justice, Post Office Box 28083, Central Station, Washington, D.C. 20005.

(c) Each requisition shall show the name, address, and registration number of the registrant and the number of books of order forms desired. Each requisition shall be signed and dated by the same person who signed the most recent application for registration or for reregistration, or by any person authorized to obtain and execute order forms by a power of attorney pursuant to §1305.07.

(d) Order forms will be serially numbered and issued with the name, address and registration number of the registrant, the authorized activity and schedules of the registrant. This information cannot be altered or changed by the registrant; any errors must be corrected by the Registration Unit of the Administration by returning the forms with notification of the error.

## SECTION 1305.06 PROCEDURE FOR EXECUTING ORDER FORMS

(a) Order forms shall be prepared and executed by the purchaser simultaneously in triplicate by means of interleaved carbon sheets which are part of the DEA Form 222. Order forms shall be prepared by use of a typewriter, pen, or indelible pencil.

(b) Only one item shall be entered on each numbered line. An item shall consist of one or more commercial or bulk containers of the same finished or bulk form and quantity of the same substance. The number of lines completed shall be noted on that form at the bottom of the form, in the space provided. Order forms for carfentanil, etorphine hydrochloride, and diprenorphine shall contain only these substances.

(c) The name and address of the supplier from whom the controlled substances are being ordered shall be entered on the form. Only one supplier may be listed on any form.

(d) Each order form shall be signed and dated by a person authorized to sign an application for registration. The name of the purchaser, if different from the individual signing the order form, shall also be inserted in the signature space. Unexecuted order forms may be kept and may be executed at a location other than the registered location printed on the form, provided that all unexecuted forms are delivered promptly to the registered location upon an inspection of such location by any officer authorized to make inspections, or to enforce, any Federal, State, or local law regarding controlled substances.

## SECTION 1305.07 POWER OF ATTORNEY

Any purchaser may authorize one or more individuals, whether or not located at the registered location of the purchaser, to obtain and execute order forms on his/her behalf by executing a power of attorney for each such individual. The power of attorney shall be signed by the same person who signed the most recent application for registration or reregistration and by the individual being authorized to obtain and execute order forms. The power of attorney shall be filed with the executed order forms of the purchaser, and shall be retained for the same period as any order form bearing the signature of the attorney. The power of attorney shall be available for inspection together with other order form records. Any power of attorney may be revoked at any time by executing a notice of revocation, signed by the person who signed (or was authorized to sign) the power of attorney or by a successor, whoever signed the most recent application for registration or reregistration, and filing it with the power of attorney being revoked. The form for the power of attorney and notice of revocation shall be similar to the following:

### POWER OF ATTORNEY FOR DEA ORDER FORMS

_____ (Name of registrant)

_____ (Address of registrant)

_____ (DEA registration number)

I, _____ (name of person granting power), the undersigned, who is authorized to sign the current application for registration of the above-named registrant under the Controlled Substances Act or Controlled Substances Import and Export Act, have made, constituted, and appointed, and by these presents, do make, constitute, and appoint _____ (name of attorney-in-fact), my true and lawful attorney for me in my name, place, and stead, to execute applications for books of official order forms and to sign such order forms in requisition for Schedule I and II controlled substances, in accordance with section 308 of the Controlled Substances Act (21 U.S.C. 828) and part 1305 of Title 21 of the Code of Federal Regulations. I hereby ratify and confirm all that said attorney shall lawfully do or cause to be done by virtue hereof.

_____

(Signature of person granting power)

I, _____ (name of attorney-in-fact), hereby affirm that I am the person named herein as attorney-in-fact and that the signature affixed hereto is my signature.

_____

(Signature of attorney-in-fact)

Witnesses:

1. _____ .

2. _____ .

Signed and dated on the _____ day of _____ , (year), at _____ .

Notice of Revocation

The foregoing power of attorney is hereby revoked by the undersigned, who is authorized to sign the current application for registration of the above-named registrant under the Controlled Substances Act of the Controlled Substances Import and Export Act.

Written notice of this revocation has been given to the attorney-in-fact _____ this same day.

_____

(Signature of person revoking power)

Witnesses:

1. _____ .

2. _____ .

Signed and dated on the _____ day of _____ , (year), at _____ .

## SECTION 1305.08 PERSONS ENTITLED TO FILL ORDER FORMS

An order form may be filled only by a person registered as a manufacturer or distributor of controlled substances listed in Schedule I or II under section 303 of the Act (21 U.S.C. 823) or as an importer of such substances under section 1008 of the Act (21 U.S.C. 958), except for the following:

(a) A person registered to dispense such substances under section 303 of the Act, or to export such substances under section 1008 of the Act, if he/she is discontinuing business or if his/her registration is expiring without reregistration, may dispose of any controlled substances listed in Schedule I or II in his/her possession pursuant to order forms in accordance with §1307.14 of this chapter;

(b) A person who has obtained any controlled substance in Schedule I or II by order form may return such substance, or portion thereof, to the person from whom he/she obtained the substance or the manufacturer of the substance pursuant to the order form of the latter person;

(c) A person registered to dispense such substances may distribute such substances to another dispenser pursuant to, and only in the circumstances described in, §1307.11 of this chapter; and

(d) A person registered or authorized to conduct chemical analysis or research with controlled substances may distribute a controlled substance listed in Schedule I or II to another person registered or authorized to conduct chemical analysis, instructional activities, or research with such substances pursuant to the order form of the latter person, if such distribution is for the purpose of furthering such chemical analysis, instructional activities, or research.

(e) A person registered as a compounder of narcotic substances for use at off-site locations in conjunction with a narcotic treatment program at the compounding location, who is authorized to handle Schedule II narcotics, is authorized to fill order forms for distribution of narcotic drugs to off-site narcotic treatment programs only.

## Section 1305.09 Procedure for Filling Order Forms

(a) The purchaser shall submit Copy 1 and Copy 2 of the order form to the supplier, and retain Copy 3 in his own files.

(b) The supplier shall fill the order, if possible and if he/she desires to do so, and record on Copies 1 and 2 the number of commercial or bulk containers furnished on each item and the date on which such containers are shipped to the purchaser. If an order cannot be filled in its entirety, it may be filled in part and the balance supplied by additional shipments within 60 days following the date of the order form. No order form shall be valid more than 60 days after its execution by the purchaser, except as specified in paragraph (f) of this section.

(c) The controlled substances shall be shipped only to the purchaser and at the location printed by the Administration on the order form, except as specified in paragraph (f) of this section.

(d) The supplier shall retain Copy 1 of the order form for his/her own files and forward Copy 2 to the Special Agent in Charge of the Drug Enforcement Administration in the area in which the supplier is located. Copy 2 shall be forwarded at the close of the month during which the order is filled; if an order is filled by partial shipments, Copy 2 shall be forwarded at the close of the month during which the final shipment is made or during which the 60-day validity period expires.

(e) The purchaser shall record on Copy 3 of the order form the number of commercial or bulk containers furnished on each item and the dates on which such containers are received by the purchaser.

(f) Order forms submitted by registered procurement officers of the Defense Personnel Support Center of Defense Supply Agency for delivery to armed services establishments within the United States may be shipped to locations other than the location printed on the order form, and in partial shipments at different times not to exceed six months from the date of the order, as designated by the procurement officer when submitting the order.

## Section 1305.10 Procedure for Endorsing Order Forms

(a) An order form made out to any supplier who cannot fill all or a part of the order within the time limitation set forth in §1305.09 may be endorsed to another supplier for filling. The endorsement shall be made only by the supplier to whom the order form was first made, shall state (in the spaces provided on the reverse sides of Copies 1 and 2 of the order form) the name and address of the second supplier, and shall be signed by a person authorized to obtain and execute order forms on behalf of the first supplier. The first supplier may not fill any part of an order on an endorsed form. The second supplier shall fill the order, if possible and if he/she desires to do so, in accordance with §1305.09 (b), (c), and (d), including shipping all substances directly to the purchaser.

(b) Distributions made on endorsed order forms shall be reported by the second supplier in the same manner as all other distributions except that where the name of the supplier is requested on the reporting form, the second supplier shall record the name, address and registration number of the first supplier.

## SECTION 1305.11 UNACCEPTED AND DEFECTIVE ORDER FORMS

(a) No order form shall be filled if it:
    (1) Is not complete, legible, or properly prepared, executed, or endorsed; or
    (2) Shows any alteration, erasure, or change of any description.
(b) If an order form cannot be filled for any reason under this section, the supplier shall return Copies 1 and 2 to the purchaser with a statement as to the reason (e.g., illegible or altered). A supplier may for any reason refuse to accept any order and if a supplier refuses to accept the order, a statement that the order is not accepted shall be sufficient for purposes of this paragraph.
(c) When received by the purchaser, Copies 1 and 2 of the order form and the statement shall be attached to Copy 3 and retained in the files of the purchaser in accordance with §1305.13. A defective order form may not be corrected; it must be replaced by a new order form in order for the order to be filled.

## SECTION 1305.12 LOST AND STOLEN ORDER FORMS

(a) If a purchaser ascertains that an unfilled order form has been lost, he shall execute another in triplicate and a statement containing the serial number and date of the lost form, and stating that the goods covered by the first order form were not received through loss of that order form. Copy 3 of the second form and a copy of the statement shall be retained with Copy 3 of the order form first executed. A copy of the statement shall be attached to Copies 1 and 2 of the second order form sent to the supplier. If the first order form is subsequently received by the supplier to whom it was directed, the supplier shall mark upon the face thereof "Not accepted" and return Copies 1 and 2 to the purchaser, who shall attach it to Copy 3 and the statement.
(b) Whenever any used or unused order forms are stolen or lost (otherwise than in the course of transmission) by any purchaser or supplier, he/she shall immediately upon discovery of such theft or loss, report the same to the Special Agent in Charge of the Drug Enforcement Administration in the Divisional Office responsible for the area in which the registrant is located, stating the serial number of each form stolen or lost. If the theft or loss includes any original order forms received from purchasers and the supplier is unable to state the serial numbers of such order forms, he/she shall report the date or approximate date of receipt thereof and the names and addresses of the purchasers. If an entire book of order forms is lost or stolen, and the purchaser is unable to state the serial numbers of the order forms contained therein, he/she shall report, in lieu of the numbers of the forms contained in such book, the date or approximate date of issuance thereof. If any unused order form reported stolen or lost is subsequently recovered or found, the Special Agent in Charge of the Drug Enforcement Administration in the Divisional Office responsible for the area in which the registrant is located shall immediately be notified.

## Section 1305.13 Preservation of Order Forms

(a) The purchaser shall retain Copy 3 of each order form which has been filled. He/she shall also retain in his files all copies of each unaccepted or defective order form and each statement attached thereto.

(b) The supplier shall retain Copy 1 of each order form which he/she has filled.

(c) Order forms must be maintained separately from all other records of the registrant. Order forms are required to be kept available for inspection for a period of 2 years. If a purchaser has several registered locations, he/she must retain Copy 3 of the executed order forms and any attached statements or other related documents (not including unexecuted order forms which may be kept elsewhere pursuant to §1305.06(d)) at the registered location printed on the order form.

(d) The supplier of carfentanil etorphine hydrochloride and diprenorphine shall maintain order forms for these substances separately from all other order forms and records required to be maintained by the registrant.

## Section 1305.14 Return of Unused Order Forms

If the registration of any purchaser terminates (because the purchaser dies, ceases legal existence, discontinues business or professional practice, or changes his name or address as shown on his registration) or is suspended or revoked pursuant to §1301.36 of this chapter as to all controlled substances listed in Schedules I and II for which he/she is registered, he/she shall return all unused order forms for such substance to the nearest office of the Administration.

## Section 1305.15 Cancellation and Voiding of Order Forms.

(a) A purchaser may cancel part or all of an order on an order form by notifying the supplier in writing of such cancellation. The supplier shall indicate the cancellation on Copies 1 and 2 of the order form by drawing a line through the canceled items and printing "canceled" in the space provided for number of items shipped.

(b) A supplier may void part or all of an order on an order form by notifying the purchaser in writing of such voiding. The supplier shall indicate the voiding in the manner prescribed for cancellation in paragraph (a) of this section.

(c) No cancellation or voiding permitted by this section shall affect in any way contract rights of either the purchaser or the supplier.

## Section 1305.16 Special Procedure for Filling Certain Order Forms

(a) The purchaser of carfentanil etorphine hydrochloride or diprenorphine shall submit copy 1 and 2 of the order form to the supplier and retain copy 3 in his own files.

(b) The supplier, if he/she determines that the purchaser is a veterinarian engaged in zoo and exotic animal practice, wildlife management programs and/or research and authorized by the Administrator to handle these substances shall fill the order in accordance with the procedures set forth in §1305.09 except that:

(1) Order forms for carfentanil etorphine hydrochloride and diprenorphine shall only contain these substances in reasonable quantities and

(2) The substances shall be shipped only to the purchaser at the location printed by the Administration upon the order form under secure conditions using substantial packaging material with no markings on the outside which would indicate the content.

# Part 1306

## Prescriptions

## CONTENTS

## SECTION 1306.01 SCOPE OF PART 1306

Rules governing the issuance, filling and filing of prescriptions pursuant to section 309 of the Act (21 U.S.C. 829) are set forth generally in that section and specifically by the sections of this part.

## SECTION 1306.02 DEFINITIONS

Any term contained in this part shall have the definition set forth in section 102 of the Act (21 U.S.C. 802) or part 1300 of this chapter.

## SECTION 1306.03 PERSONS ENTITLED TO ISSUE PRESCRIPTIONS

(a) A prescription for a controlled substance may be issued only by an individual practitioner who is:

(1) Authorized to prescribe controlled substances by the jurisdiction in which he/she is licensed to practice his/her profession and

(2) Either registered or exempted from registration pursuant to §1301.22(c) and 1301.23 of this chapter.

(b) A prescription issued by an individual practitioner may be communicated to a pharmacist by an employee or agent of the individual practitioner.

## Section 1306.04 Purpose of Issue of Prescription

(a) A prescription for a controlled substance to be effective must be issued for a legitimate medical purpose by an individual practitioner acting in the usual course of his/her professional practice. The responsibility for the proper prescribing and dispensing of controlled substances is upon the prescribing practitioner, but a corresponding responsibility rests with the pharmacist who fills the prescription. An order purporting to be a prescription issued not in the usual course of professional treatment or in legitimate and authorized research is not a prescription within the meaning and intent of section 309 of the Act (21 U.S.C. 829) and the person knowingly filling such a purported prescription, as well as the person issuing it, shall be subject to the penalties provided for violations of the provisions of law relating to controlled substances.

(b) A prescription may not be issued in order for an individual practitioner to obtain controlled substances for supplying the individual practitioner for the purpose of general dispensing to patients.

(c) A prescription may not be issued for the dispensing of narcotic drugs listed in any schedule for "detoxification treatment" or "maintenance treatment" as defined in Section 102 of the Act (21 U.S.C. 802).

## Section 1306.05 Manner of Issuance of Prescriptions

(a) All prescriptions for controlled substances shall be dated as of, and signed on, the day when issued and shall bear the full name and address of the patient, the drug name, strength, dosage form, quantity prescribed, directions for use and the name, address and registration number of the practitioner. A practitioner may sign a prescription in the same manner as he/she would sign a check or legal document (e.g., J.H. Smith or John H. Smith). Where an oral order is not permitted, prescriptions shall be written with ink or indelible pencil or typewriter and shall be manually signed by the practitioner. The prescriptions may be prepared by the secretary or agent for the signature of a practitioner, but the prescribing practitioner is responsible in case the prescription does not conform in all essential respects to the law and regulations. A corresponding liability rests upon the pharmacist who fills a prescription not prepared in the form prescribed by these regulations.

(b) An individual practitioner exempted from registration under §1301.22(c) of this chapter shall include on all prescriptions issued by him or her the registration number of the hospital or other institution and the special internal code number assigned to him or her by the hospital or other institution as provided in §1301.22(c) of this chapter, in lieu of the registration number of the practitioner required by this section. Each written prescription shall have the name of the physician stamped, typed, or handprinted on it, as well as the signature of the physician.

(c) An official exempted from registration under §1301.22(c) shall include on all prescriptions issued by him his branch of service or agency (e.g., "U.S. Army" or "Public Health Service") and his/her service identification number, in lieu of the registration number of the practitioner required by this section. The service identification number for a Public Health Service employee is his Social Security identification number. Each prescription shall have the name of the officer stamped, typed, or handprinted on it, as well as the signature of the officer.

### SECTION 1306.06 PERSONS ENTITLED TO FILL PRESCRIPTIONS

A prescription for controlled substances may be filled only by a pharmacist acting in the usual course of his/her professional practice and either registered individually or employed in a registered pharmacy or registered institutional practitioner.

### SECTION 1306.07 ADMINISTERING OR DISPENSING OF NARCOTIC DRUGS

(a) The administering or dispensing directly (but not prescribing) of narcotic drugs listed in any schedule to a narcotic drug dependent person for "detoxification treatment" or "maintenance treatment" as defined in section 102 of the Act (21 U.S.C. 802) shall be deemed to be within the meaning of the term "in the course of his professional practice or research" in section 308(e) and section 102(20) of the Act (21 U.S.C. 828 (e)): Provided, That the practitioner is separately registered with the Attorney General as required by section 303(g) of the Act (21 U.S.C. 823(g)) and then thereafter complies with the regulatory standards imposed relative to treatment qualification, security, records and unsupervised use of drugs pursuant to such Act.

(b) Nothing in this section shall prohibit a physician who is not specifically registered to conduct a narcotic treatment program from administering (but not prescribing) narcotic drugs to a person for the purpose of relieving acute withdrawal symptoms when necessary while arrangements are being made for referral for treatment. Not more than one day's medication may be administered to the person or for the person's use at one time. Such emergency treatment may be carried out for not more than three days and may not be renewed or extended.

(c) This section is not intended to impose any limitations on a physician or authorized hospital staff to administer or dispense narcotic drugs in a hospital to maintain or detoxify a person as an incidental adjunct to medical or surgical treatment of conditions other than addiction, or to administer or dispense narcotic drugs to persons with intractable pain in which no relief or cure is possible or none has been found after reasonable efforts.

## CONTROLLED SUBSTANCES LISTED IN SCHEDULE II

### SECTION 1306.11 REQUIREMENT OF PRESCRIPTION

(a) A pharmacist may dispense directly a controlled substance listed in Schedule II, which is a prescription drug as determined under the Federal Food, Drug, and Cosmetic Act, only pursuant to a written prescription signed by the practitioner, except as provided in paragraph (d) of this section. A prescription for a Schedule II controlled substance may be transmitted by the practitioner or the practitioner's agent to a pharmacy via facsimile equipment, provided that the original written, signed prescription is presented to the pharmacist for review prior to the actual dispensing of the controlled substance, except as noted in paragraph (e), (f), or (g) of this section. The original prescription shall be maintained in accordance with §1304.04(h) of this chapter.

(b) An individual practitioner may administer or dispense directly a controlled substance listed in Schedule II in the course of his professional practice without a prescription, subject to §1306.07.

(c) An institutional practitioner may administer or dispense directly (but not prescribe) a controlled substance listed in Schedule II pursuant only to a written prescription signed by the prescribing individual practitioner or to an order for medication made by an individual practitioner which is dispensed for immediate administration to the ultimate user.

(d) In the case of an emergency situation, as defined by the Secretary in §290.10 of this title, a pharmacist may dispense a controlled substance listed in Schedule II upon receiving oral authorization of a prescribing individual practitioner, provided that:

(1) The quantity prescribed and dispensed is limited to the amount adequate to treat the patient during the emergency period (dispensing beyond the emergency period must be pursuant to a written prescription signed by the prescribing individual practitioner);

(2) The prescription shall be immediately reduced to writing by the pharmacist and shall contain all information required in §1306.05, except for the signature of the prescribing individual practitioner;

(3) If the prescribing individual practitioner is not known to the pharmacist, he/she must make a reasonable effort to determine that the oral authorization came from a registered individual practitioner, which may include a callback to the prescribing individual practitioner using his/her phone number as listed in the telephone directory and/or other good faith efforts to insure his/her identity; and

(4) Within 7 days after authorizing an emergency oral prescription, the prescribing individual practitioner shall cause a written prescription for the emergency quantity prescribed to be delivered to the dispensing pharmacist. In addition to conforming to the requirements of §1306.05, the prescription shall have written on its face "Authorization for Emergency Dispensing," and the date of the oral order. The written prescription may be delivered to the pharmacist in person or by mail, but if delivered by mail it must be postmarked within the 7-day period. Upon receipt, the dispensing pharmacist shall attach this prescription to the oral emergency prescription which had earlier been reduced to writing. The pharmacist shall notify the nearest office of the Administration if the prescribing individual practitioner fails to deliver a written prescription to him; failure of the pharmacist to do so shall void the authority conferred by this paragraph to dispense without a written prescription of a prescribing individual practitioner.

(e) A prescription prepared in accordance with §1306.05 written for a Schedule II narcotic substance to be compounded for the direct administration to a patient by parenteral, intravenous, intramuscular, subcutaneous or intraspinal infusion may be transmitted by the practitioner or the practitioner's agent to the pharmacy by facsimile. The facsimile serves as the original written prescription for purposes of this paragraph (e) and it shall be maintained in accordance with §1304.04(h) of this chapter.

(f) A prescription prepared in accordance with §1306.05 written for Schedule II substance for a resident of a Long Term Care Facility may be transmitted by the practitioner or the practitioner's agent to the dispensing pharmacy by facsimile. The facsimile serves as the original written prescription for purposes of this paragraph (f) and it shall be maintained in accordance with §1304.04(h).

(g) A prescription prepared in accordance with §1306.05 written for a Schedule II narcotic substance for a patient enrolled in a hospice care program certified and/or paid for by Medicare under Title XVIII or a hospice program which is licensed by the state may be transmitted by the practitioner or the practitioner's agent to the dispensing pharmacy by facsimile. The practitioner or the practitioner's agent will note on the prescription that the patient is a hospice patient. The facsimile serves as the original written prescription for purposes of this paragraph (g) and it shall be maintained in accordance with §1304.04(h).

## Section 1306.12 Refilling Prescriptions

The refilling of a prescription for a controlled substance listed in Schedule II is prohibited.

## Section 1306.13 Partial Filling of Prescriptions

(a) The partial filling of a prescription for a controlled substance listed in Schedule II is permissible, if the pharmacist is unable to supply the full quantity called for in a written or emergency oral prescription and he/she makes a notation of the quantity supplied on the face of the written prescription (or written record of the emergency oral prescription). The remaining portion of the prescription may be filled within 72 hours of the first partial filling; however, if the remaining portion is not or cannot be filled within the 72-hour period, the pharmacist shall so notify the prescribing individual practitioner. No further quantity may be supplied beyond 72 hours without a new prescription.

(b) A prescription for a Schedule II controlled substance written for a patient in a Long Term Care Facility (LTCF) or for a patient with a medical diagnosis documenting a terminal illness may be filled in partial quantities to include individual dosage units. If there is any question whether a patient may be classified as having a terminal illness, the pharmacist must contact the practitioner prior to partially filling the prescription. Both the pharmacist and the prescribing practitioner have a corresponding responsibility to assure that the controlled substance is for a terminally ill patient. The pharmacist must record on the prescription whether the patient is "terminally ill" or an "LTCF patient." A prescription that is partially filled and does not contain the notation "terminally ill" or "LTCF patient" shall be deemed to have been filled in violation of the Act. For each partial filling, the dispensing pharmacist shall record on the back of the prescription (or on another appropriate record, uniformly maintained, and readily retrievable) the date of the partial filling, quantity dispensed, remaining quantity authorized to be dispensed, and the identification of the dispensing pharmacist. The total quantity of Schedule II controlled substances dispensed in all partial fillings must not exceed the total quantity prescribed. Schedule II prescriptions for patients in a LTCF or patients with a medical diagnosis documenting a terminal illness shall be valid for a period not to exceed 60 days from the issue date unless sooner terminated by the discontinuance of medication.

(c) Information pertaining to current Schedule II prescriptions for patients in a LTCF or for patients with a medical diagnosis documenting a terminal illness may be maintained in a computerized system if this system has the capability to permit:

(1) Output (display or printout) of the original prescription number, date of issue, identification of prescribing individual practitioner, identification of patient, address of the

LTCF or address of the hospital or residence of the patient, identification of medication authorized (to include dosage, form, strength and quantity), listing of the partial fillings that have been dispensed under each prescription and the information required in §1306.13(b).

(2) Immediate (real time) updating of the prescription record each time a partial filling of the prescription is conducted.

(3) Retrieval of partially filled Schedule II prescription information is the same as required by §1306.22(b) (4) and (5) for Schedule III and IV prescription refill information.

### SECTION 1306.14 LABELING OF SUBSTANCES AND FILLING OF PRESCRIPTIONS

(a) The pharmacist filling a written or emergency oral prescription for a controlled substance listed in Schedule II shall affix to the package a label showing date of filling, the pharmacy name and address, the serial number of the prescription, the name of the patient, the name of the prescribing practitioner, and directions for use and cautionary statements, if any, contained in such prescription or required by law.

(b) The requirements of paragraph (a) of this section do not apply when a controlled substance listed in Schedule II is prescribed for administration to an ultimate user who is institution-alized: Provided, That:

(1) Not more than 7-day supply of the controlled substance listed in Schedule II is dispensed at one time;

(2) The controlled substance listed in Schedule II is not in the possession of the ultimate user prior to the administration;

(3) The institution maintains appropriate safeguards and records regarding the proper administration, control, dispensing, and storage of the controlled substance listed in Schedule II; and

(4) The system employed by the pharmacist in filling a prescription is adequate to identify the supplier, the product, and the patient, and to set forth the directions for use and cautionary statements, if any, contained in the prescription or required by law.

(c) All written prescriptions and written records of emergency oral prescriptions shall be kept in accordance with requirements of §1304.04(h) of this chapter.

## CONTROLLED SUBSTANCES LISTED IN SCHEDULES III, IV, AND V

### SECTION 1306.21 REQUIREMENT OF PRESCRIPTION

(a) A pharmacist may dispense directly a controlled substance listed in Schedule III, IV, or V which is a prescription drug as determined under the Federal Food, Drug, and Cosmetic Act, pursuant only to either a written prescription signed by a practitioner or a facsimile of a written, signed prescription transmitted by the practitioner or the practitioner's agent to the pharmacy or pursuant to an oral prescription made by an individual practitioner and promptly reduced to writing by the pharmacist containing all information required in §1306.05, except for the signature of the practitioner.

(b) An individual practitioner may administer or dispense directly a controlled substance listed in Schedule III, IV, or V in the course of his/her professional practice without a prescription, subject to §1306.07.

(c) An institutional practitioner may administer or dispense directly (but not prescribe) a controlled substance listed in Schedule III, IV, or V pursuant only to a written prescription signed by an individual practitioner, or pursuant to a facsimile of a written prescription or order for medication transmitted by the practitioner or the practitioner's agent to the institutional practitioner-pharmacist, or pursuant to an oral prescription made by an individual practitioner and promptly reduced to writing by the pharmacist (containing all information required in §1306.05 except for the signature of the individual practitioner), or pursuant to an order for medication made by an individual practitioner which is dispensed for immediate administration to the ultimate user, subject to §1306.07.

## SECTION 1306.22 REFILLING OF PRESCRIPTIONS

(a) No prescription for a controlled substance listed in Schedule III or IV shall be filled or refilled more than six months after the date on which such prescription was issued and no such prescription authorized to be refilled may be refilled more than five times. Each refilling of a prescription shall be entered on the back of the prescription or on another appropriate document. If entered on another document, such as a medication record, the document must be uniformly maintained and readily retrievable. The following information must be retrievable by the prescription number consisting of the name and dosage form of the controlled substance, the date filled or refilled, the quantity dispensed, initials of the dispensing pharmacist for each refill, and the total number of refills for that prescription. If the pharmacist merely initials and dates the back of the prescription it shall be deemed that the full face amount of the prescription has been dispensed. The prescribing practitioner may authorize additional refills of Schedule III or IV controlled substances on the original prescription through an oral refill authorization transmitted to the pharmacist provided the following conditions are met:

(1) The total quantity authorized, including the amount of the original prescription, does not exceed five refills nor extend beyond six months from the date of issue of the original prescription.

(2) The pharmacist obtaining the oral authorization records on the reverse of the original prescription the date, quantity of refill, number of additional refills authorized, and initials the prescription showing who received the authorization from the prescribing practitioner who issued the original prescription.

(3) The quantity of each additional refill authorized is equal to or less than the quantity authorized for the initial filling of the original prescription.

(4) The prescribing practitioner must execute a new and separate prescription for any additional quantities beyond the five-refill, six-month limitation.

(b) As an alternative to the procedures provided by subsection (a), an automated data processing system may be used for the storage and retrieval of refill information for prescription orders for controlled substances in Schedule III and IV, subject to the following conditions:

(1) Any such proposed computerized system must provide on-line retrieval (via CRT display or hard-copy printout) of original prescription order information for those prescription orders which are currently authorized for refilling. This shall include, but is not limited to, data such as the original prescription number, date of issuance of the original prescription order by the practitioner, full name and address of the patient, name, address, and DEA registration number of the practitioner, and the name, strength, dosage form, quantity of the controlled substance prescribed (and quantity dispensed if different from

the quantity prescribed), and the total number of refills authorized by the prescribing practitioner.

(2) Any such proposed computerized system must also provide on-line retrieval (via CRT display or hard-copy printout) of the current refill history for Schedule III or IV controlled substance prescription orders (those authorized for refill during the past six months.) This refill history shall include, but is not limited to, the name of the controlled substance, the date of refill, the quantity dispensed, the identification code, or name or initials of the dispensing pharmacist for each refill and the total number of refills dispensed to date for that prescription order.

(3) Documentation of the fact that the refill information entered into the computer each time a pharmacist refills an original prescription order for a Schedule III or IV controlled substance is correct must be provided by the individual pharmacist who makes use of such a system. If such a system provides a hard-copy printout of each day's controlled substance prescription order refill data, that printout shall be verified, dated, and signed by the individual pharmacist who refilled such a prescription order. The individual pharmacist must verify that the data indicated is correct and then sign this document in the same manner as he/she would sign a check or legal document (e.g., J. H. Smith, or John H. Smith). This document shall be maintained in a separate file at that pharmacy for a period of two years from the dispensing date. This printout of the day's controlled substance prescription order refill data must be provided to each pharmacy using such a computerized system within 72 hours of the date on which the refill was dispensed. It must be verified and signed by each pharmacist who is involved with such dispensing. In lieu of such a printout, the pharmacy shall maintain a bound log book, or separate file, in which each individual pharmacist involved in such dispensing shall sign a statement (in the manner previously described) each day, attesting to the fact that the refill information entered into the computer that day has been reviewed by him/her and is correct as shown. Such a book or file must be maintained at the pharmacy employing such a system for a period of two years after the date of dispensing the appropriately authorized refill.

(4) Any such computerized system shall have the capability of producing a printout of any refill data which the user pharmacy is responsible for maintaining under the Act and its implementing regulations. For example, this would include a refill-by-refill audit trail for any specified strength and dosage form of any controlled substance (by either brand or generic name or both). Such a printout must include name of the prescribing practitioner, name and address of the patient, quantity dispensed on each refill, date of dispensing for each refill, name or identification code of the dispensing pharmacist, and the number of the original prescription order. In any computerized system employed by a user pharmacy the central recordkeeping location must be capable of sending the printout to the pharmacy within 48 hours, and if a DEA Special Agent or Diversion Investigator requests a copy of such printout from the user pharmacy, it must, if requested to do so by the Agent or Investigator, verify the printout transmittal capability of its system by documentation (e.g., postmark).

(5) In the event that a pharmacy which employs such a computerized system experiences system down-time, the pharmacy must have an auxiliary procedure which will be used for documentation of refills of Schedule III and IV controlled substance prescription orders. This auxiliary procedure must insure that refills are authorized by the original prescription order, that the maximum number of refills has not been exceeded, and that

all of the appropriate data is retained for on-line data entry as soon as the computer system is available for use again.

(c) When filing refill information for original prescription orders for Schedule III or IV controlled substances, a pharmacy may use only one of the two systems described in paragraphs (a) or (b) of this section.

## SECTION 1306.23 PARTIAL FILLING OF PRESCRIPTIONS

The partial filling of a prescription for a controlled substance listed in Schedule III, IV, or V is permissible, provided that:

(a) Each partial filling is recorded in the same manner as a refilling,
(b) The total quantity dispensed in all partial fillings does not exceed the total quantity prescribed, and
(c) No dispensing occurs after 6 months after the date on which the prescription was issued.

## SECTION 1306.24 LABELING OF SUBSTANCES AND FILLING OF PRESCRIPTIONS

(a) The pharmacist filling a prescription for a controlled substance listed in Schedule III, IV, or V shall affix to the package a label showing the pharmacy name and address, the serial number and date of initial filling, the name of the patient, the name of the practitioner issuing the prescription, and directions for use and cautionary statements, if any, contained in such prescription as required by law.
(b) The requirements of paragraph (a) of this section do not apply when a controlled substance listed in Schedule III, IV, or V is prescribed for administration to an ultimate user who is institutionalized: Provided, That:
    (1) Not more than a 34-day supply or 100 dosage units, whichever is less, of the controlled substance listed in Schedule III, IV, or V is dispensed at one time;
    (2) The controlled substance listed in Schedule III, IV, or V is not in the possession of the ultimate user prior to administration;
    (3) The institution maintains appropriate safeguards and records the proper administration, control, dispensing, and storage of the controlled substance listed in Schedule III, IV, or V; and
    (4) The system employed by the pharmacist in filling a prescription is adequate to identify the supplier, the product and the patient, and to set forth the directions for use and cautionary statements, if any, contained in the prescription or required by law.
(c) All prescriptions for controlled substances listed in Schedules III, IV, and V shall be kept in accordance with §1304.04(h) of this chapter.

## SECTION 1306.25 TRANSFER BETWEEN PHARMACIES OF PRESCRIPTION INFORMATION FOR SCHEDULES III, IV, AND V CONTROLLED SUBSTANCES FOR REFILL PURPOSES

(a) The transfer of original prescription information for a controlled substance listed in Schedules III, IV or V for the purpose of refill dispensing is permissible between pharmacies on a onetime basis only. However, pharmacies electronically sharing a real-time, on-line database may transfer up to the maximum refills permitted by law and the prescriber's authorization. Transfers are subject to the following requirements:

(1) The transfer is communicated directly between two licensed pharmacists and the transferring pharmacist records the following information:

   (i) Write the word **VOID** on the face of the invalidated prescription.

   (ii) Record on the reverse of the invalidated prescription the name, address and DEA registration number of the pharmacy to which it was transferred and the name of the pharmacist receiving the prescription information.

   (iii) Record the date of the transfer and the name of the pharmacist transferring the information.

(b) The pharmacist receiving the transferred prescription information shall reduce to writing the following:

   (1) Write the word **transfer** on the face of the transferred prescription.

   (2) Provide all information required to be on a prescription pursuant to 21 CFR 1306.05 and include:

      (i) Date of issuance of original prescription;

      (ii) Original number of refills authorized on original prescription;

      (iii) Date of original dispensing;

      (iv) Number of valid refills remaining and date(s) and locations of previous refill(s);

      (v) Pharmacy's name, address, DEA registration number and prescription number from which the prescription information was transferred;

      (vi) Name of pharmacist who transferred the prescription;

      (vii) Pharmacy's name, address, DEA registration number and prescription number from which the prescription was originally filled.

   (3) The original and transferred prescription(s) must be maintained for a period of two years from the date of last refill.

(c) Pharmacies electronically accessing the same prescription record must satisfy all information requirements of a manual mode for prescription transferral.

(d) The procedure allowing the transfer of prescription information for refill purposes is permissible only if allowable under existing state or other applicable law.

## SECTION 1306.26 DISPENSING WITHOUT PRESCRIPTION

A controlled substance listed in Schedules II, III, IV, or V which is not a prescription drug as determined under the Federal Food, Drug, and Cosmetic Act, may be dispensed by a pharmacist without a prescription to a purchaser at retail, provided that:

(a) Such dispensing is made only by a pharmacist (as defined in part 1300 of this chapter), and not by a nonpharmacist employee even if under the supervision of a pharmacist (although after the pharmacist has fulfilled his professional and legal responsibilities set forth in this section, the actual cash, credit transaction, or delivery, may be completed by a nonpharmacist);

(b) Not more than 240 cc (8 ounces) of any such controlled substance containing opium, nor more than 120 cc (4 ounces) of any other such controlled substance nor more than 48 dosage units of any such controlled substance containing opium, nor more than 24 dosage units of any other such controlled substance may be dispensed at retail to the same purchaser in any given 48-hour period;

(c) The purchaser is at least 18 years of age;

(d) The pharmacist requires every purchaser of a controlled substance under this section not known to him to furnish suitable identification (including proof of age where appropriate);

(e) A bound record book for dispensing of controlled substances under this section is maintained by the pharmacist, which book shall contain the name and address of the purchaser, the name and quantity of controlled substance purchased, the date of each purchase, and the name or initials of the pharmacist who dispensed the substance to the purchaser (the book shall be maintained in accordance with the recordkeeping requirement of §1304.04 of this chapter); and

(f) A prescription is not required for distribution or dispensing of the substance pursuant to any other Federal, State or local law.

# Part 1307
## Miscellaneous

## CONTENTS

### SECTION 1307.01 DEFINITIONS

Any term contained in this part shall have the definition set forth in section 102 of the Act (21 U.S.C. 802) or part 1300 of this chapter.

### SECTION 1307.02 APPLICATION OF STATE LAW AND OTHER FEDERAL LAW

Nothing in this chapter shall be construed as authorizing or permitting any person to do any act which such person is not authorized or permitted to do under other Federal laws or obligations under international treaties, conventions or protocols, or under the law of the State in which he/she desires to do such act nor shall compliance with such parts be construed as compliance with other Federal or State laws unless expressly provided in such other laws.

### SECTION 1307.03 EXCEPTIONS TO REGULATIONS

Any person may apply for an exception to the application of any provision of this chapter by filing a written request stating the reasons for such exception. Requests shall be filed with the Administrator, Drug Enforcement Administration, Department of Justice, Washington, D.C. 20537. The Administrator may grant an exception in his discretion, but in no case shall he/she be required to grant an exception to any person which is otherwise required by law or the regulations cited in this section.

# SPECIAL EXCEPTIONS FOR MANUFACTURE AND DISTRIBUTION OF CONTROLLED SUBSTANCES

## SECTION 1307.11 DISTRIBUTION BY DISPENSER TO ANOTHER PRACTITIONER

(a) A practitioner who is registered to dispense a controlled substance may distribute (without being registered to distribute) a quantity of such substance to another practitioner for the purpose of general dispensing by the practitioner to his or its patients: Provided, That:

    (1) The practitioner to whom the controlled substance is to be distributed is registered under the Act to dispense that controlled substance;

    (2) The distribution is recorded by the distributing practitioner in accordance with §1304.22(c) of this chapter and by the receiving practitioner in accordance with §1304.22(c) of this chapter;

    (3) If the substance is listed in Schedule I or II, an order form is used as required in part 1305 of this chapter;

    (4) The total number of dosage units of all controlled substances distributed by the practitioner pursuant to this section and §1301.25 of this chapter during each calendar year in which the practitioner is registered to dispense does not exceed 5 percent of the total number of dosage units of all controlled substances distributed and dispensed by the practitioner during the same calendar year.

(b) If, during any calendar year in which the practitioner is registered to dispense, the practitioner has reason to believe that the total number of dosage units of all controlled substances which will be distributed by him/her pursuant to this section and §1301.25 of this chapter will exceed 5 percent of the total number of dosage units of all controlled substances distributed and dispensed by him/her during that calendar year, the practitioner shall obtain a registration to distribute controlled substances.

## SECTION 1307.12 DISTRIBUTION TO SUPPLIER

(a) Any person lawfully in possession of a controlled substance listed in any schedule may distribute (without being registered to distribute) that substance to the person from whom he/she obtained it or to the manufacturer of the substance, provided that a written record is maintained which indicates the date of the transaction, the name, form, and quantity of the substance, the name, address, and registration number, if any, of the person making the distribution, and the name, address, and registration number, if known, of the supplier or manufacturer. In the case of returning a controlled substance in Schedule I or II, an order form shall be used in the manner prescribed in part 1305 of this chapter and be maintained as the written record of the transaction. Any person not required to register pursuant to sections 302(c) or 1007(b)(1) of the Act (21 U.S.C. 822(c) or 957(b)(1)) shall be exempt from maintaining the records required by this section.

(b) Distributions referred to in paragraph (a) may be made through a freight forwarding facility operated by the person to whom the controlled substance is being returned provided that prior arrangement has been made for the return and the person making the distribution delivers the controlled substance directly to an agent or employee of the person to whom the controlled substance is being returned.

### SECTION 1307.13 INCIDENTAL MANUFACTURE OF CONTROLLED SUBSTANCES

Any registered manufacturer who, incidentally but necessarily, manufactures a controlled substance as a result of the manufacture of a controlled substance or basic class of controlled substance for which he/she is registered and has been issued an individual manufacturing quota pursuant to part 1303 of this chapter (if such substance or class is listed in Schedule I or II) shall be exempt from the requirement of registration pursuant to part 1301 of this chapter and, if such incidentally manufactured substance is listed in Schedule I or II, shall be exempt from the requirement of an individual manufacturing quota pursuant to part 1303 of this chapter, if such substances are disposed of in accordance with §1307.21.

## DISPOSAL OF CONTROLLED SUBSTANCES

### SECTION 1307.21 PROCEDURE FOR DISPOSING OF CONTROLLED SUBSTANCES

(a) Any person in possession of any controlled substance and desiring or required to dispose of such substance may request assistance from the Special Agent in Charge of the Administration in the area in which the person is located for authority and instructions to dispose of such substance. The request should be made as follows:

    (1) If the person is a registrant, he/she shall list the controlled substance or substances which he/she desires to dispose of on DEA Form 41, and submit three copies of that form to the Special Agent in Charge in his/her area; or

    (2) If the person is not a registrant, he/she shall submit to the Special Agent in Charge a letter stating:

        (i) The name and address of the person;

        (ii) The name and quantity of each controlled substance to be disposed of;

        (iii) How the applicant obtained the substance, if known; and

        (iv) The name, address, and registration number, if known, of the person who possessed the controlled substances prior to the applicant, if known.

(b) The Special Agent in Charge shall authorize and instruct the applicant to dispose of the controlled substance in one of the following manners:

    (1) By transfer to person registered under the Act and authorized to possess the substance;

    (2) By delivery to an agent of the Administration or to the nearest office of the Administration;

    (3) By destruction in the presence of an agent of the Administration or other authorized person; or

    (4) By such other means as the Special Agent in Charge may determine to assure that the substance does not become available to unauthorized persons.

(c) In the event that a registrant is required regularly to dispose of controlled substances, the Special Agent in Charge may authorize the registrant to dispose of such substances, in accordance with paragraph (b) of this section, without prior approval of the Administration in each instance, on the condition that the registrant keep records of such disposals and file periodic reports with the Special Agent in Charge summarizing the disposals made by the registrant. In granting such authority, the Special Agent in Charge may place such conditions as he/she deems proper on the disposal of controlled substances, including the method of disposal and the frequency and detail of reports.

(d) This section shall not be construed as affecting or altering in any way the disposal of controlled substances through procedures provided in laws and regulations adopted by any State.

### SECTION 1307.22 DISPOSAL OF CONTROLLED SUBSTANCES BY THE ADMINISTRATION.

Any controlled substance delivered to the Administration under §1307.21 or forfeited pursuant to section 511 of the Act (21 U.S.C. 881) may be delivered to any department, bureau, or other agency of the United States or of any State upon proper application addressed to the Administrator, Drug Enforcement Administration, Department of Justice, Washington, D.C. 20537. The application shall show the name, address, and official title of the person or agency to whom the controlled drugs are to be delivered, including the name and quantity of the substances desired and the purpose for which intended. The delivery of such controlled drugs shall be ordered by the Administrator, if, in his opinion, there exists a medical or scientific need therefore.

## SPECIAL EXEMPT PERSONS

### SECTION 1307.31 NATIVE AMERICAN CHURCH

The listing of peyote as a controlled substance in Schedule I does not apply to the nondrug use of peyote in bona fide religious ceremonies of the Native American Church, and members of the Native American Church so using peyote are exempt from registration. Any person who manufactures peyote for or distributes peyote to the Native American Church, however, is required to obtain registration annually and to comply with all other requirements of law.

# MIL-Q-9858A

16 December 1963
Superseding
MIL-Q-9858
9 April 1959

## Military Specification
## Quality Program Requirements

*This specification has been approved by the Department of Defense and is mandatory for use by the Departments of the Army, the Navy, the Air Force and the Defense Supply Agency.*

## 1. SCOPE

### 1.1 APPLICABILITY

This specification shall apply to all supplies (including equipments, sub-systems and systems) or services when referenced in the item specification, contract or order.

### 1.2 CONTRACTUAL INTENT

This specification requires the establishment of a quality program by the contractor to assure compliance with the requirements of the contract. The program and procedures used to implement this specification shall be developed by the contractor. The quality program, including procedures, processes and product shall be documented and shall be subject to review by the Government Representative. The quality program is subject to the disapproval of the Government Representative whenever the contractor's procedures do not accomplish their objectives. The Government, at its option, may furnish written notice of the acceptability of the contractor's quality program

### 1.3 SUMMARY

An effective and economical quality program, planned and developed in consonance with the contractor's other administrative and technical programs, is required by this specification. Design of the program shall be based upon consideration of the technical and manufacturing aspects of production and related engineering design and materials. The program shall assure adequate quality throughout all areas of contract performance; for example, design, development, fabrication, processing, assembly, inspection, test, maintenance, packaging, shipping, storage and site installation.

All supplies and services under the contract, whether manufactured or performed within the contractor's plant or at any other source, shall be controlled at all points necessary to assure conformance to contractual requirements. The program shall provide for the prevention and ready detection of discrepancies and for timely and positive corrective action. The contractor shall make objective evidence of quality conformance readily available to the Government Representative. Instructions and records for quality must be controlled.

The authority and responsibility of those in charge of the design, production, testing, and inspection of quality shall be clearly stated. The program shall facilitate determinations of the effects of quality deficiencies and quality costs on price. Facilities and standards such as drawings, engineering changes, measuring equipment and the like which are necessary for the creation of the required quality shall be effectively managed. The program shall include an effective control of purchased materials and subcontracted work. Manufacturing, fabrication and assembly work conducted within the contractor's plant shall be controlled completely The quality program shall also include effective execution of responsibilities shared jointly with the Government or related to Government functions, such as control of Government property and Government source inspection.

### 1.4 Relation to Other Contract Requirements

This specification and any procedure or document executed in implementation thereof, shall be in addition to and not in derogation of other contract requirements. The quality program requirements set forth in this specification shall be satisfied in addition to all detail requirements contained in the statement of work or in other parts of the contract. The contractor is responsible for compliance with all provisions of the contract and for furnishing specified supplies and services which meet all the requirements of the contract. If any inconsistency exists between the contract schedule or its general provisions and this specification, the contract schedule and the general provisions shall control. The contractor's quality program shall be planned and used in a manner to support reliability effectively.

### 1.5 Relation to MIL-I-45208

This specification contains requirements in excess of those in specification MIL-1-45208, Inspection System Requirements, inasmuch as total conformance to contract requirements is obtained best by controlling work operations, manufacturing processes as well as inspections and tests.

## 2. SUPERSEDING, SUPPLEMENTATION AND ORDERING

### 2.1 Applicable Documents

The following documents of the issue in effect on date of the solicitation form a part of this specification to the extent specified herein.

> Specifications
> Military
> MIL-1-45208 — Inspection System Requirements
> MIL-STD-45662 — Calibration System Requirements

### 2.2 Amendments and Revisions

Whenever this specification is amended or revised subsequent to its contractually effective date, the contractor may follow or authorize his/her subcontractors to follow the amended or revised document provided no increase in price or fee is required. The contractor shall not be required to follow the amended or revised document except as a change in contract. If the contractor elects to follow the amended or revised document, he/she shall notify the Contracting Officer in writing of this election. When the contractor elects to follow the provisions of an amendment or revision, he/she must follow them in full.

## 2.3 Ordering Government Documents

Copies of specifications, standards and drawings required by contractors in connection with specific procurements may be obtained from the procuring agency, or as otherwise directed by the Contracting Officer.

# 3. QUALITY PROGRAM MANAGEMENT

## 3.1 Organization

Effective management for quality shall be clearly prescribed by the contractor. Personnel performing quality functions shall have sufficient, well-defined responsibility, authority and the organizational freedom to identify and evaluate quality problems and to initiate, recommend or provide solutions. Management regularly shall review the status and adequacy of the quality program. The term "quality program requirements" as used herein identifies the collective requirements of this specification. It does not mean that the fulfillment of the requirements of this specification is the responsibility of any single contractor's organization, function or person.

## 3.2 Initial Quality Planning

The contractor, during the earliest practical phase of contract performance, shall conduct a complete review of the requirements of the contract to identify and make timely provision for the special controls, processes, test equipments, fixtures, tooling and skills required for assuring product quality. This initial planning will recognize the need and provide for research, when necessary, to update inspection and testing techniques, instrumentation and correlation of inspection and test results with manufacturing methods and processes. This planning will also provide appropriate review and action to assure compatibility of manufacturing, inspection, testing and documentation.

## 3.3 Work Instructions

The quality program shall assure that all work affecting quality (including such things as purchasing, handling, machining, assembling, fabricating, processing, inspection, testing, modification, installation, and any other treatment of product, facilities, standards or equipment from the ordering of materials to dispatch of shipments) shall be prescribed in clear and complete documented instructions of a type appropriate to the circumstances. Such instructions shall provide the criteria for performing the work functions and they shall be compatible with acceptance criteria for workmanship. The instructions are intended also to serve for supervising, inspecting and managing work. The preparation and maintenance of and compliance with work instructions shall be monitored as a function of the quality program.

## 3.4 Records

The contractor shall maintain and use any records or data essential to the economical and effective operation of his/her quality program. These records shall be available for review by the Government

Representative and copies of individual records shall be furnished him/her upon request. Records are considered one of the principal forms of objective evidence of quality. The quality program shall assure that records are complete and reliable. Inspection and testing records shall, as a minimum, indicate the nature of the observations together with the number of observations made and the number and type of deficiencies found. Also, records for monitoring work performance and for inspection and testing shall indicate the acceptability of work or products and the action taken in connection with deficiencies. The quality program shall provide for the analysis and use of records as a basis for management action.

## 3.5 Corrective Action

The quality program shall detect promptly and correct assignable conditions adverse to quality. Design, purchasing. manufacturing, testing or other operations which could result in or have resulted in defective supplies, services, facilities, technical data, standards or other elements of contract performance which could create excessive losses or costs must be identified and changed as a result of the quality program. Corrective action will extend to the performance of all suppliers and vendors and will be responsive to data and product forwarded from users. Corrective action shall include as a minimum:

(a) Analysis of data and examination of product scrapped or reworked to determine extent and causes;

(b) Analysis of trends in processes or performance of work to prevent nonconforming product; and

(c) Introduction of required improvements and corrections, an initial review of the adequacy of such measures and monitoring of the effectiveness of corrective action taken.

## 3.6 Costs Related to Quality

The contractor shall maintain and use quality cost data as a management element of the quality program. These data shall serve the purpose of identifying the cost of both the prevention and correction of nonconforming supplies (e.g., labor and material involved in material spoilage caused by defective work, correction of defective work and for quality control exercised by the contractor at subcontractor's or vendor's facilities). The specific quality cost data to be maintained and used will be determined by the contractor. Quality cost data maintained by the contractor shall, upon request, be furnished the Government Representative for use by the Government in determining the effectiveness of the contractor's quality program.

# 4. FACILITIES AND STANDARDS

## 4.1 Drawings, Documentation and Changes

A procedure shall be maintained that concerns itself with the adequacy, the completeness and the currentness of drawings and with the control of changes in design. With respect to the currentness of drawings and changes, the contractor shall assure that requirements for the effectivity point of changes are met and that obsolete drawings and change requirements are removed from all points of issue and use. Some means of recording the effective points shall be employed and be available to the Government.

With respect to design drawings and design specifications, a procedure shall be maintained that shall provide for the evaluation of their engineering adequacy and an evaluation of the adequacy of proposed changes. The evaluation shall encompass both the adequacy in relation to standard engineering and design practices and the adequacy with respect to the design and purpose of the product to which the drawing relates.

With respect to supplemental specifications, process instructions, production engineering instructions, industrial engineering instructions and work instructions relating to a particular design, the contractor shall be responsible for a review of their adequacy, currentness and completeness. The quality program must provide complete coverage of all information necessary to produce an article in complete conformity with requirements of the design.

The quality program shall assure that there is complete compliance with contract requirements for proposing, approving, and effecting of engineering changes. The quality program shall provide for monitoring effectively compliance with contractual engineering changes requiring approval by Government design authority. The quality program shall provide for monitoring effectively the drawing changes of lesser importance not requiring approval by Government design authorities.

Delivery of correct drawings and change information to the Government in connection with data acquisition shall be an integral part of the quality program. This includes full compliance with contract requirements concerning rights and data both proprietary and other. The quality program's responsibility for drawings and changes extend to the drawings and changes provided by the subcontractors and vendors for the contract.

## 4.2 Measuring and Testing Equipment

The contractor shall provide and maintain gauges and other measuring and testing devices necessary to assure that supplies conform to technical requirements. These devices shall be calibrated against certified measurement standards which have known valid relationships to national standards at established periods to assure continued accuracy. The objective is to assure that inspection and test equipment is adjusted, replaced or repaired before it becomes inaccurate. The calibration of measuring and testing equipment shall be in conformity with military specification MIL-STD-45662. In addition, the contractor shall insure the use of only such subcontractor and vendor sources that depend upon calibration systems which effectively control the accuracy of measuring and testing equipment.

## 4.3 Production Tooling Used as Media of Inspection

When production jigs, fixtures, tooling masters, templates, patterns and such other devices are used as media of inspection, they shall be proved for accuracy prior to release for use. These devices shall be proved again for accuracy at intervals formally established in a manner to cause their timely adjustment, replacement or repair prior to becoming inaccurate.

## 4.4 Use of Contractor's Inspection Equipment

The contractor's gauges, measuring and testing devices shall be made available for use by the Government when required to determine conformance with contract requirements. If conditions warrant, contractor's personnel shall be made available for operation of such devices and for verification of their accuracy and condition.

## 4.5 Advanced Metrology Requirements

The quality program shall include timely identification and report to the Contracting Officer of any precision measurement need exceeding the known state of the art.

# 5. CONTROL OF PURCHASES

## 5.1 Responsibility

The contractor is responsible for assuring that all supplies and services procured from his/her suppliers (subcontractors and vendors) conform to the contract requirements. The selection of sources and the nature and extent of control exercised by the contractor shall be dependent upon the type of supplies, his/her supplier's demonstrated capability to perform, and the quality evidence made available. To assure an adequate and economical control of such material, the contractor shall utilize to the fullest extent objective evidence of quality furnished by his/her suppliers. When the Government elects to perform inspection at a supplier's plant, such inspection shall not be used by contractors as evidence of effective control of quality by such suppliers. The inclusion of a product on the Qualified Products List only signifies that at one time the manufacturer made a product which met specification requirements. It does not relieve the contractor of his/her responsibility for furnishing supplies that meet all specification requirements or for the performance of specified inspections and tests for such material. The effectiveness and integrity of the control of quality by his/her suppliers shall be assessed and reviewed by the contractor at intervals consistent with the complexity and quantity of product. Inspection of products upon delivery to the contractor shall be used for assessment and review to the extent necessary for adequate assurance of quality. Test reports, inspection records, certificates and other suitable evidence relating to the supplier's control of quality should be used in the contractor's assessment and review. The contractor's responsibility for the control of purchases includes the establishment of a procedure for (1) the selection of qualified suppliers, (2) the transmission of applicable design and quality requirements in the Government contracts and associated technical requirements, (3) the evaluation of the adequacy of procured items, and (4) effective provisions for early information feedback and correction of nonconformances.

## 5.2 Purchasing Data

The contractor's quality program shall not be acceptable to the Government unless the contractor requires of his/her subcontractors a quality effort achieving control of the quality of the services and supplies which they provide. The contractor shall assure that all applicable requirements are properly included or referenced in all purchase orders for products ultimately to apply on a Government contract. The purchase order shall contain a complete description of the supplies ordered including, by statement or reference, all applicable requirements for manufacturing, inspecting, testing, packaging, and any requirements for Government or contractor inspections, qualification or approvals. Technical requirements of the following nature must be included by statement or reference as a part of the required clear description: all pertinent drawings, engineering change orders, specifications (including inspection system or quality program requirements), reliability, safety, weight, or other special requirements, unusual test or inspection procedures or equipment and any special revision or model identification. The description of products ordered shall include a requirement for contractor inspection at the subcontractor or vendor source when such action is necessary to assure that the contractor's quality program effectively implements the contractor's responsibility for complete assurance of product quality. Requirements shall be included for chemical and physical testing and recording in connection with the purchase of raw materials by his/her suppliers. The purchase orders must also contain a requirement for such suppliers to notify and obtain approval from the contractor of changes in design of the products. Necessary instructions should be provided when provision is made for direct shipment from the subcontractor to Government activities.

# 6. MANUFACTURING CONTROL

## 6.1 MATERIALS AND MATERIALS CONTROL

Supplier's materials and products shall be subjected to inspection upon receipt to the extent necessary to assure conformance to technical requirements. Receiving inspection may be adjusted upon the basis of the quality assurance program exercised by suppliers. Evidence of the suppliers' satisfactory control of quality may be used to adjust the amount and kind of receiving inspection.

The quality program shall assure that raw materials to be used in fabrication or processing of products conform to the applicable physical, chemical, and other technical requirements. Laboratory testing shall be employed as necessary. Suppliers shall be required by the contractor's quality program to exercise equivalent control of the raw materials utilized in the production of the parts and items which they supply to the contractor. Raw material awaiting testing must be separately identified or segregated from already tested and approved material but can be released for initial production, providing that identification and control is maintained. Material tested and approved must be kept identified until such time as its identity is necessarily obliterated by processing. Controls will be established to prevent the inadvertent use of material failing to pass tests.

## 6.2 PRODUCTION PROCESSING AND FABRICATION

The contractor's quality program must assure that all machining, wiring, batching, shaping and all basic production operations of any type together with all processing and fabricating of any type is accomplished under controlled conditions. Controlled conditions include documented work instructions, adequate production equipment, and any special working environment. Documented work instructions are considered to be the criteria for much of the production, processing and fabrication work. These instructions are the criteria for acceptable or unacceptable "workmanship." The quality program will effectively monitor the issuance of and compliance with all of these work instructions.

Physical examination, measurement or tests of the material or products processed is necessary for each work operation and must also be conducted under controlled conditions. If physical inspection of processed material is impossible or disadvantageous, indirect control by monitoring processing methods, equipment and personnel shall be provided. Both physical inspection and process monitoring shall be provided when control is inadequate without both, or when contract or specification requires both.

Inspection and monitoring of processed material or products shall be accomplished in any suitable systematic manner selected by the contractor. Methods of inspection and monitoring shall be corrected any time their unsuitability with reasonable evidence is demonstrated. Adherence to selected methods for inspection and monitoring shall be complete and continuous. Corrective measures shall be taken when noncompliance occurs.

Inspection by machine operators, automated inspection gauges, moving line or lot sampling, setup or first piece approval, production line inspection station, inspection or test department, roving inspectors — any other type of inspection — shall be employed in any combination desired by the contractor which will adequately and efficiently protect product quality and the integrity of processing.

Criteria for approval and rejection shall be provided for all inspection of product and monitoring of methods, equipment, and personnel. Means for identifying approved and rejected product shall be provided.

Certain chemical, metallurgical, biological, sonic, electronic, and radiological processes are of so complex and specialized a nature that much more than the ordinary detailing of work documentation is required. In effect, such processing may require an entire work specification as contrasted with the normal work operation instructions established in normal plant-wide standard production control issuances such as job operation routing books and the like. For these special processes, the contractor's quality program shall assure that the process control procedures or specifications are adequate and that processing environments and the certifying, inspection, authorization and monitoring of such processes to the special degree necessary for these ultraprecise and super-complex work functions are provided.

## 6.3 COMPLETED ITEM INSPECTION AND TESTING

The quality program shall assure that there is a system for final inspection and test of completed products. Such testing shall provide a measure of the overall quality of the completed product and shall be performed so that it simulates, to a sufficient degree, product end use and functioning. Such simulation frequently involves appropriate life and endurance tests and qualification testing. Final inspection and testing shall provide for reporting to designers any unusual difficulties, deficiencies or questionable conditions. When modifications, repairs or replacements are required after final inspection or testing, there shall be reinspection and retesting of any characteristics affected.

## 6.4 HANDLING, STORAGE AND DELIVERY

The quality program shall provide for adequate work and inspection instructions for handling, storage, preservation, packaging, and shipping to protect the quality of products and prevent damage, loss, deterioration, degradation, or substitution of products. With respect to handling, the quality program shall require and monitor the use of procedures to prevent handling damage to articles. Handling procedures of this type include the use of special crates, boxes, containers, transportation vehicles and any other facilities for materials handling. Means shall be provided for any necessary protection against deterioration or damage to products in storage. Periodic inspection for the prevention and results of such deterioration or damage shall be provided. Products subject to deterioration or corrosion during fabrication or interim storage shall be cleaned and preserved by methods which will protect against such deterioration or corrosion. When necessary, packaging designing and packaging shall include means for accommodating and maintaining critical environments within packages, e.g., moisture content levels, gas pressures. The quality program shall assure that when such packaging environments must be maintained, packages are labeled to indicate this condition. The quality program shall monitor shipping work to assure that products shipped are accompanied with required shipping and technical documents and that compliance with Interstate Commerce Commission rules and other applicable shipping regulations is effected to assure safe arrival and identification at destination. In compliance with contractual requirements, the quality program shall include monitoring provisions for protection of the quality of products during transit.

## 6.5 Nonconforming Material

The contractor shall establish and maintain an effective and positive system for controlling nonconforming material, including procedures for its identification, segregation, and disposition. Repair or rework of nonconforming material shall be in accordance with documented procedures acceptable to the Government. The acceptance of nonconforming supplies is a prerogative of and shall be as prescribed by the Government and may involve a monetary adjustment. All nonconforming supplies shall be positively identified to prevent unauthorized use, shipment and intermingling with conforming supplies. Holding areas or procedures mutually agreeable to the contractor and the Government Representative shall be provided by the contractor. The contractor shall make known to the Government upon request the data associated with the costs and losses in connection with scrap and with rework necessary to reprocess nonconforming material to make it conform completely.

## 6.6 Statistical Quality Control and Analysis

In addition to statistical methods required by the contract, statistical planning, analysis, tests and quality control procedures may be utilized whenever such procedures are suitable to maintain the required control of quality. Sampling plans may be used when tests are destructive, or when the records, inherent characteristics of the product or the noncritical application of the product, indicate that a reduction in inspection or testing can be achieved without jeopardizing quality. The contractor may employ sampling inspection in accordance with applicable military standards and sampling plans (e.g., from MIL-STD-105, MIL-STD-414, or Handbooks H 106, 107 and 108). If the contractor uses other sampling plans, they shall be subject to review by the cognizant Government Representative. Any sampling plan used shall provide valid confidence and quality levels.

## 6.7 Indication of Inspection Status

The contractor shall maintain a positive system for identifying the inspection status of products. Identification may be accomplished by means of stamps, tags, routing cards, move tickets, tote box cards or other normal control devices. Such controls shall be of a design distinctly different from Government inspection identification.

# 7. COORDINATED GOVERNMENT/CONTRACTOR ACTIONS

## 7.1 Government Inspection at Subcontractor or Vendor Facilities

The Government reserves the right to inspect at source supplies or services not manufactured or performed with the contractor's facility. Government inspection shall not constitute acceptance; nor shall it in any way replace contractor inspection or otherwise relieve the contractor of his responsibility to furnish an acceptable end item. The purpose of this inspection is to assist the Government Representative at the contractor's facility to determine the conformance of supplies or services with contract requirements. Such inspection can be requested only by or under authorization of the Government Representative. When Government inspection is required, the contractor shall add to his purchasing document the following statement:

"Government inspection is required prior to shipment from your plant. Upon receipt of this order, promptly notify the Government Representative who normally services your plant so that appropriate planning for Government inspection can be accomplished."

When, under authorization of the Government Representative, copies of the purchasing document are to be furnished directly by the subcontractor or vendor to the Government Representative at his facility rather than through Government channels, the contractor shall add to his purchasing document a statement substantially as follows:

> "On receipt of this order, promptly furnish a copy to the Government Representative who normally services your plant, or, if none, to the nearest Army, Navy, Air Force, or Defense Supply Agency inspection office. In the event the representative or office cannot be located, our purchasing agent should be notified immediately."

All documents and referenced data for purchases applying to a Government contract shall be available for review by the Government Representative to determine compliance with the requirements for the control of such purchases. Copies of purchasing documents required for Government purposes shall be furnished in accordance with the instructions of the Government Representative. The contractor shall make available to the Government Representative reports of any nonconformance found on Government source inspected supplies and shall (when requested) require the supplier to coordinate with his Government Representative on corrective action.

## 7.2 Government Property

### 7.2.1 Government-Furnished Material

When material is furnished by the Government, the contractor's procedures shall include at least the following:

(a) Examination upon receipt, consistent with practicability to detect damage in transit;
(b) Inspection for completeness and proper type;
(c) Periodic inspection and precautions to assure adequate storage conditions and to guard against damage from handling and deterioration during storage;
(d) Functional testing, either prior to or after installation, or both, as required by contract to determine satisfactory operation;
(e) Identification and protection from improper use or disposition; and
(f) Verification of quantity.

### 7.2.2 Damaged Government-Furnished Material

The contractor shall report to the Government Representative any Government-furnished material found damaged, malfunctioning, or otherwise unsuitable for use. In the event of damage or malfunctioning during or after installation, the contractor shall determine and record probable cause and necessity for withholding material from use.

### 7.2.3 Bailed Property

The contractor shall, as required by the terms of the Bailment Agreement, establish procedures for the adequate storage, maintenance and inspection of bailed Government property. Records of all inspections and maintenance performed on bailed property shall be maintained. These procedures and records shall be subject to review by the Government Representative.

## 8. NOTES

(The following information is provided solely for guidance in using this specification. It has no contractual significance.)

### 8.1 INTENDED USE

This specification will apply to complex supplies, components, equipments and systems for which the requirements of MIL-I-45208 are inadequate to provide needed quality assurance. In such cases, total conformance to contract requirements cannot be obtained effectively and economically solely by controlling inspection and testing. Therefore, it is essential to control work operations and manufacturing processes as well as inspections and tests. The purpose of this control is not only to assure that particular units of hardware conform to contractual requirements, but also to assure interface compatibility among these units of hardware when they collectively comprise major equipments, sub-systems and systems.

### 8.2 EXEMPTIONS

This specification will not be applicable to types of supplies for which MIL-I-45208 applies. The following do not normally require the application of this specification:

(a) Personal services, and
(b) Research and development studies of a theoretical nature which do not require fabrication of articles.

### 8.3 ORDER DATA

Procurement documents should specify the title, number and date of this specification.

# MIL-STD-790F

1 AUGUST 1995
SUPERSEDING
MIL-STD-790E
15 DECEMBER 1989

# DEPARTMENT OF DEFENSE

# STANDARD PRACTICE FOR ESTABLISHED RELIABILITY AND HIGH RELIABILITY QUALIFIED PRODUCTS LIST (QPL) SYSTEMS FOR ELECTRICAL, ELECTRONIC, AND FIBER OPTIC PARTS SPECIFICATIONS

## FOREWORD

1. This standard is approved for use by all Departments and Agencies of the Department of Defense.
2. In implementing the Parts Specification Management for Reliability Report (PSMR-1), issued by the Department of Defense in May 1960, it was determined that a manufacturer must provide evidence of (a) adequate production and test facilities, and (b) sound procedures for process control. This standard was developed to provide guidelines.
3. Beneficial comments (recommendations, additions, deletions) and any pertinent data which may be of use in improving this document should be addressed to: Commander, Defense Electronics Supply Center, ATTN: DESC-ELDM, 1507 Wilmington Pike, Dayton, OH 45444-5765, by using the self-addressed Standardization Document Improvement Proposal (DD Form 1426) appearing at the end of this document or by letter.

# CONTENTS

# 1. SCOPE

## 1.1 SCOPE

This standard is for direct reference in established reliability and high reliability electrical, electronic, and fiber optic parts specifications and establishes the criteria for a manufacturer's qualified product system.

# 2. APPLICABLE DOCUMENTS

## 2.1 GENERAL

The documents listed in this section are specified in sections 3, 4, and 5 of this standard. This section does not include documents cited in other sections of this standard or recommended for additional information or as examples. While every effort has been made to ensure the completeness of this list, document users are cautioned that they must meet all specified requirements documents cited in sections 3, 4, and 5 of this standard, whether or not they are listed.

## 2.2 GOVERNMENT DOCUMENTS

### 2.2.1 Specifications, Standards, and Handbooks

The following specifications, standards, and handbooks form a part of this document to the extent specified herein. Unless otherwise specified, the issues of these documents are those listed in the issue of the Department of Defense Index of Specifications and Standards (DoDISS) and supplement thereto, cited in the solicitation.

> STANDARDS
> FEDERAL
> FED-STD-209  Airborne Particulate Cleanliness Classes in Clean Room and Clear Zones.

(Unless otherwise indicated, copies of the above specification, standards, and handbooks are available from the Document Automation and Production Service, Building 4D, (DPM-DODSSP), 700 Robbins Avenue, Philadelphia, PA 19111-5094.)

## 2.3 NON-GOVERNMENT PUBLICATIONS

The following documents form a part of this document to the extent specified herein. Unless otherwise specified, the issues of the documents which are DoD adopted are those listed in the issue of the DoDISS cited in the solicitation. Unless otherwise specified, the issues of documents not listed in the DoDISS are the issues of the documents cited in the solicitation.

> AMERICAN NATIONAL STANDARDS INSTITUTE (ANSI)
> ANSI/EIA 557  Statistical Process Control Systems.
> ANSI/NCSL-Z540-1  Calibration Laboratories and Measuring and Test Equipment — General Requirements.
> ISO 10012-1  Quality Assurance Requirements for Measuring Equipment — Part 1: Metrological Confirmation System for Measuring Equipment.

(Application for copies should be addressed to the American National Standards Institute, 11 West 42nd Street, New York, NY 10036-8002.)

## 2.4 ORDER OF PRECEDENCE

In the event of a conflict between the text of this document and the references cited herein, the text of this document takes precedence. Nothing in this document, however, supersedes applicable laws and regulations unless a specific exemption has been obtained.

# 3. DEFINITIONS

## 3.1 RELIABILITY TERMS

The definitions of reliability terms used herein are as follows:

a. *Assembly plant.* A plant established by a manufacturer or operated by a distributor authorized by the manufacturer to perform specified functions pertaining to the manufacturer's identified qualified products in accordance with specified assembly procedures, test methods, processes, controls, and storage, handling, and packaging techniques.

b. *Burn-in (pre-conditioning).* The operation of an item under stress to stabilize its characteristics.

c. *Calibration.* A comparison of a measuring device with a known standard.

d. *Corrective action.* A documented design, process, procedure, or materials change implemented and validated to correct the cause of failure or design deficiency.

e. *Criticality.* A relative measure of the consequence of a failure mode and its frequency of occurrence.

f. *Defect analysis.* The process of examining technical or management (nontechnical) data, manufacturing techniques, processes, or materials to determine the cause of variations of electrical, mechanical, optical, or physical characteristics outside the established limitations.

g. *Degradation.* A gradual impairment in ability to perform.

h. *Demonstrated.* That which has been measured by the use of objective evidence gathered under specified conditions.

i. *Electrical, electronic, and fiber optic parts.* Basic circuit elements which cannot be dissembled and still perform their intended function, such as capacitors, connects, filters, resistors, switches, relays, transformers, crystals, electron tubes, semiconductor, and fiber optic devices.

j. *Environmental.* The aggregate of all external and internal conditions (such as temperature, humidity, radiation, magnetic and electric fields, and shock vibration) either natural or man made, or self-induced, that influences the form, performance, reliability or survival of an item.

k. *Established reliability.* A quantitative maximum failure rate demonstrated under controlled conditions specified in a military specification and usually expressed as percent failures per thousand hours of test.

l. *Failure.* The event, or inoperable state, in which any item or part of an item does not, or would not, perform as previously specified.

m. *Failure activating cause.* The stresses or forces, thermal, electrical shocks, or vibration, which induce or activate a failure mechanism.

n. *Failure analysis.* The process of examining electrical, electronic, or fiber optic parts to determine the cause of variation of performance characteristics outside of previously established limits with the end result that failure modes, failure mechanisms, and failure activating causes will be identified.

o. *Failure mechanism.* The process of degradation or chain of events which results in a particular failure mode.

p. *Failure mode.* The abnormality of an electrical, electronic, or fiber optic parts performance which causes the part to be classified as failed.

q. *Failure rate.* The total number of failures within an item population, divided by the total number of life units expended by that population, during a particular measurement interval under stated conditions.

r. *Inspection lot.* A group of electrical, electronic, and fiber optic parts offered for inspection at one time and in combinations authorized by the applicable specification.

s. *Item.* A non-specific term used to denote any product, including system, material parts, subassemblies, sets, or accessories.

t. *Manufacturer.* The actual producer of electrical, electronic, and fiber optic parts.

u. *Production lot.* A group of electrical, electronic, or fiber optic parts manufactured during the same period from the same raw materials processes under the same specifications and procedures, produced with the same type equipment, and identified by the documentation defined in the manufacturer's qualified product system through all significant manufacturing operations, including final assembly operations. Final assembly operations shall be considered the last major assembly operations such as casing, hermetic sealing, or lead attachment rather than painting or marking.

v. *Qualification.* The entire procedure by which electrical, electronic, and fiber optic parts are processed, examined, and tested to obtain and maintain approval for listing.

w. *Qualifying activity.* The military preparing activity or its government agent delegated to administer the qualification program.

x. *Quality assurance.* Quality assurance is a planned and systematic pattern of all actions necessary to provide adequate technical requirements are established; products and services conform to established technical requirements; and satisfactory performance is achieved.

y. *Reliability.*
   (1) The duration or probability of failure-free performance under stated conditions.
   (2) The probability that an item can perform its intended function for a specified interval under stated conditions. (For non-redundant items this is equivalent to definition (1). For redundant items this is equivalent to definition of mission reliability.)

z. *Screening.* A process for inspecting items to remove those that are unsatisfactory or those likely to exhibit early failure. Inspection includes visual examination, physical dimension measurement and functional performance measurement under specified environmental conditions.

aa. *Sub-assembly facilities.* A facility authorized, by both the manufacturer and the qualifying activity, to perform manufacturing steps in accordance with processing contained in the qualified product system.

bb. *Self-assessment.* The performance of periodic review by the manufacturer's designated personnel to verify that the requirements of this standard are being met.

cc. *System.* General — A composite of equipment and skills, and techniques capable of performing or supporting an operational role, or both. A complete system includes all equipment, related facilities, material, software, services, and personnel required for its operation and support to the degree that it can be considered self-sufficient in its intended operational environmental.

dd. *Technology Review Board (TRB).* A board established by the manufacturer that is given authority and responsibility to oversee the MIL-STD-790 qualified product system as described herein. The TRB consists of designated manufacturers representatives that have the knowledge and expertise to administer the system.

ee. *Time.* The universal measure of duration. The general word "time" will be modified by an additional term when used in reference to operating time, mission time, and test time.

ff. *Traveler.* The production and raw material process routing sheet.

## 4. GENERAL REQUIREMENTS

### 4.1 GENERAL

Manufacturers of established reliability and high reliability electrical, electronic, and fiber optic components shall demonstrate to the qualifying activity that a system is in place to integrate all design, planning, manufacturing, inspection, and test functions as described herein.

## 4.2 Validation

The qualifying activity is responsible for determining if the manufacturer meets the requirements of this standard. Validation is required as part of the qualification and retention of qualification to the individual product specification. The qualifying activity shall perform a review of the manufacturing facility as part of the validation effort. Revalidations are required to maintain qualification and shall be performed within 24 months of the last review. This validation period may be extended by the qualifying activity if the manufacturer can demonstrate adequate controls of the system through Statistical Process Control (SPC), self-assessment, or Technology Review Boards (TRBs).

## 4.3 Elements

The manufacturer shall demonstrate a system for established reliability and high reliability parts that includes the specific elements as defined in the detailed requirements of this standard (see section 5).

# 5. DETAILED REQUIREMENTS

## 5.1 General

The detailed requirements for meeting this standard are described in this section. It is not intended that the manufacturer create a military unique system in order to meet these requirements. Manufacturers may use existing internal systems in meeting these requirements provided they are validated by the qualifying activity.

### 5.1.1 Key Personnel and Organizations

The responsibility and authority of key personnel and organizations associated with the qualified products shall be identified. The manufacturer shall identify changes affecting key organizations and personnel. The qualifying activity shall be informed of any changes within 30 days after such an occurrence.

### 5.1.2 Test Facilities

The manufacturer shall identify the test facilities and equipment used for qualification and conformance inspection of the electrical, electronic, and fiber optic parts.

### 5.1.3 GIDEP Alerts

The manufacturer shall notify the qualifying activity of all pending GIDEP alerts prior to issuance.

### 5.1.4 Sub-Assembly Facilities

Manufacturers validated to this standard may utilize sub-assembly facilities to perform specific manufacturing steps in accordance with the authorized qualification system.

### 5.1.5 Distributors

Manufacturers validated to this standard may authorize distributors to perform additional functions and operation on the qualified products. The manufacturer is responsible for validation of these distributors to the requirements of this standard as applicable. In case of dispute or quality related problems, the qualifying activity reserves the right to perform a validation of the distributor. The controls and requirements shall be such as to assure the product sold by distributor is of the same quality and performance as parts supplied directly from the manufacturer. The manufacturer is responsible for ensuring that all products sold through these distributors meet the requirements of the applicable product specification. The manufacturer shall identify each distributor and the functions that they are authorized to perform according to the following categories:

a. Category A distributor. This category of distributor is authorized to store, pack, handle, and distribute qualified products.
b. Category B distributor. This category of distributor is authorized to perform additional operation, tests, and inspection in addition to responsibilities of a Category A distributor. If the distributor is authorized to mark the parts, a code symbol is to be added to the modified part to identify the distributor (in accordance with agreement with original manufacturer) in addition to the original part marking and lot identification by the manufacturer.
c. Category C distributor. This category is authorized to perform assembly of the qualified products in addition to the responsibilities of a Category B distributor including part marking requirements.

## 5.2 QPL System Elements

The manufacturer's system shall address, as a minimum, the elements described herein. This system shall be maintained by the manufacturer such that the qualifying activity can verify and validate these elements (e.g., internet documentation and control systems).

### 5.2.1 Training

The manufacturer shall maintain a training program to cover all phases of their activity involved in producing electrical, electronic, and fiber optic parts. The type and extent of training shall be determined by the manufacturer.

### 5.2.2 Calibration

Each instrument used to measure or control production process or to measure the acceptability of parts under test shall be calibrated in accordance with ANSI/NCSL Z540-1, ISO 10012-1, or equivalent system as approved by the qualifying activity.

### 5.2.3 Proprietary Processes and Procedures

The qualifying activity shall have access to all areas of the manufacturer's plant for the purpose of verifying implementation of this standard.

### 5.2.4 Failure and Defect Analysis System

The manufacturer shall maintain a failure and defect analysis system. Failure analysis of parts are required when failures exceed the number allowed by the specification in qualification and conformance inspections or which have failed during field use (either at equipment contractor or military field activities).

## 5.2.4.1 Failure Reporting

The manufacturer shall maintain a failure recording and reporting system for parts which have failed during qualification or conformance inspections or while in use in equipment. The system shall provide for at least the following:

a. The operating or test conditions under which the part failed, including environmental exposure levels, if known.
b. The source from which the failed part was received.
c. Verification of the reported condition of the failed part by the manufacturer's personnel responsible for production, inspection, quality, or engineering.
d. The length of time the part has been operating if it failed in life testing. Compliance with failure rate levels shall be calculated in accordance with the governing applicable product specification.
e For field failures, review and corrective action (as applicable) shall be within 30 days after receipt of parts and supporting information.

## 5.2.4.2 Failure and Defect Analysis

The manufacturer shall maintain a system to retain the results of failure and defect analysis. This system shall provide for at least the following:

a. The results of analysis.
b. The probable failure activating cause when possible.
c. Recommended corrective action, if any.
d. Include approval by responsible authority.

## 5.2.4.3 Failure and Defect Analysis Capabilities and Facilities

The manufacturer shall have, either as part of their facilities or an arrangement with suitable laboratories outside of their facilities, proper capabilities.

## 5.2.5 Corrective Action

Where failures or defects are greater than the prescribed limits, the manufacturer shall prepare a recommendation for corrective action. Corrective action recommendations for performance failures shall include failure mode information when established and shall be supported by verifying information, or a proposed evaluation test plan. Corrective actions on parts covered by the specification shall not be made without approval from the qualifying activity, except those actions which consist only of improvements in control procedures. Corrective action affecting control procedures shall not be implemented for production until approved by qualified personnel.

## 5.2.5.1 Production of Prototype Parts for Evaluation

Prototype parts for change evaluation shall be produced on the controlled production line to the point at which the proposed corrective change must be made; the change shall then be effected; and the changed prototype parts shall then be continued through the balance of the normal series of production operations. Shipment of product incorporating the change shall not occur until approved by the qualifying activity.

## 5.2.6 Clean Rooms

When process control includes the requirements of a clean room, airborne particulate class limits shall be defined. The proper class shall be specified by the manufacturer's design activity in the process specification (FED-STD-209 may be used as a guideline). The manufacturer shall establish action and absolute control limits (at which point work stops until corrective action is completed) based on historical data and criticality of the process in each particular area.

## 5.2.7 Description of Production Processes and Controls

The manufacturer shall maintain a system that details the production processes, steps, and controls applied to parts currently produced and proposed for inclusion in this program. Requirements and tolerances shall be specified for all critical environments and utilities which come in contact with the production and test of electrical, electronic, and fiber optic parts. When applicable, the system shall include such items as:

a. List of process control equipment and periodic calibrations.
b. Control of chemical purity and ionization of water.
c. Known composition of all gases and chemicals, including degree and type of contamination, used in the processes and control of fabrication.
d. Definition of maximum permissible variations in voltage used in the processes or supplied to the test equipment which may introduce errors or variations in the performance or inaccuracies in test data.
e. Definition of clean rooms or other controlled atmospheric requirements.
f. Process specifications showing process tolerances.
g. Detailed engineering specification requirements covering specific types of parts.
h. Identification of each inspection operation for receiving inspection, inspection during manufacture, inspection of completed parts including related sampling plans, and inspection tolerances.
i. Procedures for forming conformance inspection lots which will comply with part specification criteria.
j. Procedure for identification of each production lot through all significant manufacturing operations, including final assembly operations such as casing, hermetic sealing, or lead attachment. Alternately, where this procedure is impractical (e.g., where a part cannot be identified until after final assembly and determination of its performance characteristics), the manufacturer shall as a minimum be able to identify the time period during which the final production operation was performed on each item of product prior to final test. The date or lot code marked on each part shall be identified to a production lot.
k. The manufacturer listed must notify the qualifying activity of any major change affecting the design or process of the qualified product.

## 5.2.8 Acquisition and Production Control System

The manufacturer shall maintain an acquisition and production control system that identifies pertinent internal documents relating to acquisition and processing of materials, production of parts and methods of product assurance (e.g., name, number, release date, and latest revision).

## 5.2.9 Statistical Process Control

When specified in the individual component specification, a statistical process control (SPC) program system in accordance with ANSI/EIA-557 shall be established.

## 5.2.10 Acceptance Criteria for Incoming Materials and Work in-Process

Acceptance criteria shall be identified including type of inspection, the materials group inspected, the sampling and test procedures, the date of completion of inspection, the amount of material tested, acceptance rejection criteria and frequency of inspection.

## 5.2.11 Handling and Packaging Procedures

Handling procedures shall be established to provide physical protection of material during all sequences of production and inspection. Assembled parts shall be physically protected during testing and conformance inspections. Handling and packaging procedures shall be prepared to cover storage of parts in a controlled storage area, their removal from the area, and their preparation for shipment.

## 5.2.12 Materials

### 5.2.12.1 Incoming, in-Process, and Outgoing Inventory Control

The methods and procedures shall be identified which are used to control storage and handling of incoming materials, work in-process, and warehoused and outgoing product in order to achieve such factors as age control of limited-life materials, and prevent inadvertent mixing of conforming and nonconforming materials, work, or finished product. Procedures shall be maintained for controlling the receipt of acquired materials and supplies. The procedures shall provide the following:

a. Withholding received materials or supplies from use pending completion of the required inspection or tests, or the receipt of necessary reports.
b. Segregation and identification of nonconforming materials and supplies from conforming materials.
c. Identification and control of limited-life materials and supplies.
d. Identification and control of raw materials.
e. Assurance that the required test reports, certification, etc., have been received.
f. Clear identification of materials released from receiving inspection and test to clearly indicate acceptance or rejection status of material pending review action.

### 5.2.12.2 Conforming Materials

The manufacturer shall maintain a positive system of identifying the inspection status by means of stamps, tags, routing cards, or other control devices. In controlling the status of materials, the manufacturer shall establish suitable controls to assure that identification of status is applied under the jurisdiction of authorized inspection personnel.

### 5.2.12.3 Nonconforming Materials

Nonconforming materials shall be controlled by a positive system of identification to prevent their inadvertent use or intermingling with conforming materials.

### 5.2.12.4 Material Traceability

Conforming materials shall be identified upon receipt and, where possible, throughout the production process to the accepted product. Where another basis of part production lot identification (e.g., the time period during which certain operations are performed) is used, the accepted product shall be identified to the appropriate production lot, and records of conforming material batches or lots used in each production lot shall be maintained. Completed parts shall be identified to permit positive correlation to the production lot.

## 5.2.13 Product Traceability

The traceability system shall be maintained such that the qualifying activity can trace and determine that the qualified product passed the applicable screening, qualification, and conformance inspections as well as be able to trace and determine the exact processes (includes machines, operators, equipment, etc.), piece parts, and raw materials used in the actual manufacture of the qualified product.

## 5.2.14 Controlled Storage Area

The manufacturers shall describe the procedures and controls which will be used to maintain a separate storage area (e.g., specially marked containers, special cabinets, or stockroom) for parts that have passed the specification conformance inspections. Such an area shall be maintained and no other parts shall be permitted in this area.

## 5.2.15 Quality Assurance Operations

Quality assurance operations shall be identified as to type, procedures, equipment, judgment and action criteria, records, and frequency of use.

## 5.2.16 Manufacturer's Self-Assessment System

The manufacturer shall establish a self-assessment system to verify that the requirements of this standard are being met. Appendix A offers guidance for a self-assessment system. Self-assessments shall be performed at least annually.

## 5.2.17 Technology Review Board (TRB)

The manufacturer may request the approval and use of an internal TRB. Requirements for TRBs are specified in appendix B. Establishment of a TRB at any level is subject to approval/withdrawal by the qualifying activity. The TRB allows manufacturers to assume more responsibility and authority for meeting the requirements of this standard.

## 6. NOTES

(This section contains information of a general or explanatory nature that may be helpful, but is not mandatory.)

## 6.1 INTENDED USE

This standard specifies general requirements for qualified product systems for established reliability and high reliability electrical, electronic, and fiber optic part specifications.

## 6.2 Subject Term (Key Word) Listing

Assessment
Calibration
GIDEP
Process control
Production
Qualification
Technology Review Boards
Traceability

## 6.3 Changes from Previous Issue

Marginal notations are not used in this revision to identify changes with respect to the previous issue due to the extensiveness of the changes.

# APPENDIX A
# SELF-ASSESSMENT SYSTEM

## A.1 Scope

### A.1.1 Scope

This appendix contains guidance for use in the manufacturer's self-assessment system. The intent of this system is to assure continued conformance to specification requirements. The information contained herein is intended for guidance.

## A.2 Applicable Documents

This section is not applicable to this appendix.

## A.3 General

### A.3.1 Self-Assessment Program

The manufacturer shall have an independent self-assessment system to assess the effectiveness of the manufacturer's quality assurance system. This system shall identify any deficiencies for resolution in the processing, testing, or deviations from specification requirements.

### A.3.2 Self-Assessment Representatives

The manufacturer's quality assurance representative or their designated appointees shall perform all self-assessments. The representatives shall be independent from the areas reviewed.

### A.3.3 Deficiencies

All deficiencies shall be identified and submitted to the department head for corrective actions. All corrective actions shall be agreed to by the manufacturer's quality department prior to implementation.

### A.3.4 Follow Up

The quality department shall establish a procedure to follow up on all deficiencies to assure the corrective actions have been implemented in a timely manner.

### A.3.5 Schedules

The original self-assessment frequency shall be established by the quality department but in no case exceed 12 months for each area unless authorized by the qualifying activity. Changes to the frequency of self-assessment shall require approval of the quality department.

### A.3.6 Self-Assessment Results

The results of the self-assessment shall be made available to the qualifying activity prior to validation. The manufacturer shall make available to the qualifying activity, during validation, all corrective actions taken as a result of the self-assessment.

### A.3.7 Self-Assessment Requirements

The following is an explanation of the typical requirements that every manufacturer should meet.

a. Manufacturer should identify key personnel and organizations (see 5.1.1).
b. Manufacturer should have a system that details the manufacturing flow processes performed, quality control stations, and the internal document control number pertaining to each (see 5.2.7).
c. It should be verified that the manufacturer's manufacturing and quality documentation control system is being adequately maintained. This means tracing document status all the way from responsible authority to its location on the production line (see 5.2.8).
d. Incoming inspection area should be examined to determine that the conforming and non-conforming materials are segregated. Traceability should be from the finished product traveler to the incoming inspection (see 5.2.12). Adherence to applicable material specifications and standards in section 2 of the specification should be verifiable.
e. Manufacturing travelers should be checked to determine that they are being filled out and signed off at every step of the production and test stages. A sample review of past travelers is recommended. Significant steps of the process should be included on the traveler (see 5.2.7j). If part of group A is performed in production testing, the manufacturer should be able to produce data for group A tests for each inspection lot.
f. Logs on voltage and temperature checks in ovens and chambers, life and burn-in start and stop times, etc., should be in place and filled in.
g. Voltages and temperatures should be checked at least once a week on life test ovens.
h. Overvoltage and thermal runaway protections on reliability life test chambers should be utilized (see 5.2.7).
i. Environmental controls should be maintained and monitored as required in the specification or standard that is applicable.
j. All instructions (e.g., setting equipment, testing, and handling) should be signed, dated, and part of documentation control system. Operators should follow instructions for procedures (see 5.2.8).

k. Process control records should be reviewed to verify that they are being utilized and that process corrections are implemented when a need is indicated by the control charts (e.g., X bar and R charts) (see 5.2.9).

l. Manufacturer should describe and maintain failure and defect analysis systems which should result in corrective actions to reduce part failures and defects to acceptable level (see 5.2.4).

m. A sample of production processes and controls should be reviewed to determine that operators are following the steps outlined in the production and test documents (see 5.2.8).

n. It should be determined that distributors are being controlled to assure that the product sold by the distributor is of the same quality and performance as parts acquired directly from the manufacturer (see 5.1.5).

o. The calibration system should always be checked (see 5.2.2).

p. Compliance to specification test requirements should be clearly shown on internal control documents.

q. The manufacturer should demonstrate the ability to perform the tests required by the specification.

r. The manufacturer should describe, conduct, and maintain a training program producing qualified parts (see 5.2.1).

s. Both the original and altered data should be readable. When changes are made the entry should be initialed and all entries should be permanent (ink).

# APPENDIX B
# ESTABLISHMENT OF TECHNOLOGY REVIEW BOARDS (TRBS)

## B.1 Scope

### B.1.1 Scope

This appendix is not a mandatory part of this standard. The information contained herein provides for manufacturer selection of a TRB operation at stated levels. However once a TRB option is selected and designated as a requirement, this appendix becomes mandatory. This appendix establishes requirements for a TRB system which provides for a phased assignment of responsibility to the TRB for the development of design, procurement, manufacturing, test, reliability, and quality assurance standard procedures. A TRB shall be established as noted in table II at either level 1 or 2 which defines the extent of control authorized by the qualifying activity. A prerequisite for a TRB shall be an approved Mil-STD-790 system.

## B.2 Applicable Documents

This section is not applicable to this appendix.

## B.3 Requirements

### B.3.1 TRB Levels of Responsibilities

Establishment of a TRB at any level is subject to approval/withdrawal by the qualifying activity at any time. Upon initiation of a TRB, the level of responsibility (see B.3.4) will be assigned by the qualifying activity. The qualifying activity shall provide an overview and determination of the level of authority based upon the experience demonstrated by the TRB. Continual review and monitoring of the TRB operation and procedures will allow the qualifying activity to determine the assignment of the TRB to the appropriate level, 1 or 2. TRBs are not authorized for class S (space level) programs unless specified in the individual class S product specification. In addition, TRB level 2 is not authorized unless specified in the individual product level specification.

### B.3.1.1 Purpose of TRB

The TRB shall assess the impact of proposed changes upon the reliability, form, fit, and function of the product, and oversee review to assure proper implementation of concepts to improve the operational procedures and requirements.

### B.3.1.2 Level 1 TRB

This level provides for a mature program and provides authority for changes by a TRB. The qualifying activity will authorize responsibilities as noted in B.3.4 herein.

### B.3.1.3 Level 2 TRB

When authorized by the individual product specification, this level is for TRBs that have matured sufficiently to attain the confidence of the qualifying activity. This level TRB is accorded major authority as designated in B.3.4. TRB level 2 is not authorized unless specified in the individual product level specification.

## B.3.2 TRB

The manufacturer electing to establish a technology review board shall develop the necessary system to govern its operation. The manufacturer shall be responsible for ensuring that the actions of the TRB result in products that meet all applicable specification requirements. In the event of disputes, the referee point shall be the original specification requirement as defined by the preparing/qualifying activity. As a minimum, these procedures shall address the following:

a. Record retention.
b. Minimum organizational membership by function.
c. Responsibilities.
d. System for recovery of data used in TRB decisions.
e. TRB operating structure.
f. Decision making/approval procedures.
g. MIL-STD-790 oversight.

**TABLE II**
**TRB Authority Levels\***

| TRB Authority | TRB Level |
|---|---|
| Process/material changes, traceability (B.3.4a) | 1 and 2 |
| Alternate methods/equipment, procedures (B.3.4b) | 1 and 2 |
| Evaluation/corrective action on quality reliability (B.3.4c) | 2 |
| Deletion/alternative tests (B.3.4d) | 2 |

\* Table II defines the minimum authority of each TRB level. Actual responsibility and authority will be determined by the qualifying activity.

### B.3.3 TRB Organizational Structure

The following functions, as a minimum, shall be considered for the manufacturers TRB: design and construction, material procurement, manufacturing, test, reliability, and quality assurance. Other personnel with decision making responsibilities affecting the product, its processes, or its production facility may participate as required. The manufacturer shall identify those organizations that must be represented on the TRB. A responsible technical representative within each of these organizations shall be identified to the qualifying activity. Any changes to either permanent participating organizations or their corresponding technical representatives must be reported, within 30 days, to the qualifying activity.

### B.3.4 TRB Responsibilities

The TRB authority may extend to review and oversight of the manufacturer's entire process. Under no circumstances shall the establishment of a TRB relieve the manufacturer of the responsibility to report nonconformance to specification requirements to the qualifying activity. The life test requirements of the individual product specification are not within the spectrum of authority or responsibilities of the TRB. The TRB may be responsible for the following as authorized by the qualifying activity:

a. Managing and approving design/construction process/material confirmation and change control activities including traceability.
b. Approving alternative equipment/methods/procedures that modify, delete, or substitute for existing equipment methods/procedures detailed in MIL-STD-790 (other than those required by the individual product specification).
c. When performance or reliability of shipped components, including tests, is called into question, the TRB shall provide quick evaluation, approve appropriate corrective action, and provide prompt notification of all problems to the qualifying activity.
d. Approving alternative methods that modify, substitute for, or delete existing tests required by the product specification (e.g., tests within groups A, B, C, etc.).

## CONCLUDING MATERIAL

Custodians:
    Army — ER
    Navy — EC
    Air Force — 85
    NASA — NA
Review activities:
    Army — AR, CR, MI
    Navy — AS, CG, MC, OS, SH
    Air Force — 17, 19, 99
    DLA — DH
Preparing activity:
    DLA — ES
(Project 59GP–0138)

STANDARDIZATION DOCUMENT IMPROVEMENT PROPOSAL

*Instructions*

1. The preparing activity must complete blocks 1, 2, 3, and 8. In block 1, both the document number and revision letter should be given.

2. The submitter of this form must complete blocks 4, 5, 6, and 7.

3. The preparing activity must provide a reply within 30 days from receipt of the form.

NOTE: This form may not be used to request copies of documents, nor to request waivers, or clarification of requirements on current contracts. Comments submitted on this form do not constitute or imply authorization to waive any portion of the referenced document(s) or to amend contractual requirements.

| I RECOMMEND A CHANGE | 1. DOCUMENT NUMBER<br>MIL-STD-790F | 2. DOCUMENT DATE<br>(YYMMDD)<br>1 August 1995 |
|---|---|---|
| 3. DOCUMENT TITLE<br>STANDARD PRACTICE FOR ESTABLISHED RELIABILITY AND HIGH RELIABILITY QUALIFIED PRODUCTS LIST (QPL) SYSTEMS FOR ELECTRICAL, ELECTRONIC, AND FIBER OPTIC PARTS SPECIFICATIONS | | |
| 4. NATURE OF CHANGE<br>(Identify paragraph number and include proposed rewrite, if possible. Attach extra sheets as needed.) | | |
| 5. REASON FOR RECOMMENDATION | | |
| 6. SUBMITTER | | |
| a. NAME | B. ORGANIZATION | |
| c. ADDRESS (Include Zip Code) | d. TELEPHONE (include Area Code)<br>(1) Commercial<br>(2) AUTOVON (If applicable) | 7. DATE SUBMITTED<br>(YYMMDD) |

| 8. PREPARING ACTIVITY | |
|---|---|
| a. NAME<br>Defense Electronics Supply Center | b. TELEPHONE (include Area Code)<br><br>(1) Commercial<br>(513) 296-5255<br><br>(2) AUTOVON<br>986-5255 |
| ADDRESS (Include Zip Code)<br>ATTN: ELDM<br>1507 Wilmington Pike<br>Dayton, OH 45444-5765 | IF YOU DO NOT RECEIVE A REPLY WITHIN<br>45 DAYS, CONTACT:<br><br>Defense Quality and Standardization Office<br>5203 Leesburg Pike, Suite 1403,<br>Falls Church, VA 22041-3466<br>Telephone (703) 756-2340<br>AUTOVON 289-2340 |

DD Form 1426, OCT 89 Previous editions are obsolete 198/290

# The Rules Governing Medicinal Products in the European Union

## Volume 4
## Good Manufacturing Practices: Medicinal Products for Human and Veterinary Use

1998 Edition
European Commission
Directorate General III — Industry
Pharmaceuticals and Cosmetics

## THE RULES GOVERNING MEDICINAL PRODUCTS IN THE EUROPEAN UNION

## FOREWORD

The Pharmaceutical Industry of the European Community maintains high standards of Quality Assurance in the development, manufacture and control of medicinal products. A system of Marketing Authorisations ensures that all medicinal products are assessed by a Competent Authority to ensure compliance with contemporary requirements of safety, quality and efficacy. A system of Manufacturing Authorisations ensures that all products authorised on the European market are manufactured only by authorised manufacturers, whose activities are regularly inspected by the Competent Authorities. Manufacturing Authorisations are required by all pharmaceutical manufacturers in the European Community whether the products are sold within or outside of the Community.

Two directives laying down principles and guidelines of good manufacturing practice (GMP) for medicinal products were adopted by the Commission in 1991, the first for medicinal products for human use (Directive 91/356/EEC), the second one for veterinary use (Directive 91/412/EEC). Detailed guidelines in accordance with those principles are published in the Guide to Good Manufacturing Practice which will be used in assessing applications for Manufacturing authorisations and as a basis for inspection of manufacturers of medicinal products.

The principles of GMP and the detailed guidelines are applicable to all operations which require the authorisation referred to in Article 16 of Directive 75/319/EEC and in Article 24 of Directive 81/851/EEC as modified. They are also relevant for all other large scale pharmaceutical manufacturing processes, such as that undertaken in hospitals, and for the preparation of products for use in clinical trials.

All Member States and the Industry itself are agreed that the GMP requirements applicable to the manufacture of veterinary medicinal products are the same as those applicable to the manufacture of medicinal products for human use. Certain detailed adjustments to the GMP guidelines are set out in two annexes specific to veterinary medicinal products and to immunological veterinary medicinal products.

The Guide is presented in chapters, each headed by a principle. Chapter 1 on Quality Management outlines the fundamental concept of Quality Assurance as applied to the manufacture of medicinal products. Thereafter each chapter has a principle outlining the Quality Assurance objectives of that chapter and a text which provides sufficient detail for manufacturers to be made aware of the essential matters to be considered when implementing the principle.

In addition to the general matters of Good Manufacturing Practice outlined in the 9 chapters of this guide, a series of annexes providing detail about specific areas of activity is included. For some manufacturing processes, different annexes will apply simultaneously (e.g. annex on sterile preparations and on radiopharmaceuticals and/or on biological medicinal products).

A glossary of some terms used in the Guide has been incorporated after the annexes.

The first edition of the Guide was published in 1989, including an annex on the manufacture of sterile medicinal products.

The second edition was published in January 1992; including the Commission Directives 91/356 of 13 June 1991 and 91/412 of 23 July 1991 laying down the principles and guidelines on good manufacturing practice for medicinal products for human use as well as for veterinary medicinal products. The second edition also included 12 additional annexes.

The basic requirements in the main guide have not been modified. 14 annexes on the manufacture of medicinal products have been included in this third edition.

Annex 1 on the manufacture of sterile medicinal products has been modified. Annex 13 on the manufacture of investigational medicinal products, which was not included in the second edition of the Guide, has been modified and included in this version. Annex 14 on the manufacture of products derived from human blood or human plasma, which was not included In the second edition of the Guide, has been included in this version and a revision is scheduled for 1998.

The Guide is not intended to cover security aspects for the personnel engaged in manufacture. This may be particularly important in the manufacture of certain medicinal products such as highly active, biological and radioactive medicinal products, but they are governed by other provisions of Community or national law.

Throughout the Guide it is assumed that the requirements of the Marketing Authorisation relating to the safety, quality and efficacy of the products, are systematically incorporated into all the manufacturing, control and release for sale arrangements of the holder of the Manufacturing Authorisation.

The manufacture of medicinal products has for many years taken place in accordance with guidelines for Good Manufacturing Practice and the manufacture of medicinal products is not governed by CEN/ISO standards. Harmonised standards as adopted by the European standardisation organisations CEN/ISO may be used at industry's discretion as a tool for implementing a quality system in the pharmaceutical sector. The CEN/ISO standards have been considered but the terminology of these standards has not been implemented in this third edition of the Guide.

It is recognised that there are acceptable methods, other than those described in the Guide, which are capable of achieving the principles of Quality Assurance. The Guide is not intended to place any restraint upon the development of any new concepts or new technologies which have been validated and which provide a level of Quality Assurance at least equivalent to those set out in this Guide.

It will be regularly revised.

## CONTENTS

# Chapter 1: Quality Management

## PRINCIPLE

The holder of a Manufacturing Authorisation must manufacture medicinal products so as to ensure that they are fit for their intended use, comply with the requirements of the Marketing Authorisation and do not place patients at risk due to inadequate safety, quality or efficacy. The attainment of this quality objective is the responsibility of senior management and requires the participation and commitment by staff in many different departments and at all levels within the company, by the company's suppliers and by the distributors. To achieve the quality objective reliably there must be a comprehensively designed and correctly implemented system of Quality Assurance incorporating Good Manufacturing Practice and thus Quality Control. It should be fully documented and its effectiveness monitored. All parts of the Quality Assurance system should be adequately resourced with competent personnel, and suitable and sufficient premises, equipment and facilities. There are additional legal responsibilities for the holder of the Manufacturing Authorisation and for the Qualified Person(s).

1.1 The basic concepts of Quality Assurance, Good Manufacturing Practice and Quality Control are inter-related. They are described here in order to emphasise their relationships and their fundamental importance to the production and control of medicinal products.

## QUALITY ASSURANCE

1.2 Quality Assurance is a wide ranging concept that covers all matters which individually or collectively influence the quality of a product. It is the sum total of the organised arrangements made with the object of ensuring that medicinal products are of the quality required for their intended use. Quality Assurance therefore incorporates Good Manufacturing Practice plus other factors outside the scope of this Guide.

The system of Quality Assurance appropriate for the manufacture of medicinal products should ensure that:

i. Medicinal products are designed and developed in a way that takes account of the requirements of Good Manufacturing Practice and Good Laboratory Practice;
ii. Production and control operations are clearly specified and Good Manufacturing Practice adopted;
iii. Managerial responsibilities are clearly specified;
iv. Arrangements are made for the manufacture, supply and use of the correct starting and packaging materials;
v. All necessary controls on intermediate products, and any other in-process controls and validations are carried out;
vi. The finished product is correctly processed and checked, according to the defined procedures;
vii. Medicinal products are not sold or supplied before a Qualified Person has certified that each production batch has been produced and controlled in accordance with the requirements of the Marketing Authorisation and any other regulations relevant to the production, control and release of medicinal products;

viii. Satisfactory arrangements exist to ensure, as far as possible, that the medicinal products are stored, distributed and subsequently handled so that quality is maintained throughout their shelf life;

ix. There is a procedure for Self-Inspection and/or quality audit which regularly appraises the effectiveness and applicability of the Quality Assurance system.

## GOOD MANUFACTURING PRACTICE FOR MEDICINAL PRODUCTS (GMP)

1.3 Good Manufacturing Practice is that part of Quality Assurance which ensures that products are consistently produced and controlled to the quality standards appropriate to their intended use and as required by the Marketing Authorisation or product specification.

Good Manufacturing Practice is concerned with both production and quality control. The basic requirements of GMP are that:

i. All manufacturing processes are clearly defined, systematically reviewed in the light of experience and shown to be capable of consistently manufacturing medicinal products of the required quality and complying with their specifications;

ii. Critical steps of manufacturing processes and significant changes to the process are validated;

iii. All necessary facilities for GMP are provided including:
   a. appropriately qualified and trained personnel;
   b. adequate premises and space;
   c. suitable equipment and services;
   d. correct materials, containers and labels;
   e. approved procedures and instructions;
   f. suitable storage and transport;

iv. Instructions and procedures are written in an instructional form in clear and unambiguous language, specifically applicable to the facilities provided;

v. Operators are trained to carry out procedures correctly;

vi. Records are made, manually and/or by recording instruments, during manufacture which demonstrate that all the steps required by the defined procedures and instructions were in fact taken and that the quantity and quality of the product was as expected. Any significant deviations are fully recorded and investigated;

vii. Records of manufacture including distribution which enable the complete history of a batch to be traced, are retained in a comprehensible and accessible form;

viii. The distribution (wholesaling) of the products minimises any risk to their quality;

ix. A system is available to recall any batch of product, from sale or supply;

x. Complaints about marketed products are examined, the causes of quality defects investigated and appropriate measures taken in respect of the defective products and to prevent reoccurrence.

## QUALITY CONTROL

1.4 Quality Control is that part of Good Manufacturing Practice which is concerned with sampling, specifications and testing, and with the organisation, documentation and release procedures which ensure that the necessary and relevant tests are actually carried out and that materials are not released for use, nor products released for sale or supply, until their quality has been judged to be satisfactory.

The basic requirements of Quality Control are that:

i. Adequate facilities, trained personnel and approved procedures are available for sampling, inspecting and testing starting materials, packaging materials, intermediate, bulk, and finished products, and where appropriate for monitoring environmental conditions for GMP purposes;

ii. Samples of starting materials, packaging materials, intermediate products, bulk products and finished products are taken by personnel and by methods approved by Quality Control;

iii. Test methods are validated;

iv. Records are made, manually and/or by recording instruments, which demonstrate that all the required sampling, inspecting and testing procedures were actually carried out. Any deviations are fully recorded and investigated;

v. The finished products contain active ingredients complying with the qualitative and quantitative composition of the Marketing Authorisation, are of the purity required, and are enclosed within their proper containers and correctly labelled;

vi. Records are made of the results of inspection and that testing of materials, intermediate, bulk, and finished products is formally assessed against specification. Product assessment includes a review and evaluation of relevant production documentation and an assessment of deviations from specified procedures;

vii. No batch of product is released for sale or supply prior to certification by a Qualified Person that it is in accordance with the requirements of the Marketing Authorisation;

viii. Sufficient reference samples of starting materials and products are retained to permit future examination of the product if necessary and that the product is retained in its final pack unless exceptionally large packs are produced.

# Chapter 2: Personnel

## PRINCIPLE

The establishment and maintenance of a satisfactory system of quality assurance and the correct manufacture of medicinal products relies upon people. For this reason there must be sufficient qualified personnel to carry out all the tasks which are the responsibility of the manufacturer. Individual responsibilities should be clearly understood by the individuals and recorded. All personnel should be aware of the principles of Good Manufacturing Practice that affect them and receive initial and continuing training, including hygiene instructions, relevant to their needs.

## GENERAL

2.1  The manufacturer should have an adequate number of personnel with the necessary qualifications and practical experience. The responsibilities placed on any one individual should not be so extensive as to present any risk to quality.

2.2  The manufacturer must have an organisation chart. People in responsible positions should have specific duties recorded in written job descriptions and addequate authority to carry out their responsibilities. Their duties may be delegated to designated deputies of a satisfactory qualification level. There should be no gaps or unexplained overlaps in the responsibilities of those personnel concerned with the application of Good Manufacturing Practice.

## KEY PERSONNEL

2.3  Key Personnel include the head of Production, the head of Quality Control, and if at least one of these persons is not responsible for the duties described in Article 22 of Directive 75/319/EEC, the Qualified Person(s) designated for the purpose. Normally key posts should be occupied by full-time personnel. The heads of Production and Quality Control must be independent from each other. In large organisations, it may be necessary to delegate some of the functions listed in 2.5, 2.6 and 2.7.

2.4  The duties of the Qualified Person(s) are fully described in Article 22 of Directive 75/319/EEC, and can be summarised as follows:

(a) For medicinal products manufactured within the European Community, a Qualified Person must ensure that each batch has been produced and tested/checked in accordance with the directives and the marketing authorisation*;
(b) For medicinal products manufactured outside the European Community, a Qualified Person must ensure that each imparted batch has undergone, in the importing country, the testing specified in paragraph 1 (b) of Article 22;
(c) A Qualified Person must certify in a register or equivalent document, as operations are carried out and before any release, that each production batch satisfies the provisions of Article 22.

---

* According to Directive 75/319/EEC and the Ruling (Case 247/81) of the Court of Justice of the European Communities, medicinal products which have been properly controlled in the EU by a Qualified Person do not have to be recontrolled or rechecked in any other Member State of the Community.

The persons responsible for these duties must meet the qualification requirements laid down in Article 23 of the same Directive, they shall be permanently and continuously at the disposal of the holder of the Manufacturing Authorisation to carry out their responsibilities. Their responsibilities may be delegated, but only to other Qualified Person(s).

2.5 The head of the Production Department generally has the following responsibilities:

   i. To ensure that products are produced and stored according to the appropriate documentation in order to obtain the required quality;
   ii. To approve the instructions relating to production operations and to ensure their strict implementation;
   iii. To ensure that the production records are evaluated and signed by an authorised person before they are sent to the Quality Control Department;
   iv. To check the maintenance of his/her department, premises and equipment;
   v. To ensure that the appropriate validations are done;
   vi. To ensure that the required initial and continuing training of his/her department personnel is carried out and adapted according to need.

2.6 The head of the Quality Control Department generally has the following responsibilities:

   i. To approve or reject, as he/she sees fit, starting materials, packaging materials, and intermediate, bulk and finished products;
   ii. To evaluate batch records;
   iii. To ensure that all necessary testing is carried out;
   iv. To approve specifications, sampling instructions, test methods and other Quality Control procedures;
   v. To approve and monitor any contract analysts;
   vi. To check the maintenance of his department, premises and equipment;
   vii. To ensure that the appropriate validations are done;
   viii. To ensure that the required initial and continuing training of his department personnel is carried out and adapted according to need.

Other duties of the Quality Control Department are summarised in Chapter 6.

2.7 The heads of Production and Quality Control generally have some shared, or jointly exercised, responsibilities relating to quality. These may include, subject to any national regulations:

   • The authorisation of written procedures and other documents, including amendments;
   • The monitoring and control of the manufacturing environment;
   • Plant hygiene;
   • Process validation;
   • Training;
   • The approval and monitoring of suppliers of materials;
   • The approval and monitoring of contract manufacturers;
   • The designation and monitoring of storage conditions for materials and products;
   • The retention of records;

- The monitoring of compliance with the requirements of Good Manufacturing Practice;
- The inspection, investigation, and taking of samples, in order to monitor factors which may affect product quality.

## TRAINING

2.8  The manufacturer should provide training for all the personnel whose duties take them into production areas or into control laboratories (including the technical, maintenance and cleaning personnel), and for other personnel whose activities could affect the quality of the product.

2.9  Besides the basic training on the theory and practice of Good Manufacturing Practice, newly recruited personnel should receive training appropriate to the duties assigned to them. Continuing training should also be given, and its practical effectiveness should be periodically assessed. Training programmes should be available, approved by either the head of Production or the head of Quality Control, as appropriate. Training records should be kept.

2.10  Personnel working in areas where contamination is a hazard, e.g., clean areas or areas where highly active, toxic, infectious or sensitising materials are handled, should be given specific training.

2.11  Visitors or untrained personnel should, preferably, not be taken into the production and quality control areas. If this is unavoidable, they should be given information in advance, particularly about personal hygiene and the prescribed protective clothing. They should be closely supervised.

2.12  The concept of Quality Assurance and all the measures capable of improving its understanding and implementation should be fully discussed during the training sessions.

## PERSONNEL HYGIENE

2.13  Detailed hygiene programmes should be established and adapted to the different needs within the factory. They should include procedures relating to the health, hygiene practices and clothing of personnel. These procedures should be understood and followed in a very strict way by every person whose duties take him into the production and control areas. Hygiene programmes should be promoted by management and widely discussed during training sessions.

2.14  All personnel should receive medical examination upon recruitment. It must be the manufacturer's responsibility that there are instructions ensuring that health conditions that can be of relevance to the quality of products come to the manufacturer's knowledge. After the first medical examination, examinations should be carried out when necessary for the work and personal health.

2.15  Steps should be taken to ensure as far as is practicable that no person affected by an infectious disease or having open lesions on the exposed surface of the body is engaged in the manufacture of medicinal products.

2.16  Every person entering the manufacturing areas should wear protective garments appropriate to the operations to be carried out.

2.17  Eating, drinking, chewing or smoking, or the storage of food, drink, smoking materials or personal medication in the production and storage areas should be prohibited. In general, any unhygienic practice within the manufacturing areas or in any other area where the product might be adversely affected, should be forbidden.

2.18 Direct contact should be avoided between the operator's hands and the exposed product as well as with any part of the equipment that comes into contact with the products.

2.19 Personnel should be instructed to use the hand-washing facilities.

2.20 Any specific requirements for the manufacture of special groups of products, for example sterile preparations, are covered in the annexes.

# Chapter 3: Premises and Equipment

## PRINCIPLE

Premises and equipment must be located, designed, constructed, adapted and maintained to suit the operations to be carried out. Their layout and design must aim to minimise the risk of errors and permit effective cleaning and maintenance in order to avoid cross-contamination, build up of dust or dirt and, in general, any adverse effect on the quality of products.

## PREMISES

### GENERAL

3.1 Premises should be situated in an environment which, when considered together with measures to protect the manufacture, presents minimal risk of causing contamination of materials or products.

3.2 Premises should be carefully maintained, ensuring that repair and maintenance operations do not present any hazard to the quality of products. They should be cleaned and, where applicable, disinfected according to detailed written procedures.

3.3 Lighting, temperature, humidity and ventilation should be appropriate and such that they do not adversely affect, directly or indirectly, either the medicinal products during their manufacture and storage, or the accurate functioning of equipment.

3.4 Premises should be designed and equipped so as to afford maximum protection against the entry of insects or other animals.

3.5 Steps should be taken in order to prevent the entry of unauthorised people. Production, storage and quality control areas should not be used as a right of way by personnel who do not work in them.

### PRODUCTION AREA

3.6 In order to minimise the risk of a serious medical hazard due to cross-contamination, dedicated and self contained facilities must be available for the production of particular medicinal products, such as highly sensitising materials (e.g. penicillins) or biological preparations (e.g. from live micro-organisms). The production of certain additional products, such as certain antibiotics, certain hormones, certain cytotoxics, certain highly active drugs and non-medicinal products should not be conducted in the same facilities. For those products, in exceptional cases, the principle of campaign working in the same facilities can be accepted provided that specific precautions are taken and the necessary validations are made. The manufacture of technical poisons, such as pesticides and herbicides, should not be allowed in premises used for the manufacture of medicinal products.

3.7 Premises should preferably be laid out in such a way as to allow the production to take place in areas connected in a logical order corresponding to the sequence of the operations and to the requisite cleanliness levels.

3.8 The adequacy of the working and in-process storage space should permit the orderly and logical positioning of equipment and materials so as to minimise the risk of confusion between different medicinal products or their components, to avoid cross-contamination and to minimise the risk of omission or wrong application of any of the manufacturing or control steps.

3.9 Where starting and primary packaging materials, intermediate or bulk products are exposed to the environment, interior surfaces (walls, floors and ceilings) should be smooth, free from cracks and open joints, and should not shed particulate matter and should permit easy and effective cleaning and, if necessary, disinfection.

3.10 Pipework, light fittings, ventilation points and other services should be designed and sited to avoid the creation of recesses which are difficult to clean. As far as possible, for maintenance purposes, they should be accessible from outside the manufacturing areas.

3.11 Drains should be of adequate size, and have trapped gullies. Open channels should be avoided where possible, but if necessary, they should be shallow to facilitate cleaning and disinfection.

3.12 Production areas should be effectively ventilated, with air control facilities (including temperature and, where necessary, humidity and filtration) appropriate both to the products handled, to the operations undertaken within them and to the external environment.

3.13 Weighing of starting materials usually should be carried out in a separate weighing room designed for that use.

3.14 In cases where dust is generated (e.g. during sampling, weighing, mixing and processing operations, packaging of dry products), specific provisions should be taken to avoid cross-contamination and facilitate cleaning.

3.15 Premises for the packaging of medicinal products should be specifically designed and laid out so as to avoid mix-ups or cross-contamination.

3.16 Production areas should be well lit, particularly where visual on-line controls are carried out.

3.17 In-process controls may be carried out within the production area provided they do not carry any risk for the production.

## STORAGE AREAS

3.18 Storage areas should be of sufficient capacity to allow orderly storage of the various categories of materials and products: starting and packaging materials, intermediate, bulk and finished products, products in quarantine, released, rejected, returned or recalled.

3.19 Storage areas should be designed or adapted to ensure good storage conditions. In particular, they should be clean and dry and maintained within acceptable temperature limits. Where special storage conditions are required (e.g. temperature, humidity) these should be provided, checked and monitored.

3.20 Receiving and dispatch bays should protect materials and products from the weather. Reception areas should be designed and equipped to allow containers of incoming materials to be cleaned where necessary before storage.

3.21 Where quarantine status is ensured by storage in separate areas, these areas must be clearly marked and their access restricted to authorised personnel. Any system replacing the physical quarantine should give equivalent security.

3.22 There should normally be a separate sampling area for starting materials. If sampling is performed in the storage area, it should be conducted in such a way as to prevent contamination or cross-contamination.

3.23 Segregated areas should be provided for the storage of rejected, recalled or returned materials or products.

3.24 Highly active materials or products should be stored in safe and secure areas.

3.25 Printed packaging materials are considered critical to the conformity of the medicinal product and special attention should be paid to the safe and secure storage of these materials.

## Quality Control Areas

3.26 Normally, Quality Control laboratories should be separated from production areas. This is particularly important for laboratories for the control of biologicals, microbiologicals and radioisotopes, which should also be separated from each other.

3.27 Control laboratories should be designed to suit the operations to be carried out in them. Sufficient space should be given to avoid mix-ups and cross-contamination. There should be adequate suitable storage space for samples and records.

3.28 Separate rooms may be necessary to protect sensitive instruments from vibration, electrical interference, humidity, etc.

3.29 Special requirements are needed in laboratories handling particular substances, such as biological or radioactive samples.

## Ancillary Areas

3.30 Rest and refreshment rooms should be separate from other areas.

3.31 Facilities for changing clothes, and for washing and toilet purposes should be easily accessible and appropriate for the number of users. Toilets should not directly communicate with production or storage areas.

3.32 Maintenance workshops should as far as possible be separated from production areas. Whenever parts and tools are stored in the production area, they should be kept in rooms or lockers reserved for that use.

3.33 Animal houses should be well isolated from other areas, with separate entrance (animal access) and air handling facilities.

# EQUIPMENT

3.34 Manufacturing equipment should be designed, located and maintained to suit its intended purpose.

3.35 Repair and maintenance operations should not present any hazard to the quality of the products.

3.36 Manufacturing equipment should be designed so that it can be easily and thoroughly cleaned. It should be cleaned according to detailed and written procedures and stored only in a clean and dry condition.

3.37 Washing and cleaning equipment should be chosen and used in order not to be a source of contamination.

3.38 Equipment should be installed in such a way as to prevent any risk of error or of contamination.

3.39 Production equipment should not present any hazard to the products. The parts of the production equipment that come into contact with the product must not be reactive, additive or absorptive to such an extent that it will affect the quality of the product and thus present any hazard.

3.40 Balances and measuring equipment of an appropriate range and precision should be available for production and control operations.

3.41 Measuring, weighing, recording and control equipment should be calibrated and checked at defined intervals by appropriate methods. Adequate records of such tests should be maintained.

3.42 Fixed pipework should be clearly labelled to indicate the contents and, where applicable, the direction of flow.

3.43 Distilled, deionized and, where appropriate, other water pipes should be sanitised according to written procedures that detail the action limits for microbiological contamination and the measures to be taken.

3.44 Defective equipment should, if possible, be removed from production and quality control areas, or at least be clearly labelled as defective.

# Chapter 4: Documentation

## PRINCIPLE

Good documentation constitutes an essential part of the quality assurance system. Clearly written documentation prevents errors from spoken communication and permits tracing of batch history. Specifications, Manufacturing Formulae and instructions, procedures, and records must be free from errors and available in writing. The legibility of documents is of paramount importance.

## GENERAL

4.1 *Specifications* describe in detail the requirements with which the products or materials used or obtained during manufacture have to conform. They serve as a basis for quality evaluation.

*Manufacturing Formulae, Processing and Packaging Instructions* state all the starting materials used and lay down all processing and packaging operations.

*Procedures* give directions for performing certain operations e.g., cleaning, clothing, environmental control, sampling, testing, equipment operation.

*Records* provide a history of each batch of product, including its distribution, and also of all other relevant circumstances pertinent to the quality of the final product.

4.2 Documents should be designed, prepared, reviewed and distributed with care. They should comply with the relevant parts of the manufacturing and marketing authorisation dossiers.

4.3 Documents should be approved, signed and dated by appropriate and authorised persons.

4.4 Documents should have unambiguous contents; title, nature and purpose should be clearly stated. They should be laid out in an orderly fashion and be easy to check. Reproduced documents should be clear and legible. The reproduction of working documents from master documents must not allow any error to be introduced through the reproduction process.

4.5 Documents should be regularly reviewed and kept up-to-date. When a document has been revised, systems should be operated to prevent inadvertent use of superseded documents.

4.6 Documents should not be handwritten; although, where documents require the entry of data, these entries may be made in clear, legible, indelible handwriting. Sufficient space should be provided for such entries.

4.7 Any alteration made to the entry on a document should be signed and dated; the alteration should permit the reading of the original information. Where appropriate, the reason for the alteration should be recorded.

4.8 The records should be made or completed at the time each action is taken and in such a way that all significant activities concerning the manufacture of medicinal products are traceable. They should be retained for at least one year after the expiry date of the finished product.

4.9 Data may be recorded by electronic data processing systems, photographic or other reliable means, but detailed procedures relating to the system in use should be available and the accuracy of the records should be checked. If documentation is handled by electronic data processing methods, only authorised persons should be able to enter or modify data in the computer and there should be a record of changes and deletions; access should be restricted by passwords or other means and the result of entry of critical data should be independently checked. Batch records electronically stored should be protected by back-up transfer on magnetic tape, microfilm, paper or other means. It is particularly important that the data are readily available throughout the period of retention.

## DOCUMENTS REQUIRED

### SPECIFICATIONS

4.10 There should be appropriately authorised and dated specifications for starting and packaging materials, and finished products; where appropriate, they should be also available for intermediate or bulk products.

### SPECIFICATIONS FOR STARTING AND PACKAGING MATERIALS

4.11 Specifications for starting and primary or printed packaging materials should include, if applicable:

  a) A description of the materials, including:
    • The designated name and the internal code reference;
    • The reference, if any, to a pharmacopoeial monograph;
    • The approved suppliers and, if possible, the original producer of the products,
    • A specimen of printed materials;
  b) Directions for sampling and testing or reference to procedures;
  c) Qualitative and quantitative requirements with acceptance limits,
  d) Storage conditions and precautions;
  e) The maximum period of storage before re-examination.

### SPECIFICATIONS FOR INTERMEDIATE AND BULK PRODUCTS

4.12 Specifications for intermediate and bulk products should be available if these are purchased or dispatched, or if data obtained from intermediate products are used for the evaluation of the finished product. The specifications should be similar to specifications for starting materials or for finished products, as appropriate.

### SPECIFICATIONS FOR FINISHED PRODUCTS

4.13 Specifications for finished products should include:

  a) The designated name of the product and the code reference where applicable;
  b) The formula or a reference to;
  c) A description of the pharmaceutical form and package details;
  d) Directions for sampling and testing or a reference to procedures;
  e) The qualitative and quantitative requirements, with the acceptance limits;
  f) The storage conditions and any special handling precautions, where applicable;
  g) The shelf-life.

## MANUFACTURING FORMULA AND PROCESSING INSTRUCTIONS

Formally authorised Manufacturing Formula and Processing Instructions should exist for each product and batch size to be manufactured. They are often combined in one document.

4.14 The Manufacturing Formula should include:

a) The name of the product, with a product reference code relating to its specification;
b) A description of the pharmaceutical form, strength of the product and batch size;
c) A list of all starting materials to be used, with the amount of each, described using the designated name and a reference which is unique to that material; mention should be made of any substance that may disappear in the course of processing;
d) A statement of the expected final yield with the acceptable limits, and of relevant intermediate yields, where applicable.

4.15 The Processing Instructions should include:

a) A statement of the processing location and the principal equipment to be used;
b) The methods, or reference to the methods, to be used for preparing the critical equipment (e.g. cleaning, assembling, calibrating, sterilising);
c) Detailed stepwise processing instructions (e.g. checks on materials, pre-treatments, sequence for adding materials, mixing times, temperatures);
d) The instructions for any in-process controls with their limits;
e) Where necessary, the requirements for bulk storage of the products; including the container, labelling and special storage conditions where applicable;
f) Any special precautions to be observed.

## PACKAGING INSTRUCTIONS

4.16 There should be formally authorised Packaging Instructions for each product, pack size and type. These should normally include, or have a reference to, the following:

a) Name of the product;
b) Description of its pharmaceutical form, and strength where applicable;
c) The pack size expressed in terms of the number, weight or volume of the product in the final container;
d) A complete list of all the packaging materials required for a standard batch size, including quantities, sizes and types, with the code or reference number relating to the specifications of each packaging material;
e) Where appropriate, an example or reproduction of the relevant printed packaging materials, and specimens indicating where to apply batch number references, and shelf life of the product;
f) Special precautions to be observed, including a careful examination of the area and equipment in order to ascertain the line clearance before operations begin;
g) A description of the packaging operation, including any significant subsidiary operations, and equipment to be used;
h) Details of in-process controls with instructions for sampling and acceptance limits.

## BATCH PROCESSING RECORDS

4.17 A Batch Processing Record should be kept for each batch processed. It should be based on the relevant parts of the currently approved Manufacturing Formula and Processing Instructions. The method of preparation of such records should be designed to avoid transcription errors. The record should carry the number of the batch being manufactured.

Before any processing begins, there should be recorded checks that the equipment and work station are clear of previous products, documents or materials not required for the planned process, and that equipment is clean and suitable for use.

During processing, the following information should be recorded at the time each action is taken and, after completion, the record should be dated and signed in agreement by the person responsible for the processing operations:

a) The name of the product;
b) Dates and times of commencement, of significant intermediate stages and of completion of production;
c) Name of the person responsible for each stage of production;
d) Initials of the operator of different significant steps of production and, where appropriate, of the person who checked each of these operations (e.g. weighing);
e) The batch number and/or analytical control number as well as the quantities of each starting material actually weighed (including the batch number and amount of any recovered or reprocessed material added);
f) Any relevant processing operation or event and major equipment used;
g) A record of the in-process controls and the initials of the person(s) carrying them out, and the results obtained;
h) The product yield obtained at different and pertinent stages of manufacture;
i) Notes on special problems including details, with signed authorisation for any deviation from the Manufacturing Formula and Processing Instructions.

## BATCH PACKAGING RECORDS

4.18 A Batch Packaging Record should be kept for each batch or part batch processed. It should be based on the relevant parts of the Packaging Instructions and the method of preparation of such records should be designed to avoid transcription errors. The record should carry the batch number and the quantity of bulk product to be packed, as well as the batch number and the planned quantity of finished product that will be obtained.

Before any packaging operation begins, there should be recorded checks that the equipment and work station are clear of previous products, documents or materials not required for the planned packaging operations, and that equipment is clean and suitable for use.

The following information should be entered at the time each action is taken and, after completion, the record should be dated and signed in agreement by the person(s) responsible for the packaging operations:

a) The name of the product;
b) The date(s) and times of the packaging operations;
c) The name of the responsible person carrying out the packaging operation;

d) The initials of the operators of the different significant steps;

e) Records of checks for identity and conformity with the packaging instructions including the results of in-process controls;

f) Details of the packaging operations carried out, including references to equipment and the packaging lines used;

g) Whenever possible, samples of printed packaging materials used, including specimens of the batch coding, expiry dating and any additional overprinting;

h) Notes on any special problems or unusual events including details, with signed authorisation for any deviation from the Manufacturing Formula and Processing Instructions;

i) The quantities and reference number or identification of all printed packaging materials and bulk product issued, used, destroyed or returned to stock and the quantities of obtained product, in order to provide for an adequate reconciliation.

## PROCEDURES AND RECORDS

### RECEIPT

4.19  There should be written procedures and records for the receipt of each delivery of each starting and primary and printed packaging material.

4.20  The records of the receipts should include:

a) The name of the material on the delivery note and the containers;

b) The "in-house" name and/or code of material (if different from a);

c) Date of receipt;

d) Supplier's name and, if possible, manufacturer's name;

e) Manufacturer's batch or reference number;

f) Total quantity, and number of containers received;

g) The batch number assigned after receipt;

h) Any relevant comment (e.g. state of the containers).

4.21  There should be written procedures for the internal labelling, quarantine and storage of starting materials, packaging materials and other materials, as appropriate.

### SAMPLING

4.22  There should be written procedures for sampling, which include the person(s) authorised to take samples, the methods and equipment to be used, the amounts to be taken and any precautions to be observed to avoid contamination of the material or any deterioration in its quality (see Chapter 6, item 13).

### TESTING

4.23  There should be written procedures for testing materials and products at different stages of manufacture, describing the methods and equipment to be used. The tests performed should be recorded (see Chapter 6, item 17).

## OTHER

4.24 Written release and rejection procedures should be available for materials and products, and in particular for the release for sale of the finished product by the Qualified Person(s) in accordance with the requirements of Article 22 of Directive 75/319/EEC.

4.25 Records should be maintained of the distribution of each batch of a product in order to facilitate the recall of the batch if necessary (see Chapter 8).

4.26 There should be written procedures and the associated records of actions taken or conclusions reached, where appropriate, for:

- Validation;
- Equipment assembly and calibration;
- Maintenance, cleaning and sanitation;
- Personnel matters including training, clothing, hygiene;
- Environmental monitoring;
- Pest control;
- Complaints;
- Recalls;
- Returns.

4.27 Clear operating procedures should be available for major items of manufacturing and test equipment.

4.28 Log books should be kept for major or critical equipment recording, as appropriate, any validations, calibrations, maintenance, cleaning or repair operations, including the dates and identity of people who carried these operations out.

4.29 Log books should also record in chronological order the use of major or critical equipment and the areas where the products have been processed.

# Chapter 5: Production

## PRINCIPLE

Production operations must follow clearly defined procedures; they must comply with the principles of Good Manufacturing Practice in order to obtain products of the requisite quality and be in accordance with the relevant manufacturing and marketing authorisations.

## GENERAL

5.1  Production should be performed and supervised by competent people.

5.2  All handling of materials and products, such as receipt and quarantine, sampling, storage, labelling, dispensing, processing, packaging and distribution should be done in accordance with written procedures or instructions and, where necessary, recorded.

5.3  All incoming materials should be checked to ensure that the consignment corresponds to the order. Containers should be cleaned where necessary and labelled with the prescribed data.

5.4  Damage to containers and any other problem which might adversely affect the quality of a material should be investigated, recorded and reported to the Quality Control Department.

5.5  Incoming materials and finished products should be physically or administratively quarantined immediately after receipt or processing, until they have been released for use or distribution.

5.6  Intermediate and bulk products purchased as such should be handled on receipt as though they were starting materials.

5.7  All materials and products should be stored under the appropriate conditions established by the manufacturer and in an orderly fashion to permit batch segregation and stock rotation.

5.8  Checks on yields, and reconciliation of quantities, should be carried out as necessary to ensure that there are no discrepancies outside acceptable limits.

5.9  Operations on different products should not be carried out simultaneously or consecutively in the same room unless there is no risk of mix-up or cross-contamination.

5.10  At every stage of processing, products and materials should be protected from microbial and other contamination.

5.11  When working with dry materials and products, special precautions should be taken to prevent the generation and dissemination of dust. This applies particularly to the handling of highly active or sensitising materials.

5.12  At all times during processing, all materials, bulk containers, major items of equipment and where appropriate rooms used should be labelled or otherwise identified with an indication of the product or material being processed, its strength (where applicable) and batch number. Where applicable, this indication should also mention the stage of production.

5.13 Labels applied to containers, equipment or premises should be clear, unambiguous and in the company's agreed format. It is often helpful in addition to the wording on the labels to use colours to indicate status (for example, quarantined, accepted, rejected, clean,...).

5.14 Checks should be carried out to ensure that pipelines and other pieces of equipment used for the transportation of products from one area to another are connected in a correct manner.

5.15 Any deviation from instructions or procedures should be avoided as far as possible. If a deviation occurs, it should be approved in writing by a competent person, with the involvement of the Quality Control Department when appropriate.

5.16 Access to production premises should be restricted to authorised personnel.

5.17 Normally, the production of non-medicinal products should be avoided in areas and with the equipment destined for the production of medicinal products.

## PREVENTION OF CROSS-CONTAMINATION IN PRODUCTION

5.18 Contamination of a starting material or of a product by another material or product must be avoided. This risk of accidental cross-contamination arises from the uncontrolled release of dust, gases, vapours, sprays or organisms from materials and products in process, from residues on equipment, and from operators' clothing. The significance of this risk varies with the type of contaminant and of product being contaminated. Amongst the most hazardous contaminants are highly sensitising materials, biological preparations containing living organisms, certain hormones, cytotoxics, and other highly active materials. Products in which contamination is likely to be most significant are those administered by injection, those given in large doses and/or over a long time.

5.19 Cross-contamination should be avoided by appropriate technical or organisational measures, for example:

a) Production in segregated areas (required for products such as penicillins, live vaccines, live bacterial preparations and some other biologicals), or by campaign (separation in time) followed by appropriate cleaning;
b) Providing appropriate air-locks and air extraction;
c) Minimising the risk of contamination caused by recirculation or re-entry of untreated or insufficiently treated air;
d) Keeping protective clothing inside areas where products with special risk of cross-contamination are processed;
e) Using cleaning and decontamination procedures of known effectiveness, as ineffective cleaning of equipment is a common source of cross-contamination;
f) Using "closed systems" of production;
g) Testing for residues and use of cleaning status labels on equipment.

5.20 Measures to prevent cross-contamination and their effectiveness should be checked periodically according to set procedures.

## VALIDATION

5.21 Validation studies should reinforce Good Manufacturing Practice and be conducted in accordance with defined procedures. Results and conclusions should be recorded.

5.22 When any new manufacturing formula or method of preparation is adopted, steps should be taken to demonstrate its suitability for routine processing. The defined process, using the materials and equipment specified, should be shown to yield a product consistently of the required quality.

5.23 Significant amendments to the manufacturing process, including any change in equipment or materials, which may affect product quality and/or the reproducibility of the process should be validated.

5.24 Processes and procedures should undergo periodic critical re-validation to ensure that they remain capable of achieving the intended results.

## STARTING MATERIALS

5.25 The purchase of starting materials is an important operation which should involve staff who have a particular and thorough knowledge of the suppliers.

5.26 Starting materials should only be purchased from approved suppliers named in the relevant specification and, where possible, directly from the producer. It is recommended that the specifications established by the manufacturer for the starting materials be discussed with the suppliers. It is of benefit that all aspects of the production and control of the starting material in question, including handling, labelling and packaging requirements, as well as complaints and rejection procedures are discussed with the manufacturer and the supplier.

5.27 For each delivery, the containers should be checked for integrity of package and seal and for correspondence between the delivery note and the supplier's labels.

5.28 If one material delivery is made up of different batches, each batch must be considered as separate for sampling, testing and release.

5.29 Starting materials in the storage area should be appropriately labelled (see Chapter 5, item 13). Labels should bear at least the following information:

- The designated name of the product and the internal code reference where applicable;
- A batch number given at receipt;
- Where appropriate, the status of the contents (e.g. in quarantine, on test, released, rejected);
- Where appropriate, an expiry date or a date beyond which retesting is necessary.

When fully computerised storage systems are used, all the above information need not necessarily be in a legible form on the label.

5.30 There should be appropriate procedures or measures to assure the identity of the contents of each container of starting material. Bulk containers from which samples have been drawn should be identified (see Chapter 6, item 13).

5.31 Only starting materials which have been released by the Quality Control Department and which are within their shelf life should be used.

5.32 Starting materials should be dispensed only by designated persons, following a written procedure, to ensure that the correct materials are accurately weighed or measured into clean and properly labelled containers.

5.33 Each dispensed material and its weight or volume should be independently checked and the check recorded.

5.34 Materials dispensed for each batch should be kept together and conspicuously labelled as such.

## PROCESSING OPERATIONS: INTERMEDIATE AND BULK PRODUCTS

5.35 Before any processing operation is started, steps should be taken to ensure that the work area and equipment are clean and free from any starting materials, products, product residues or documents not required for the current operation.

5.36 Intermediate and bulk products should be kept under appropriate conditions.

5.37 Critical processes should be validated (see "VALIDATION" in this Chapter).

5.38 Any necessary in-process controls and environmental controls should be carried out and recorded.

5.39 Any significant deviation from the expected yield should be recorded and investigated.

## PACKAGING MATERIALS

5.40 The purchase, handling and control of primary and printed packaging materials shall be accorded attention similar to that given to starting materials.

5.41 Particular attention should be paid to printed materials. They should be stored in adequately secure conditions such as to exclude unauthorised access. Cut labels and other loose printed materials should be stored and transported in separate closed containers so as to avoid mix-ups. Packaging materials should be issued for use only by authorised personnel following an approved and documented procedure.

5.42 Each delivery or batch of printed or primary packaging material should be given a specific reference number or identification mark.

5.43 Outdated or obsolete primary packaging material or printed packaging material should be destroyed and this disposal recorded.

## PACKAGING OPERATIONS

5.44 When setting up a programme for the packaging operations, particular attention should be given to minimising the risk of cross-contamination, mix-ups or substitutions. Different products should not be packaged in close proximity unless there is physical segregation.

5.45 Before packaging operations are begun, steps should be taken to ensure that the work area, packaging lines, printing machines and other equipment are clean and free from any products, materials or documents previously used, if these are not required for the current operation. The line-clearance should be performed according to an appropriate check-list.

5.46 The name and batch number of the product being handled should be displayed at each packaging station or line.

5.47 All products and packaging materials to be used should be checked on delivery to the packaging department for quantity, identity and conformity with the Packaging Instructions.

5.48 Containers for filling should be clean before filling. Attention should be given to avoiding and removing any contaminants such as glass fragments and metal particles.

5.49 Normally, filling and sealing should be followed as quickly as possible by labelling. If it is not the case, appropriate procedures should be applied to ensure that no mix-ups or mislabelling can occur.

5.50 The correct performance of any printing operation (for example code numbers, expiry dates) to be done separately or in the course of the packaging should be checked and recorded. Attention should be paid to printing by hand which should be re-checked at regular intervals.

5.51 Special care should be taken when using cut-labels and when over-printing is carried out off-line. Roll-feed labels are normally preferable to cut-labels, in helping to avoid mix-ups.

5.52 Checks should be made to ensure that any electronic code readers, label counters or similar devices are operating correctly.

5.53 Printed and embossed information on packaging materials should be distinct and resistant to fading or erasing.

5.54 On-line control of the product during packaging should include at least checking the following:

   a) General appearance of the packages;
   b) Whether the packages are complete;
   c) Whether the correct products and packaging materials are used,
   d) Whether any over-printing is correct;
   e) Correct functioning of line monitors.

   Samples taken away from the packaging line should not be returned.

5.55 Products which have been involved in an unusual event should be reintroduced into the process only after special inspection, investigation and approval by authorised personnel. Detailed record should be kept of this operation.

5.56 Any significant or unusual discrepancy observed during reconciliation of the amount of bulk product and printed packaging materials and the number of units produced should be investigated and satisfactorily accounted for before release.

5.57 Upon completion of a packaging operation, any unused batch-coded packaging materials should be destroyed and the destruction recorded. A documented procedure should be followed if uncoded printed materials are returned to stock.

## FINISHED PRODUCTS

5.58 Finished products should be held in quarantine until their final release under conditions established by the manufacturer.

5.59 The evaluation of finished products and documentation which is necessary before release of product for sale are described in Chapter 6 (Quality Control).

5.60 After release, finished products should be stored as usable stock under conditions established by the manufacturer.

## REJECTED, RECOVERED AND RETURNED MATERIALS

5.61 Rejected materials and products should be clearly marked as such and stored separately in restricted areas. They should either be returned to the suppliers or, where appropriate, reprocessed or destroyed. Whatever action is taken should be approved and recorded by authorised personnel.

5.62 The reprocessing of rejected products should be exceptional. It is permitted only if the quality of the final product is not affected, if the specifications are met and if it is done in accordance with a defined and authorised procedure after evaluation of the risks involved. Record should be kept of the reprocessing.

5.63 The recovery of all or part of earlier batches which conform to the required quality by incorporation into a batch of the same product at a defined stage of manufacture should be authorised beforehand. This recovery should be carried out in accordance with a defined procedure after evaluation of the risks involved, including any possible effect on shelf life. The recovery should be recorded.

5.64 The need for additional testing of any finished product which has been reprocessed, or into which a recovered product has been incorporated, should be considered by the Quality Control Department.

5.65 Products returned from the market and which have left the control of the manufacturer should be destroyed unless without doubt their quality is satisfactory; they may be considered for re-sale, re-labelling or recovery in a subsequent batch only after they have been critically assessed by the Quality Control Department in accordance with a written procedure. The nature of the product, any special storage conditions it requires, its condition and history, and the time elapsed since it was issued should all be taken into account in this assessment. Where any doubt arises over the quality of the product, it should not be considered suitable for re-issue or re-use, although basic chemical re-processing to recover active ingredient may be possible. Any action taken should be appropriately recorded.

# Chapter 6: Quality Control

## PRINCIPLE

Quality Control is concerned with sampling, specifications and testing as well as the organisation, documentation and release procedures which ensure that the necessary and relevant tests are carried out, and that materials are not released for use, nor products released for sale or supply, until their quality has been judged satisfactory. Quality Control is not confined to laboratory operations, but must be involved in all decisions which may concern the quality of the product. The independence of Quality Control from Production is considered fundamental to the satisfactory operation of Quality Control.

(See also Chapter 1.)

## GENERAL

6.1 Each holder of a manufacturing authorisation should have a Quality Control Department. This department should be independent from other departments, and under the authority of a person with appropriate qualifications and experience, who has one or several control laboratories at his disposal. Adequate resources must be available to ensure that all the Quality Control arrangements are effectively and reliably carried out.

6.2 The principal duties of the head of Quality Control are summarised in Chapter 2. The Quality Control Department as a whole will also have other duties, such as to establish, validate and implement all quality control procedures, keep the reference samples of materials and products, ensure the correct labelling of containers of materials and products, ensure the monitoring of the stability of the products, participate in the investigation of complaints related to the quality of the product, etc. All these operations should be carried out in accordance with written procedures and, where necessary, recorded.

6.3 Finished product assessment should embrace all relevant factors, including production conditions, results of in-process testing, a review of manufacturing (including packaging) documentation, compliance with Finished Product Specification and examination of the final finished pack.

6.4 Quality Control personnel should have access to production areas for sampling and investigation as appropriate.

## GOOD QUALITY CONTROL LABORATORY PRACTICE

6.5 Control laboratory premises and equipment should meet the general and specific requirements for Quality Control areas given in Chapter 3.

6.6 The personnel, premises, and equipment in the laboratories should be appropriate to the tasks imposed by the nature and the scale of the manufacturing operations. The use of outside laboratories, in conformity with the principles detailed in Chapter 7, Contract Analysis, can be accepted for particular reasons, but this should be stated in the Quality Control records.

## DOCUMENTATION

6.7 Laboratory documentation should follow the principles given in Chapter 4. An important part of this documentation deals with Quality Control and the following details should be readily available to the Quality Control Department:

- Specifications;
- Sampling procedures;
- Testing procedures and records (including analytical worksheets and/or laboratory notebooks);
- Analytical reports and/or certificates;
- Data from environmental monitoring, where required;
- Validation records of test methods, where applicable;
- Procedures for and records of the calibration of instruments and maintenance of equipment.

6.8 Any Quality Control documentation relating to a batch record should be retained for one year after the expiry date of the batch and at least 5 years after the certification referred to in article 22.2 of Directive 75/319/EEC.

6.9 For some kinds of data (e.g. analytical tests results, yields, environmental controls,…) it is recommended that records be kept in a manner permitting trend evaluation.

6.10 In addition to the information which is part of the batch record, other original data such as laboratory notebooks and/or records should be retained and readily available.

## SAMPLING

6.11 The sample taking should be done in accordance with approved written procedures that describe:

- The method of sampling;
- The equipment to be used;
- The amount of the sample to be taken;
- Instructions for any required sub-division of the sample;
- The type and condition of the sample container to be used;
- The identification of containers sampled;
- Any special precautions to be observed, especially with regard to the sampling of sterile or noxious materials;
- The storage conditions;
- Instructions for the cleaning and storage of sampling equipment.

6.12 Reference samples should be representative of the batch of materials or products from which they are taken. Other samples may also be taken to monitor the most stressed part of a process (e.g. beginning or end of a process).

6.13 Sample containers should bear a label indicating the contents, with the batch number, the date of sampling and the containers from which samples have been drawn.

6.14 Reference samples from each batch of finished products should be retained till one year after the expiry date. Finished products should usually be kept in their final packaging and stored under the recommended conditions. Samples of starting materials (other than solvents, gases and water) should be retained for at least two years* after the release of the product if their stability allows. This period may be shortened if their stability, as mentioned in the relevant specification, is shorter. Reference samples of materials and products should be of a size sufficient to permit at least a full re-examination.

## Testing

6.15 Analytical methods should be validated. All testing operations described in the marketing authorisation should be carried out according to the approved methods.

6.16 The results obtained should be recorded and checked to make sure that they are consistent with each other. Any calculations should be critically examined.

6.17 The tests performed should be recorded and the records should include at least the following data:

a) Name of the material or product and, where applicable, dosage form;
b) Batch number and, where appropriate, the manufacturer and/or supplier;
c) References to the relevant specifications and testing procedures;
d) Test results, including observations and calculations, and reference to any certificates of analysis;
e) Dates of testing;
f) Initials of the persons who performed the testing;
g) Initials of the persons who verified the testing and the calculations, where appropriate;
h) A clear statement of release or rejection (or other status decision) and the dated signature of the designated responsible person.

6.18 All the in-process controls, including those made in the production area by production personnel, should be performed according to methods approved by Quality Control and the results recorded.

6.19 Special attention should be given to the quality of laboratory reagents, volumetric glassware and solutions, reference standards and culture media. They should be prepared in accordance with written procedures.

6.20 Laboratory reagents intended for prolonged use should be marked with the preparation date and the signature of the person who prepared them. The expiry date of unstable reagents and culture media should be indicated on the label, together with specific storage conditions. In addition, for volumetric solutions, the last date of standardisation and the last current factor should be indicated.

6.21 Where necessary, the date of receipt of any substance used for testing operations (e.g. reagents and reference standards) should be indicated on the container. Instructions for use and storage should

---

* In Federal Republic of Germany, France, Belgium and Greece, samples of starting materials should be retained for as long as the corresponding finished product.

be followed. In certain cases it may be necessary to carry out an identification test and/or other testing of reagent materials upon receipt or before use.

6.22 Animals used for testing components, materials or products, should, where appropriate, be quarantined before use. They should be maintained and controlled in a manner that assures their suitability for the intended use. They should be identified, and adequate records should be maintained, showing the history of their use.

# Chapter 7:
# Contract Manufacture and Analysis

## PRINCIPLE

Contract manufacture and analysis must be correctly defined, agreed and controlled in order to avoid misunderstandings which could result in a product or work of unsatisfactory quality. There must be a written contract between the Contract Giver and the Contract Acceptor which clearly establishes the duties of each party. The contract must clearly state the way in which the Qualified Person releasing each batch of product for sale exercises his full responsibility.

*Note:* This Chapter deals with the responsibilities of manufacturers towards the Competent Authorities of the Member States with respect to the granting of marketing and manufacturing authorisations. It is not intended in any way to affect the respective liability of contract acceptors and contract givers to consumers, this is governed by other provisions of Community and national law.

## GENERAL

7.1 There should be a written contract covering the manufacture and/or analysis arranged under contract and any technical arrangements made in connection with it.

7.2 All arrangements for contract manufacture and analysis including any proposed changes in technical or other arrangements should be in accordance with the marketing authorisation for the product concerned.

## THE CONTRACT GIVER

7.3 The Contract Giver is responsible for assessing the competence of the Contract Acceptor to carry out successfully the work required and for ensuring by means of the contract that the principles and guidelines of GMP as interpreted in this Guide are followed.

7.4 The Contract Giver should provide the Contract Acceptor with all the information necessary to carry out the contracted operations correctly in accordance with the marketing authorisation and any other legal requirements. The Contract Giver should ensure that the Contract Acceptor is fully aware of any problems associated with the product or the work which might pose a hazard to his premises, equipment, personnel, other materials or other products.

7.5 The Contract Giver should ensure that all processed products and materials delivered to him by the Contract Acceptor comply with their specifications or that the products have been released by a Qualified Person.

## THE CONTRACT ACCEPTOR

7.6 The Contract Acceptor must have adequate premises and equipment, knowledge and experience, and competent personnel to carry out satisfactorily the work ordered by the Contract Giver. Contract manufacture may be undertaken only by a manufacturer who is the holder of a manufacturing authorisation.

7.7  The Contract Acceptor should ensure that all products or materials delivered to him are suitable for their intended purpose.

7.8  The Contract Acceptor should not pass to a third party any of the work entrusted to him under the contract without the Contract Giver's prior evaluation and approval of the arrangements. Arrangements made between the Contract Acceptor and any third party should ensure that the manufacturing and analytical information is made available in the same way as between the original Contract Giver and Contract Acceptor.

7.9  The Contract Acceptor should refrain from any activity which may adversely affect the quality of the product manufactured and/or analysed for the Contract Giver.

## THE CONTRACT

7.10  A contract should be drawn up between the Contract Giver and the Contract Acceptor which specifies their respective responsibilities relating to the manufacture and control of the product. Technical aspects of the contract should be drawn up by competent persons suitably knowledgeable in pharmaceutical technology, analysis and Good Manufacturing Practice. All arrangements for manufacture and analysis must be in accordance with the marketing authorisation and agreed by both parties.

7.11  The contract should specify the way in which the Qualified Person releasing the batch for sale ensures that each batch has been manufactured and checked for compliance with the requirements of Marketing Authorisation.

7.12  The contract should describe clearly who is responsible for purchasing materials, testing and releasing materials, undertaking production and quality controls, including in-process controls, and who has responsibility for sampling and analysis. In the case of contract analysis, the contract should state whether or not the Contract Acceptor should take samples at the premises of the manufacturer.

7.13  Manufacturing, analytical and distribution records, and reference samples should be kept by, or be available to, the Contract Giver. Any records relevant to assessing the quality of a product in the event of complaints or a suspected defect must be accessible and specified in the defect/recall procedures of the Contract Giver.

7.14  The contract should permit the Contract Giver to visit the facilities of the Contract Acceptor.

7.15  In the case of contract analysis, the Contract Acceptor should understand that he is subject to Inspection by the competent Authorities.

# Chapter 8: Complaints and Product Recall

## PRINCIPLE

All complaints and other information concerning potentially defective products must be reviewed carefully according to written procedures. In order to provide for all contingencies, and in accordance with Article 28 of Directive 75/319/EEC, a system should be designed to recall, if necessary, promptly and effectively products known or suspected to be defective from the market.

## COMPLAINTS

8.1 A person should be designated responsible for handling the complaints and deciding the measures to be taken together with sufficient supporting staff to assist him. If this person is not the Qualified Person, the latter should be made aware of any complaint, investigation or recall.

8.2 There should be written procedures describing the action to be taken, including the need to consider a recall, in the case of a complaint concerning a possible product defect.

8.3 Any complaint concerning a product defect should be recorded with all the original details and thoroughly investigated. The person responsible for Quality Control should normally be involved in the study of such problems.

8.4 If a product defect is discovered or suspected in a batch, consideration should be given to checking other batches in order to determine whether they are also affected. In particular, other batches which may contain reworks of the defective batch should be investigated.

8.5 All the decisions and measures taken as a result of a complaint should be recorded and referenced to the corresponding batch records.

8.6 Complaints records should be reviewed regularly for any indication of specific or recurring problems requiring attention and possibly the recall of marketed products.

8.7 The Competent Authorities should be informed if a manufacturer is considering action following possibly faulty manufacture, product deterioration, or any other serious quality problems with a product.

## RECALLS

8.8 A person should be designated as responsible for execution and co-ordination of recalls and should be supported by sufficient staff to handle all the aspects of the recalls with the appropriate degree of urgency. This responsible person should normally be independent of the sales and marketing organisation. If this person is not the Qualified Person, the latter should be made aware of any recall operation.

8.9 There should be established written procedures, regularly checked and updated when necessary, in order to organise any recall activity.

8.10 Recall operations should be capable of being initiated promptly and at any time.

8.11 All Competent Authorities of all countries to which products may have been distributed should be informed promptly if products are intended to be recalled because they are, or are suspected of being defective.

8.12 The distribution records should be readily available to the person(s) responsible for recalls, and should contain sufficient information on wholesalers and directly supplied customers (with addresses, phone and/or fax numbers inside and outside working hours, batches and amounts delivered), including those for exported products and medical samples.

8.13 Recalled products should be identified and stored separately in a secure area while awaiting a decision on their fate.

8.14 The progress of the recall process should be recorded and a final report issued, including a reconciliation between the delivered and recovered quantities of the products.

8.15 The effectiveness of the arrangements for recalls should be evaluated from time to time.

# Chapter 9: Self Inspection

## PRINCIPLE

Self inspections should be conducted in order to monitor the implementation and compliance with Good Manufacturing Practice principles and to propose necessary corrective measures.

9.1  Personnel matters, premises, equipment, documentation, production, quality control, distribution of the medicinal products, arrangements for dealing with complaints and recalls, and self inspection, should be examined at intervals following a pre-arranged programme in order to verify their conformity with the principles of Quality Assurance.

9.2  Self inspections should be conducted in an independent and detailed way by designated competent person(s) from the company. Independent audits by external experts may also be useful.

9.3  All self inspections should be recorded. Reports should contain all the observations made during the inspections and, where applicable, proposals for corrective measures. Statements on the actions subsequently taken should also be recorded.

**EUROPEAN COMMISSION**
**ENTERPRISE DIRECTORATE-GENERAL**

Single market: management & legislation for consumer goods
Pharmaceuticals: regulatory framework and market authorisations
Brussels, 30 May 2003

## AD HOC GMP INSPECTIONS SERVICES GROUP

## EC GUIDE TO GOOD MANUFACTURING PRACTICE
## REVISION TO ANNEX 1

### TITLE: MANUFACTURE OF STERILE MEDICINAL PRODUCTS

| | |
|---|---|
| 1st draft adopted by ad hoc GMP Inspectors Group | October 2002 |
| Released for public consultation | November 2002–January 2003 |
| Final draft adopted by ad hoc GMP Inspectors Group | April 2003 |
| Adopted by Pharmaceutical Committee | May 2003 |
| Date for coming into operation | September 2003 |

Note:

Annex 1 of the EC Guide to Good Manufacturing Practice (GMP) provides supplementary guidance on the application of the principles and guidelines of GMP to sterile medicinal products. The guidance includes recommendations on standards of environmental cleanliness for clean rooms. The guidance has been reviewed in the light of the international standard EN/ISO 14644-1 and amended in the interests of harmonisation but taking into account specific concerns unique to the production of sterile medicinal products.

The changes affect section 3 of the annex together with a minor change to section 20. The remainder of the annex remains unchanged.

# Annex 1:
# Manufacture of Sterile Medicinal Products

## PRINCIPLE

The manufacture of sterile products is subject to special requirements in order to minimise risks of microbiological contamination, and of particulate and pyrogen contamination. Much depends on the skill, training and attitudes of the personnel involved. Quality Assurance is particularly important, and this type of manufacture must strictly follow carefully established and validated methods of preparation and procedure. Sole reliance for sterility or other quality aspects must not be placed on any terminal process or finished product test.

Note:

This guidance does not lay down detailed methods for determining the microbiological and particulate cleanliness of air, surfaces etc. Reference should be made to other documents such as the EN/ISO Standards.

## GENERAL

1. The manufacture of sterile products should be carried out in clean areas entry to which should be through airlocks for personnel and/or for equipment and materials. Clean areas should be maintained to an appropriate cleanliness standard and supplied with air which has passed through filters of an appropriate efficiency.

2. The various operations of component preparation, product preparation and filling should be carried out in separate areas within the clean area.

Manufacturing operations are divided into two categories; First those where the product is terminally sterilised, and Second those which are conducted aseptically at some or all stages.

3. Clean areas for the manufacture of sterile products are classified according to the required characteristics of the environment. Each manufacturing operation requires an appropriate environmental cleanliness level in the operational state in order to minimise the risks of particulate or microbial contamination of the product or materials being handled.

In order to meet "in operation" conditions these areas should be designed to reach certain specified air-cleanliness levels in the "at rest" occupancy state. The "at rest" state is the condition where the installation is installed and operating, complete with production equipment but with no operating personnel present. The "in operation" state is the condition where the installation is functioning in the defined operating mode with the specified number of personnel working.

The "in operation" and "at rest" states should be defined for each clean room or suite of clean rooms.

For the manufacture of sterile medicinal products 4 grades can be distinguished.

Grade A: The local zone for high risk operations, e.g., filling zone, stopper bowls, open ampoules and vials, making aseptic connections. Normally such conditions are provided by a laminar air flow work station. Laminar air flow systems should provide a homogeneous

air speed in a range of 0.36–0.54 m/s (guidance value) at the working position in open clean room applications. The maintenance of laminarity should be demonstrated and validated. A uni-directional air flow and lower velocities may be used in closed isolators and glove boxes.
Grade B: For aseptic preparation and filling, this is the background environment for the grade A zone.
Grades C and D: Clean areas for carrying out less critical stages in the manufacture of sterile products.

The airborne articulate classification for these grades is given in the following table.

| | At rest[b] | | In operation[b] | |
|---|---|---|---|---|
| | Maximum permitted number of particles/m³ equal to or above[a] | | | |
| Grade | 0.5 µm[d] | 5 µm | 0.5 µm[d] | 5 µm |
| A | 3,500 | 1[e] | 3,500 | 1[e] |
| B[c] | 3,500 | 1[e] | 350,000 | 2,000 |
| C[c] | 350,000 | 2,000 | 3,500,000 | 20,000 |
| D[c] | 3,500,000 | 20,000 | Not defined[f] | Not defined[f] |

Notes:

(a) Particle measurement based on the use of a discrete airborne particle counter to measure the concentration of particles at designated sizes equal to or greater than the threshold stated.
   A continuous measurement system should be used for monitoring the concentration of particles in the grade A zone, and is recommended for the surrounding grade B areas.
   For routine testing the total sample volume should not be less than 1 m³ for grade A and B areas and preferably also in grade C areas.
(b) The particulate conditions given in the table for the "at rest" state should be achieved after a short "clean up" period of 15–20 minutes (guidance value) in an unmanned state after completion of operations. The particulate conditions for grade A "in operation" given in the table should be maintained in the zone immediately surrounding the product whenever the product or open container is exposed to the environment. It is accepted that it may not always be possible to demonstrate conformity with particulate standards at the point of fill when filling is in progress, due to the generation of particles or droplets from the product itself.
(c) In order to reach the B, C and D air grades, the number of air changes should be related to the size of the room and the equipment and personnel present in the room. The air system should be provided with appropriate terminal filters such as HEPA for grades A, B and C.
(d) The guidance given for the maximum permitted number of particles in the "at rest" and "in operation" conditions correspond approximately to the cleanliness classes in the EN/ISO 14644-1 at a particle size of 0.5 µm.
(e) These areas are expected to be completely free from particles of size greater than or equal to 5 µm. As it is impossible to demonstrate the absence of particles with any statistical significance the limits are set to 1 particle/m³. During the clean room qualification it should be shown that the areas can be maintained within the defined limits.
(f) The requirements and limits will depend on the nature of the operations carried out.

Other characteristics such as temperature and relative humidity depend on the product and nature of the operations carried out. These parameters should not interfere with the defined cleanliness standard.

Examples of operations to be carried out in the various grades are given in the table below (see also par. 11 and 12).

| Grade | Examples of operations for terminally sterilised products (see par. 11) |
|---|---|
| A | Filling of products, when unusually at risk |
| C | Preparation of solutions, when unusually at risk. Filling of products |
| D | Preparation of solutions and components for subsequent filling |

| Grade | Examples of operations for aseptic preparations (see par. 12) |
|-------|---------------------------------------------------------------|
| A | Aseptic preparation and filling |
| C | Preparation of solutions to be filtered |
| D | Handling of components after washing |

The particulate conditions given in the table for the "at rest" state should be achieved in the unmanned state after a short "clean-up" period of 15–20 minutes (guidance value), after completion of operations. The particulate conditions for grade A in operation given in the table should be maintained in the zone immediately surrounding the product whenever the product or open container is exposed to the environment. It is accepted that it may not always be possible to demonstrate conformity with particulate standards at the point of fill when filling is in progress, due to the generation of particles or droplets from the product itself.

4. The areas should be monitored during operation, in order to control the particulate cleanliness of the various grades.

5. Where aseptic operations are performed monitoring should be frequent using methods such as settle plates, volumetric air and surface sampling (e.g. swabs and contact plates). Sampling methods used in operation should not interfere with zone protection. Results from monitoring should be considered when reviewing batch documentation for finished product release. Surfaces and personnel should be monitored after critical operations.

Additional microbiological monitoring is also required outside production operations, e.g., after validation of systems, cleaning and sanitisation.

### Recommended Limits for Microbiological Monitoring of Clean Areas during Operation

| Grade | Recommended limits for microbial contamination[a] | | | |
|-------|--------------------------|-----------------------------------------------|-------------------------------------------|------------------------------------|
| | Air sample (cfu/m³) | Settle plates (diam. 9.0 mm; cfu/4 hours[b]) | Contact plates (diam. 55 mm; cfu/plate) | Glove print 5 fingers (cfu/glove) |
| A | <1 | <1 | <1 | <1 |
| B | 10 | 5 | 5 | 5 |
| C | 100 | 50 | 25 | – |
| D | 200 | 100 | 50 | – |

Notes
(a) These are average values.
(b) Individual settle plates may be exposed for less than 4 hours.

6. Appropriate alert and action limits should be set for the results of particulate and microbiological monitoring. If these limits are exceeded operating procedures should prescribe corrective action.

## ISOLATOR TECHNOLOGY

7. The utilisation of isolator technology to minimise human interventions in processing areas may result in a significant decrease in the risk of microbiological contamination of aseptically manufactured products from the environment. There are many possible designs of isolators and transfer devices. The isolator and the background environment should be designed so that the required air quality for the respective zones can be realised. Isolators are constructed of various materials more

or less prone to puncture and leakage. Transfer devices may vary from a single door to double door designs to fully sealed systems incorporating sterilisation mechanisms.

The transfer of materials into and out of the unit is one of the greatest potential sources of contamination. In general the area inside the isolator is the local zone for high risk manipulations, although it is recognised that laminar air flow may not exist in the working zone of all such devices.

The air classification required for the background environment depends on the design of the isolator and its application. It should be controlled and for aseptic processing it should be at least grade D.

8. Isolators should be introduced only after appropriate validation. Validation should take into account all critical factors of isolator technology, for example the quality of the air inside and outside (background) the isolator, sanitisation of the isolator, the transfer process and isolator integrity.

9. Monitoring should be carried out routinely and should include frequent leak testing of the isolator and glove/sleeve system.

## BLOW/FILL/SEAL TECHNOLOGY

10. Blow/fill/seal units are purpose built machines in which, in one continuous operation, containers are formed from a thermoplastic granulate, filled and then sealed, all by the one automatic machine. Blow/fill/seal equipment used for aseptic production which is fitted with an effective grade A air shower may be installed in at least a grade C environment, provided that grade AS clothing is used. The environment should comply with the viable and non viable limits at rest and the viable limit only when in operation. Blow/fill/seal equipment used for the production of products which are terminally sterilised should be installed in at least a grade D environment.

Because of this special technology particular attention should be paid to, at least the following: equipment design and qualification, validation and reproducibility of cleaning-in-place and sterilisation-in-place, background cleanroom environment in which the equipment is located, operator training and clothing, and interventions in the critical zone of the equipment including any aseptic assembly prior to the commencement of filling.

## TERMINALLY STERILISED PRODUCTS

11. Preparation of components and most products should be done in at least a grade D environment in order to give low risk of microbial and particulate contamination, suitable for filtration and sterilisation. Where the product is at a high or unusual risk of microbial contamination, (for example, because the product actively supports microbial growth or must be held for a long period before sterilisation or is necessarily processed not mainly in closed vessels), then preparation should be carried out in a grade C environment.

Filling of products for terminal sterilisation should be carried out in at least a grade C environment.

Where the product is at unusual risk of contamination from the environment, for example because the filling operation is slow or the containers are wide-necked or are necessarily exposed for more than a few seconds before sealing, the filling should be done in a grade A zone with at least a grade C background. Preparation and filling of ointments, creams, suspensions and emulsions should generally be carried out in a grade C environment before terminal sterilisation.

## ASEPTIC PREPARATION

12. Components after washing should be handled in at least a grade D environment. Handling of sterile starting materials and components, unless subjected to sterilisation or filtration through a micro-organism-retaining filter later in the process, should be done in a grade A environment with grade B background.

Preparation of solutions which are to be sterile filtered during the process should be done in a grade C environment; if not filtered, the preparation of materials and products should be done in a grade A environment with a grade B background.

Handling and filling of aseptically prepared products should be done in a grade A environment with a grade B background.

Prior to the completion of stoppering, transfer of partially closed containers, as used in freeze drying should be done either in a grade A environment with grade B background or in sealed transfer trays in a grade B environment.

Preparation and filling of sterile ointments, creams, suspensions and emulsions should be done in a grade A environment, with a grade B background, when the product is exposed and is not subsequently filtered.

## PERSONNEL

13. Only the minimum number of personnel required should be present in clean areas; this is particularly important during aseptic processing. Inspections and controls should be conducted outside the clean areas as far as possible.

14. All personnel (including those concerned with cleaning and maintenance) employed in such areas should receive regular training in disciplines relevant to the correct manufacture of sterile products. This training should include reference to hygiene and to the basic elements of microbiology. When outside staff who have not received such training (e.g. building or maintenance contractors) need to be brought in, particular care should be taken over their instruction and supervision.

15. Staff who have been engaged in the processing of animal tissue materials or of cultures of micro-organisms other than those used in the current manufacturing process should not enter sterile-product areas unless rigorous and clearly defined entry procedures have been followed.

16. High standards of personal hygiene and cleanliness are essential. Personnel involved in the manufacture of sterile preparations should be instructed to report any condition which may cause the shedding of abnormal numbers or types of contaminants; periodic health checks for such conditions are desirable. Actions to be taken about personnel who could be introducing undue microbiological hazard should be decided by a designated competent person.

17. Changing and washing should follow a written procedure designed to minimise contamination of clean area clothing or carry-through of contaminants to the clean areas.

18. Wristwatches, make-up and jewelery should not be worn in clean areas.

19. The clothing and its quality should be appropriate for the process and the grade of the working area. It should be worn in such a way as to protect the product from contamination.

The description of clothing required for each grade is given below:

Grade D: Hair and, where relevant, beard should be covered. A general protective suit and appropriate shoes or overshoes should be worn. Appropriate measures should be taken to avoid any contamination coming from outside the clean area.

Grade C: Hair and where relevant beard and moustache should be covered. A single or two-piece trouser suit, gathered at the wrists and with high neck and appropriate shoes or overshoes should be worn. They should shed virtually no fibres or particulate matter.

Grade A/B: Headgear should totally enclose hair and, where relevant, beard and moustache; it should be tucked into the neck of the suit; a face mask should be worn to prevent the shedding of droplets. Appropriate sterilised, non-powdered rubber or plastic gloves and sterilised or disinfected footwear should be worn. Trouser-legs should be tucked inside the footwear and garment sleeves into the gloves. The protective clothing should shed virtually no fibres or particulate matter and retain particles shed by the body.

20. Outdoor clothing should not be brought into changing rooms leading to grade B and C rooms. For every worker in a grade A/B area, clean sterile (sterilised or adequately sanitised) protective garments should be provided at each work session. Gloves should be regularly disinfected during operations. Masks and gloves should be changed at least for every working session.

21. Clean area clothing should be cleaned and handled in such a way that it does not gather additional contaminants which can later be shed. These operations should follow written procedures. Separate laundry facilities for such clothing are desirable. Inappropriate treatment of clothing will damage fibres and may increase the risk of shedding of particles.

## PREMISES

22. In clean areas, all exposed surfaces should be smooth, impervious and unbroken in order to minimise the shedding or accumulation of particles or micro-organisms and to permit the repeated application of cleaning agents, and disinfectants where used.

23. To reduce accumulation of dust and to facilitate cleaning there should be no uncleanable recesses and a minimum of projecting ledges, shelves, cupboards and equipment. Doors should be designed to avoid those uncleanable recesses; sliding doors may be undesirable for this reason.

24. False ceilings should be sealed to prevent contamination from the space above them.

25. Pipes and ducts and other utilities should be installed so that they do not create recesses, unsealed openings and surfaces which are difficult to clean.

26. Sinks and drains should be prohibited in grade A/B areas used for aseptic manufacture. In other areas air breaks should be fitted between the machine or sink and the drains. Floor drains in lower grade clean rooms should be fitted with traps or water seals to prevent back-flow.

27. Changing rooms should be designed as airlocks and used to provide physical separation of the different stages of changing and so minimise microbial and particulate contamination of protective clothing. They should be flushed effectively with filtered air. The final stage of the changing room should, in the at-rest state, be the same grade as the area into which it leads. The use of separate changing rooms for entering and leaving clean areas is sometimes desirable. In general hand washing facilities should be provided only in the first stage of the changing rooms.

28. Both airlock doors should not be opened simultaneously. An interlocking system or a visual and/or audible warning system should be operated to prevent the opening of more than one door at a time.

29. A filtered air supply should maintain a positive pressure and an air flow relative to surrounding areas of a lower grade under all operational conditions and should flush the area effectively. Adjacent rooms of different grades should have a pressure differential of 10–15 pascals (guidance values). Particular attention should be paid to the protection of the zone of greatest risk, that is, the immediate environment to which a product and cleaned components which contact the product are exposed. The various recommendations regarding air supplies and pressure differentials may need to be modified where it becomes necessary to contain some materials, e.g., pathogenic, highly toxic, radioactive or live viral or bacterial materials or products. Decontamination of facilities and treatment of air leaving a clean area may be necessary for some operations.

30. It should be demonstrated that air-flow patterns do not present a contamination risk, e.g., care should be taken to ensure that air flows do not distribute particles from a particle-generating person, operation or machine to a zone of higher product risk.

31. A warning system should be provided to indicate failure in the air supply. Indicators of pressure differences should be fitted between areas where these differences are important. These pressure differences should be recorded regularly or otherwise documented.

## EQUIPMENT

32. A conveyor belt should not pass through a partition between a grade A or B area and a processing area of lower air cleanliness, unless the belt itself is continually sterilised (e.g. in a sterilising tunnel).

33. As far as practicable equipment, fittings and services should be designed and installed so that operations, maintenance and repairs can be carried out outside the clean area. If sterilisation is required, it should be carried out, wherever possible, after complete reassembly.

34. When equipment maintenance has been carried out within the clean area, the area should be cleaned, disinfected and/or sterilised where appropriate, before processing recommences if the required standards of cleanliness and/or asepsis have not been maintained during the work.

35. Water treatment plants and distribution systems should be designed, constructed and maintained so as to ensure a reliable source of water of an appropriate quality. They should not be operated beyond their designed capacity. Water for injections should be produced, stored and distributed in a manner which prevents microbial growth, for example by constant circulation at a temperature above 70°C.

36. All equipment such as sterilisers, air handling and filtration systems, air vent and gas filters, water treatment, generation, storage and distribution systems should be subject to validation and planned maintenance; their return to use should be approved.

## SANITATION

37. The sanitation of clean areas is particularly important. They should be cleaned thoroughly in accordance with a written programme. Where disinfectants are used, more than one type should be

employed. Monitoring should be undertaken regularly in order to detect the development of resistant strains.

38. Disinfectants and detergents should be monitored for microbial contamination; dilutions should be kept in previously cleaned containers and should only be stored for defined periods unless sterilised. Disinfectants and detergents used in Grades A and B areas should be sterile prior to use.

39. Fumigation of clean areas may be useful for reducing microbiological contamination in inaccessible places.

## PROCESSING

40. Precautions to minimise contamination should be taken during all processing stages including the stages before sterilisation.

41. Preparations of microbiological origin should not be made or filled in areas used for the processing of other medicinal products; however, vaccines of dead organisms or of bacterial extracts may be filled, after inactivation, in the same premises as other sterile medicinal products.

42. Validation of aseptic processing should include a process simulation test using a nutrient medium (media fill). Selection of the nutrient medium should be made based on dosage form of the product and selectivity, clarity, concentration and suitability for sterilisation of the nutrient medium. The process simulation test should imitate as closely as possible the routine aseptic manufacturing process and include all the critical subsequent manufacturing steps. It should also take into account various interventions known to occur during normal production as well as worst case situations. Process simulation tests should be performed as initial validation with three consecutive satisfactory simulation tests per shift and repeated at defined intervals and after any significant modification to the HVAC-system, equipment, process and number of shifts. Normally process simulation tests should be repeated twice a year per shift and process. The number of containers used for media fills should be sufficient to enable a valid evaluation. For small batches, the number of containers for media fills should at least equal the size of the product batch. The target should be zero growth but a contamination rate of less than 0.1% with 95% confidence limit is acceptable. The manufacturer should establish alert and action limits. Any contamination should be investigated.

43. Care should be taken that any validation does not compromise the processes.

44. Water sources, water treatment equipment and treated water should be monitored regularly for chemical and biological contamination and, as appropriate, for endotoxins. Records should be maintained of the results of the monitoring and of any action taken.

45. Activities in clean areas and especially when aseptic operations are in progress should be kept to a minimum and movement of personnel should be controlled and methodical, to avoid excessive shedding of particles and organisms due to over-vigorous activity. The ambient temperature and humidity should not be uncomfortably high because of the nature of the garments worn.

46. Microbiological contamination of starting materials should be minimal. Specifications should include requirements for microbiological quality when the need for this has been indicated by monitoring.

47. Containers and materials liable to generate fibres should be minimised in clean areas.

48. Where appropriate, measures should be taken to minimise the particulate contamination of the end product.

49. Components, containers and equipment should be handled after the final cleaning process in such a way that they are not recontaminated.

50. The interval between the washing and drying and the sterilisation of components, containers and equipment as well as between their sterilisation and use should be minimised and subject to a time limit appropriate to the storage conditions.

51. The time between the start of the preparation of a solution and its sterilisation or filtration through a micro-organism-retaining filter should be minimised. There should be a set maximum permissible time for each product that takes into account its composition and the prescribed method of storage.

52. The bioburden should be monitored before sterilisation. There should be working limits on contamination immediately before sterilisation which are related to the efficiency of the method to be used. Where appropriate the absence of pyrogens should be monitored. All solutions, in particular large volume infusion fluids, should be passed through a micro-organism-retaining filter, if possible sited immediately before filling.

53. Components, containers, equipment and any other article required in a clean area where aseptic work takes place should be sterilised and passed into the area through double-ended sterilisers sealed into the wall, or by a procedure which achieves the same objective of not introducing contamination. Non-combustible gases should be passed through micro-organism retentive fitters.

54. The efficacy of any new procedure should be validated, and the validation verified at scheduled intervals based on performance history or when any significant change is made in the process or equipment.

## STERILISATION

55. All sterilisation processes should be validated. Particular attention should be given when the adopted sterilisation method is not described in the current edition of the European Pharmacopoeia, or when it is used for a product which is not a simple aqueous or oily solution. Where possible, heat sterilisation is the method of choice. In any case, the sterilisation process must be in accordance with the marketing and manufacturing authorisations.

56. Before any sterilisation process is adopted its suitability for the product and its efficacy in achieving the desired sterilising conditions in all parts of each type of load to be processed should be demonstrated by physical measurements and by biological indicators where appropriate. The validity of the process should be verified at scheduled intervals, at least annually, and whenever significant modifications have been made to the equipment. Records should be kept of the results.

57. For effective sterilisation the whole of the material must be subjected to the required treatment and the process should be designed to ensure that this is achieved.

58. Validated loading patterns should be established for all sterilisation processes.

59. Biological indicators should be considered as an additional method for monitoring the sterilisation. They should be stored and used according to the manufacturers instructions, and their quality checked by positive controls.

If biological indicators are used, strict precautions should be taken to avoid transferring microbial contamination from them.

60. There should be a clear means of differentiating products which have not been sterilised from those which have. Each basket, tray or other carrier of products or components should be clearly labelled with the material name, its batch number and an indication of whether or not it has been sterilised. Indicators such as autoclave tape may be used, where appropriate, to indicate whether or not a batch (or sub-batch) has passed through a sterilisation process, but they do not give a reliable indication that the lot is, in fact, sterile.

61. Sterilisation records should be available for each sterilisation run. They should be approved as part of the batch release procedure.

## STERILISATION BY HEAT

62. Each heat sterilisation cycle should be recorded on a time/temperature chart with a sufficiently large scale or by other appropriate equipment with suitable accuracy and precision. The position of the temperature probes used for controlling and/or recording should have been determined during the validation, and where applicable also checked against a second independent temperature probe located at the same position.

63. Chemical or biological indicators may also be used, but should not take the place of physical measurements.

64. Sufficient time must be allowed for the whole of the load to reach the required temperature before measurement of the sterilising time-period is commenced. This time must be determined for each type of load to be processed.

65. After the high temperature phase of a heat sterilisation cycle, precautions should be taken against contamination of a sterilised load during cooling. Any cooling fluid or gas in contact with the product should be sterilised unless it can be shown that any leaking container would not be approved for use.

## MOIST HEAT

66. Both temperature and pressure should be used to monitor the process. Control instrumentation should normally be independent of monitoring instrumentation and recording charts. Where automated control and monitoring systems are used for these applications they should be validated to ensure that critical process requirements are met. System and cycle faults should be registered by the system and observed by the operator. The reading of the independent temperature indicator should be routinely checked against the chart recorder during the sterilisation period. For sterilisers fitted with a drain at the bottom of the chamber, it may also be necessary to record the temperature at this position, throughout the sterilisation period. There should be frequent leak tests on the chamber when a vacuum phase is part of the cycle.

67. The items to be sterilised, other than products in sealed containers, should be wrapped in a material which allows removal of air and penetration of steam but which prevents recontamination after sterilisation. All parts of the load should be in contact with the sterilising agent at the required temperature for the required time.

68. Care should be taken to ensure that steam used for sterilisation is of suitable quality and does not contain additives at a level which could cause contamination of product or equipment.

## DRY HEAT

69. The process used should include air circulation within the chamber and the maintenance of a positive pressure to prevent the entry of non-sterile air. Any air admitted should be passed through a HEPA filter. Where this process is also intended to remove pyrogens, challenge tests using endotoxins should be used as part of the validation.

## STERILISATION BY RADIATION

70. Radiation sterilisation is used mainly for the sterilisation of heat sensitive materials and products. Many medicinal products and some packaging materials are radiation-sensitive, so this method is permissible only when the absence of deleterious effects on the product has been confirmed experimentally. Ultraviolet irradiation is not normally an acceptable method of sterilisation.

71. During the sterilisation procedure the radiation dose should be measured. For this purpose, dosimetry indicators which are independent of dose rate should be used, giving a quantitative measurement of the dose received by the product itself. Dosimeters should be inserted in the load in sufficient number and close enough together to ensure that there is always a dosimeter in the irradiator. Where plastic dosimeters are used they should be used within the time limit of their calibration. Dosimeter absorbances should be read within a short period after exposure to radiation.

72. Biological indicators may be used as an additional control.

73. Validation procedures should ensure that the effects of variations in density of the packages are considered.

74. Materials handling procedures should prevent mix-up between irradiated and non-irradiated materials. Radiation sensitive colour disks should also be used on each package to differentiate between packages which have been subjected to irradiation and those which have not.

75. The total radiation dose should be administered within a predetermined time span.

## STERILISATION WITH ETHYLENE OXIDE

76. This method should be used only when no other method is practicable. During process validation it should be shown that there is no damaging effect on the product and that the conditions and time allowed for degassing are such as to reduce any residual gas and reaction products to defined acceptable limits for the type of product or material.

77. Direct contact between gas and microbial cells is essential; precautions should be taken to avoid the presence of organisms likely to be enclosed in material such as crystals or dried protein. The nature and quantity of packaging materials can significantly affect the process.

78. Before exposure to the gas, materials should be brought into equilibrium with the humidity and temperature required by the process. The time required for this should be balanced against the opposing need to minimise the time before sterilisation.

79. Each sterilisation cycle should be monitored with suitable biological indicators, using the appropriate number of test pieces distributed throughout the load. The information so obtained should form part of the batch record.

80. For each sterilisation cycle, records should be made of the time taken to complete the cycle, of the pressure, temperature and humidity within the chamber during the process and of the gas concentration and of the total amount of gas used. The pressure and temperature should be recorded throughout the cycle on a chart. The record(s) should form part of the batch record.

81. After sterilisation, the load should be stored in a controlled manner under ventilated conditions to allow residual gas and reaction products to reduce to the defined level. This process should be validated.

## FILTRATION OF MEDICINAL PRODUCTS WHICH CANNOT BE STERILISED IN THEIR FINAL CONTAINER

82. Filtration alone is not considered sufficient when sterilisation in the final container is possible. With regard to methods currently available, steam sterilisation is to be preferred. If the product cannot be sterilised in the final container, solutions or liquids can be filtered through a sterile filter of nominal pore size of 0.22 micron (or less), or with at least equivalent micro-organism retaining properties, into a previously sterilised container. Such fitters can remove most bacteria and moulds, but not all viruses or mycoplasmas. Consideration should be given to complementing the filtration process with some degree of heat treatment.

83. Due to the potential additional risks of the filtration method as compared with other sterilisation processes, a second filtration via a further sterilised micro-organism retaining filter, immediately prior to filling, may be advisable. The final sterile filtration should be carried out as dose as possible to the filling point.

84. Fibre shedding characteristics of filters should be minimal.

85. The integrity of the sterilised filter should be verified before use and should be confirmed immediately after use by an appropriate method such as a bubble point, diffusive flow or pressure hold test. The time taken to filter a known volume of bulk solution and the pressure difference to be used across the filter should be determined during validation and any significant differences from this during routine manufacturing, should be noted and investigated. Results of these checks should be included in the batch record. The integrity of critical gas and air vent filters should be confirmed after use. The integrity of other filters should be confirmed at appropriate intervals.

86. The same filter should not be used for more than one working day unless such use has been validated.

87. The filter should not affect the product by removal of ingredients from it or by release of substances into it.

## FINISHING OF STERILE PRODUCTS

88. Containers should be closed by appropriately validated methods. Containers closed by fusion, e.g., glass or plastic ampoules should be subject to 100% integrity testing. Samples of other containers should be checked for integrity according to appropriate procedures.

89. Containers sealed under vacuum should be tested for maintenance of that vacuum after an appropriate, pre-determined period.

90. Filled containers of parenteral products should be inspected individually for extraneous contamination or other defects. When inspection is done visually, it should be done under suitable and controlled conditions of illumination and background. Operators doing the inspection should pass regular eye-sight checks, with spectacles if worn, and be allowed frequent breaks from inspection. Where other methods of inspection are used, the process should be validated and the performance of the equipment checked at intervals. Results should be recorded.

## QUALITY CONTROL

91. The sterility test applied to the finished product should be regarded only as the last in a series of control measures by which sterility is assured. The test should be validated for the product(s) concerned.

92. In those cases where parametric release has been authorised, special attention should be paid to the validation and the monitoring of the entire manufacturing process.

93. Samples taken for sterility testing should be representative of the whole of the batch, but should in particular include samples taken from parts of the batch considered to be most at risk of contamination, e.g.:

    a. for products which have been filled aseptically, samples should include containers filled at the beginning and end of the batch and after any significant intervention,
    b. or products which have been heat sterilised in their final containers, consideration should be given to taking samples from the potentially coolest part of the load.

**EUROPEAN COMMISSION**
**ENTERPRISE DIRECTORATE-GENERAL**

Single market, regulatory environment, industries under vertical legislation
Pharmaceuticals and cosmetics
Brussels, July 2001

## WORKING PARTY ON CONTROL OF MEDICINES AND INSPECTIONS

## FINAL VERSION OF ANNEX 15 TO THE EU GUIDE TO GOOD MANUFACTURING PRACTICE

TITLE: QUALIFICATION AND VALIDATION

| First discussion in drafting group | |
|---|---|
| Discussion at the working Party on Control of Medicines and Inspection for release for consultation | 16 September 1999 |
| Pharmaceutical Committee | 28 September 1999 |
| Released for consultation | 30 October 1999 |
| Deadline for comments | 28 February 2000 |
| Final approval by Inspector's working party | December 2000 |
| Pharmaceutical Committee (for information) | April 2001 |
| Date for coming into operation | September 2001 |

Note that this document is based in the PICS/S recommendations.


# QUALIFICATION AND VALIDATION

## PRINCIPLE

1. This Annex describes the principles of qualification and validation which are applicable to the manufacture of medicinal products. It is a requirement of GMP that manufacturers identify what validation work is needed to prove control of the critical aspects of their particular operations. Significant changes to the facilities, the equipment and the processes, which may affect the quality of the product, should be validated. A risk assessment approach should be used to determine the scope and extent of validation.

## PLANNING FOR VALIDATION

2. All validation activities should be planned. The key elements of a validation programme should be clearly defined and documented in a validation master plan (VMP) or equivalent documents.

3. The VMP should be a summary document which is brief, concise and clear.

4. The VMP should contain data on at least the following:

(a) validation policy;
(b) organisational structure of validation activities;
(c) summary of facilities, systems, equipment and processes to be validated;
(d) documentation format: the format to be used for protocols and reports;
(e) planning and scheduling;
(f) change control;
(g) reference to existing documents.

5. In case of large projects, it may be necessary to create separate validation master plans.

## DOCUMENTATION

6. A written protocol should be established that specifies how qualification and validation will be conducted. The protocol should be reviewed and approved. The protocol should specify critical steps and acceptance criteria.

7. A report that cross-references the qualification and/or validation protocol should be prepared, summarising the results obtained, commenting on any deviations observed, and drawing the necessary conclusions, including recommending changes necessary to correct deficiencies. Any changes to the plan as defined in the protocol should be documented with appropriate justification.

8. After completion of a satisfactory qualification, a formal release for the next step in qualification and validation should be made as a written authorisation.

## QUALIFICATION

### DESIGN QUALIFICATION

9. The first element of the validation of new facilities, systems or equipment could be design qualification (DQ).

10. The compliance of the design with GMP should be demonstrated and documented.

### INSTALLATION QUALIFICATION

11. Installation qualification (IQ) should be performed on new or modified facilities, systems and equipment.

12. IQ should include, but not be limited to the following:

(a) installation of equipment, piping, services and instrumentation checked to current engineering drawings and specifications;
(b) collection and collation of supplier operating and working instructions and maintenance requirements;
(c) calibration requirements;
(d) verification of materials of construction.

## OPERATIONAL QUALIFICATION

13. Operational qualification (OQ) should follow Installation qualification.

14. OQ should include, but not be limited to the following:

(a) tests that have been developed from knowledge of processes, systems and equipment;
(b) tests to include a condition or a set of conditions encompassing upper and lower operating limits, sometimes referred to as "worst case" conditions.

15. The completion of a successful Operational qualification should allow the finalisation of calibration, operating and cleaning procedures, operator training and preventative maintenance requirements. It should permit a formal "release" of the facilities, systems and equipment.

## PERFORMANCE QUALIFICATION

16. Performance qualification (PQ) should follow successful completion of Installation qualification and Operational qualification.

17. PQ should include, but not be limited to the following:

(a) tests, using production materials, qualified substitutes or simulated product, that have been developed from knowledge of the process and the facilities, systems or equipment;
(b) tests to include a condition or set of conditions encompassing upper and lower operating limits.

18. Although PQ is described as a separate activity, it may in some cases be appropriate to perform it in conjunction with OQ.

## QUALIFICATION OF ESTABLISHED (IN-USE) FACILITIES, SYSTEMS AND EQUIPMENT

19. Evidence should be available to support and verify the operating parameters and limits for the critical variables of the operating equipment. Additionally, the calibration, cleaning, preventative maintenance, operating procedures and operator training procedures and records should be documented.

# PROCESS VALIDATION

## GENERAL

20. The requirements and principles outlined in this chapter are applicable to the manufacture of pharmaceutical dosage forms. They cover the initial validation of new processes, subsequent validation of modified processes and re-validation.

21. Process validation should normally be completed prior to the distribution and sale of the medicinal product (prospective validation). In exceptional circumstances, where this is not possible, it may be necessary to validate processes during routine production (concurrent validation). Processes in use for some time should also be validated (retrospective validation).

22. Facilities, systems and equipment to be used should have been qualified and analytical testing methods should be validated. Staff taking part in the validation work should have been appropriately trained.

23. Facilities, systems, equipment and processes should be periodically evaluated to verify that they are still operating in a valid manner.

## PROSPECTIVE VALIDATION

24. Prospective validation should include, but not be limited to the following:

(a) short description of the process;
(b) summary of the critical processing steps to be investigated;
(c) list of the equipment/facilities to be used (including measuring/monitoring/recording equipment) together with its calibration status;
(d) finished product specifications for release;
(e) list of analytical methods, as appropriate;
(f) proposed in-process controls with acceptance criteria;
(g) additional testing to be carried out, with acceptance criteria and analytical validation, as appropriate;
(h) sampling plan;
(i) methods for recording and evaluating results;
(j) functions and responsibilities;
(k) proposed timetable.

25. Using this defined process (including specified components) a series of batches of the final product may be produced under routine conditions. In theory the number of process runs carried out and observations made should be sufficient to allow the normal extent of variation and trends to be established and to provide sufficient data for evaluation. It is generally considered acceptable that three consecutive batches/runs within the finally agreed parameters, would constitute a validation of the process.

26. Batches made for process validation should be the same size as the intended industrial scale batches.

27. If it is intended that validation batches be sold or supplied, the conditions under which they are produced should comply fully with the requirements of Good Manufacturing Practice, including the satisfactory outcome of the validation exercise, and with the marketing authorisation.

## CONCURRENT VALIDATION

28. In exceptional circumstances it may be acceptable not to complete a validation programme before routine production starts.

29. The decision to carry out concurrent validation must be justified, documented and approved by authorised personnel.

30. Documentation requirements for concurrent validation are the same as specified for prospective validation.

## RETROSPECTIVE VALIDATION

31. Retrospective validation is acceptable only for well-established processes and will be inappropriate where there have been recent changes in the composition of the product, operating procedures or equipment.

32. Validation of such processes should be based on historical data. The steps involved require the preparation of a specific protocol and the reporting of the results of the data review, leading to a conclusion and a recommendation.

33. The source of data for this validation should include, but not be limited to batch processing and packaging records, process control charts, maintenance log books, records of personnel changes, process capability studies, finished product data, including trend cards and storage stability results.

34. Batches selected for retrospective validation should be representative of all batches made during the review period, including any batches that failed to meet specifications, and should be sufficient in number to demonstrate process consistency. Additional testing of retained samples may be needed to obtain the necessary amount or type of data to retrospectively validate the process.

35. For retrospective validation, generally data from ten to thirty consecutive batches should be examined to assess process consistency, but fewer batches may be examined if justified.

## CLEANING VALIDATION

36. Cleaning validation should be performed in order to confirm the effectiveness of a cleaning procedure. The rationale for selecting limits of carry over of product residues, cleaning agents and microbial contamination should be logically based on the materials involved. The limits should be achievable and verifiable.

37. Validated analytical methods having sensitivity to detect residues or contaminants should be used. The detection limit for each analytical method should be sufficiently sensitive to detect the established acceptable level of the residue or contaminant.

38. Normally only cleaning procedures for product contact surfaces of the equipment need to be validated. Consideration should be given to noncontact parts. The intervals between use and cleaning as well as cleaning and reuse should be validated. Cleaning intervals and methods should be determined.

39. For cleaning procedures for products and processes which are similar, it is considered acceptable to select a representative range of similar products and processes. A single validation study utilising a "worst case" approach can be carried out which takes account of the critical issues.

40. Typically three consecutive applications of the cleaning procedure should be performed and shown to be successful in order to prove that the method is validated.

41. "Test until clean" is not considered an appropriate alternative to cleaning validation.

42. Products which simulate the physicochemical properties of the substances to be removed may exceptionally be used instead of the substances themselves, where such substances are either toxic or hazardous.

## CHANGE CONTROL

43. Written procedures should be in place to describe the actions to be taken if a change is proposed to a starting material, product component, process equipment, process environment (or site), method of production or testing or any other change that may affect product quality or reproducibility of the process. Change control procedures should ensure that sufficient supporting data are generated to demonstrate that the revised process will result in a product of the desired quality, consistent with the approved specifications.

44. All changes that may affect product quality or reproducibility of the process should be formally requested, documented and accepted. The likely impact of the change of facilities, systems and equipment on the product should be evaluated, including risk analysis. The need for, and the extent of, requalification and re-validation should be determined.

## REVALIDATION

45. Facilities, systems, equipment and processes, including cleaning, should be periodically evaluated to confirm that they remain valid. Where no significant changes have been made to the validated status, a review with evidence that facilities, systems, equipment and processes meet the prescribed requirements fulfils the need for revalidation.

## GLOSSARY

Definitions of terms relating to qualification and validation which are not given in the glossary of the current EC Guide to GMP, but which are used in this Annex, are given below.

**Change Control:** A formal system by which qualified representatives of appropriate disciplines review proposed or actual changes that might affect the validated status of facilities, systems, equipment or processes. The intent is to determine the need for action that would ensure and document that the system is maintained in a validated state.

**Cleaning Validation:** Cleaning validation is documented evidence that an approved cleaning procedure will provide equipment which is suitable for processing medicinal products.

**Concurrent Validation:** Validation carried out during routine production of products intended for sale.

**Design Qualification (DQ):** The documented verification that the proposed design of the facilities, systems and equipment is suitable for the intended purpose.

**Installation Qualification (IQ):** The documented verification that the facilities, systems and equipment, as installed or modified, comply with the approved design and the manufacturer's recommendations.

**Operational Qualification (OQ):** The documented verification that the facilities, systems and equipment, as installed or modified, perform as intended throughout the anticipated operating ranges.

**Performance Qualification (PQ):** The documented verification that the facilities, systems and equipment, as connected together, can perform effectively and reproducibly, based on the approved process method and product specification.

**Process Validation:** The documented evidence that the process, operated within established parameters, can perform effectively and reproducibly to produce a medicinal product meeting its predetermined specifications and quality attributes.

**Prospective Validation:** Validation carried out before routine production of products intended for sale.

**Retrospective Validation:** Validation of a process for a product which has been marketed based upon accumulated manufacturing, testing and control batch data.

**Re-Validation:** A repeat of the process validation to provide an assurance that changes in the process/equipment introduced in accordance with change control procedures do not adversely affect process characteristics and product quality.

**Risk Analysis:** Method to assess and characterise the critical parameters in the functionality of an equipment or process.

**Simulated Product:** A material that closely approximates the physical and, where practical, the chemical characteristics (e.g. viscosity, particle size, pH, etc.) of the product under validation. In many cases, these characteristics may be satisfied by a placebo product batch.

**System:** A group of equipment with a common purpose.

**Worst Case:** A condition or set of conditions encompassing upper and lower processing limits and circumstances, within standard operating procedures, which pose the greatest chance of product or process failure when compared to ideal conditions. Such conditions do not necessarily induce product or process failure.

# INTERNATIONAL STANDARD ISO 9000-3*

# Quality Management and Quality Assurance Standards — Part 3:

## GUIDELINES FOR THE APPLICATION OF ISO 9001:1994 TO THE DEVELOPMENT, SUPPLY, INSTALLATION AND MAINTENANCE OF COMPUTER SOFTWARE**

## CONTENTS

\*   Second edition
   1997-12-15
\*\* Reference number ISO 9000-3:1997(E)

ISO (the International Organization for Standardization) is a worldwide federation of national standards bodies (ISO member bodies). The work of preparing International Standards is normally carried out through ISO technical committees. Each member body interested in a subject for which a technical committee has been established has the right to be represented on that committee. International organizations, governmental and nongovernmental, in liaison with ISO, also take part in the work. ISO collaborates closely with the International Electrotechnical Commission (IEC) on all matters of electrotechnical standardization.

Draft International Standards adopted by the technical committees are circulated to the member bodies for voting. Publication as an International Standard requires approval by at least 75 % of the member bodies casting a vote.

International Standard ISO 9000-3 was prepared by Technical Committee ISOITC 176, *Quality management and quality assurance*, Subcommittee SC 2, *Quality systems.*

This second edition cancels and replaces the first edition (ISO 9000-3:1991), which has been technically revised.

ISO 9000 consists of the following parts, under the general title *Quality management and quality assurance standards*:

Part 1: Guidelines for selection and use
Part 2: Generic guidelines for the application of ISO 9001, ISO 9002 and ISO 9003
Part 3: Guidelines for the application of ISO 9001:1994 to the development, supply, installation and maintenance of computer software
Part 4: Guide to dependability programme management.

Annexes A and B of this part of ISO 9000 are for information only.

## INTRODUCTION

This part of ISO 9000 provides guidance in applying the requirements of ISO 9001:1994 where computer software design, development, installation and maintenance are an element of the business of a supplier:

a) as part of a commercial contract with an external organization;
b) as a product available for a market sector;
c) in support of the business processes of the supplier;
d) as software embedded in a hardware product.

It identifies the issues which need to be addressed and is independent of the technology, life cycle models, development processes, sequence of activities, or organization structure used by a supplier.

Where the scope of an organization's activities includes areas other than computer software development, the relationship between the computer software elements of that organization's quality system and the remaining aspects should be clearly documented within the quality system as a whole.

This part of ISO 9000 provides guidelines for the application of ISO 9001:1994. **Where text has been quoted from ISO 9001:1994, that text is enclosed in a box, for ease of identification.**

Throughout this part of ISO 9000, "shall" is used to express a provision that is binding between two or more parties; "will" to express a declaration of purpose, or intent, of one party; "should" to express a recommendation among possibilities; and "may" to indicate a course of action permissible within the limits of this parts of ISO 9000.

## QUALITY MANAGEMENT AND QUALITY ASSURANCE STANDARDS — PART 3:

## GUIDELINES FOR THE APPLICATION OF ISO 9001:1994 TO THE DEVELOPMENT, SUPPLY, INSTALLATION AND MAINTENANCE OF COMPUTER SOFTWARE

### 1 SCOPE

This part of ISO 9000 sets out guidelines to facilitate the application of ISO 9001:1994 for organizations developing, supplying, installing and maintaining computer software. It does not add to, or otherwise change, the requirements of ISO 9001.

This part of ISO 9000 is not intended to be used as assessment criteria in quality system registration/certification.

### 2 NORMATIVE REFERENCES

The following standards contain provisions which, through reference in this text, constitute provisions of this part of ISO 9000. At the time of publication, the editions indicated were valid. All standards are subject to revision, and parties to agreements based on this part of ISO 9000 are encouraged to investigate the possibility of applying the most recent edition of the standards indicated below. Members of IEC and ISO maintain registers of currently valid International Standards.

ISO 8402:1994, Quality management and quality assurance - Vocabulary.
ISO 9001:1994, Quality systems - Model for quality assurance in design, development, production, installation and servicing.

### 3 DEFINITIONS

For the purposes of this part of ISO 9000, the definitions given in ISO 8402 and the following definitions apply.

### 3.1 Product

Result of activities or processes.

NOTE 1 A product may include service, hardware, processed materials, software or a combination thereof.

NOTE 2 A product can be tangible (e.g. assemblies or processed materials) or intangible (e.g. knowledge or concepts), or a combination thereof.

NOTE 3 For the purposes of this International Standard, the term "product" applies to the intended product offering only and not to unintended "by-products" affecting the environment. This differs from the definition given in ISO 8402.

[ISO 90011]

### 3.2 Tender

Offer made by a supplier in response to an invitation to satisfy a contract award to provide product.
[ISO 9001]

### 3.3 Contract

Agreed requirements between a supplier and customer transmitted by any means.
[ISO 9001]

### 3.4 Baseline

A formally approved version of a configuration item, regardless of media, formally designated and fixed at a specific time during the configuration item's life cycle.
[ISO/IEC 12207]

### 3.5 Development

Software life cycle process that comprises the activities of requirements analysis, design, coding, integration, testing, installation and support for acceptance of software products.

### 3.6 Life Cycle Model

A framework containing the processes, activities, and tasks involved in the development, operation, and maintenance of a software product, spanning the life of the system from the definition of its requirements to the termination of its use.
[ISO/IEC 12207]

### 3.7 Phase

Defined segment of work.
NOTE: A phase does not imply the use of any specific life cycle model.

### 3.8 Regression Testing

Testing to determine that changes made in order to correct defects have not introduced additional defects.

### 3.9 Replication

Copying a software product from one medium to another.

### 3.10 Software

See software product (3.11).
NOTE: In this part of ISO 9000, the term "software" is confined to computer software.

### 3.11 Software Product

The set of computer programs, procedures, and possibly associated documentation and data.
[ISO/IEC 12207]
NOTE: A software product may be designated for delivery, an integral part of another product, or used in the development process.

### 3.12 Software Item

Any identifiable part of a software product.

## 4   QUALITY SYSTEM REQUIREMENTS

### 4.1 Management Responsibility

#### 4.1.1 Quality policy

> The supplier's management with executive responsibility shall define and document its policy for quality including objectives for quality and its commitment to quality. The quality policy shall be relevant to the supplier's organizational goals and the expectations and needs of its customers. The supplier shall ensure that this policy is understood, implemented and maintained at all levels of the organization.

No further software-related guidance is provided.

#### 4.1.2 Organization

#### 4.1.2.1 Responsibility and authority

> The responsibility, authority and the interrelation of personnel who manage, perform and verify work affecting quality shall be defined and documented, particularly for personnel who need the organizational freedom and authority to:
>
> a) initiate action to prevent the occurrence of any nonconformities relating to product, process and quality system;
> b) identify and record any problems relating to the product, process and quality system;
> c) initiate, recommend or provide solutions through designated channels;
> d) verify the implementation of solutions;
> e) control further processing, delivery or installation of nonconforming product until the deficiency or unsatisfactory condition has been corrected.

No further software-related guidance is provided.

#### 4.1.2.2 Resources

> The supplier shall identify resource requirements and provide adequate resources, including the assignment of trained personnel (see 4.18), for management, performance of work and verification activities including internal quality audits.

No further software-related guidance is provided.

NOTE: For further information see ISO/EC 12207:1995, subclause 7.2.

### 4.1.2.3   Management representative

> The supplier's management with executive responsibility shall appoint a member of the supplier's own management who, irrespective of other responsibilities, shall have defined authority for:
>
> a) ensuring that a quality system is established, implemented and maintained in accordance with this International Standard, and;
> b) reporting on the performance of the quality system to the supplier's management for review and as a basis for improvement of the quality system.
>
> NOTE 5 The responsibility of a management representative may also include liaison with external parties on matters relating to the supplier's quality system.

No further software-related guidance is provided.
NOTE: For further information, see ISO/IEC 12207:1995, subclause 6.3.1.6.

### 4.1.3   Management review

> The supplier's management with executive responsibility shall review the quality system at defined intervals sufficient to ensure its continuing suitability and effectiveness in satisfying the requirements of this International Standard and the supplier's stated quality policy and objectives (see 4.1.1). Records of such reviews shall be maintained (see 4.16)

No further software-related guidance is provided.
NOTE: For further information, see ISOIIEC 12207:1995, subclause 7.1.4.

## 4.2   Quality System

### 4.2.1   General

> The supplier shall establish, document and maintain a quality system as a means of ensuring that product conforms to specified requirements. The supplier shall prepare a quality manual covering the requirements of this International Standard. The quality manual shall include or make reference to the quality system procedures and outline the structure of the documentation used in the quality system.
>
> NOTE 6 Guidance on quality manuals is given in ISO 10013.

No further software-related guidance is provided.

### 4.2.2   Quality system procedures

> The supplier shall:
>
> a) prepare documented procedures consistent with the requirements of this International Standard and the supplier's stated quality policy, and;
> b) effectively implement the quality system and its documented procedures.
>
> For the purpose of this International Standard, the range and detail of the procedures that form part of the quality system shall be dependent upon the complexity of the work, the methods used, and the skills and training needed by personnel involved in carrying out the activity.
>
> NOTE 7 Documented procedures may make reference to work instructions that define how an activity is performed.

No further software-related guidance is provided.

## 4.2.3   Quality planning

> The supplier shall define and document how the requirements for quality will be met. Quality planning shall be consistent with all other requirements of a supplier's quality system and shall be documented in a format to suit the supplier's method of operation. The supplier shall give consideration to the following activities, as appropriate, in meeting the specified requirements for products, projects or contracts:
>
> a) the preparation of quality plans;
> b) the identification and acquisition of any controls, processes, equipment (including inspection and test equipment), fixtures, resources and skills that may be needed to achieve the required quality;
> c) ensuring the compatibility of the design, the production process, installation, servicing, inspection and test procedures and the applicable documentation.
> d) the updating, as necessary, of quality control, inspection and testing techniques, including the development of new instrumentation;
> e) the identification of any measurement requirement involving capability that exceeds the known state of the art, in sufficient time for the needed capability to be developed;
> f) the identification of suitable verification at appropriate stages in the realization of product;
> g) the clarification of standards of acceptability for all features and requirements, including those which contain a subjective element;
> h) the identification and preparation of quality records (see 4.16).
>
> NOTE 8 The quality plans referred to [see 4.2.3a)] may be in the form of a reference to the appropriate documented procedures that form an integral part of the supplier's quality system.

Quality planning should address the following items, as appropriate:

a) quality requirements, expressed in measurable terms, where appropriate;
b) the life cycle model to be used for software development;
c) defined criteria for starting and ending each project phase;
d) identification of types of reviews, tests and other verification and validation activities to be carried out;
e) identification of configuration management procedures to be carried out;
f) detailed planning (including schedules, procedures, resources and approval) and specific responsibilities and authorities for:
   - configuration management,
   - verification and validation of developed products, verification and validation of purchased products, verification of customer-supplied products,
   - control of nonconforming product and corrective action,
   - assuring that activities described in the quality plan are carried out.

A quality plan provides the means for tailoring the application of the quality system to a specific project, product or contract.

A quality plan may include or reference generic and/or project/product/contract-specific procedures, as appropriate. A quality plan should be updated along with the progress of the development, and items concerned with each phase should be completely defined when starting that phase.

A quality plan should be reviewed and agreed by all organizations concerned in its implementation.

The document that describes a quality plan may be an independent document (entitled quality plan), a part of another document, or composed of several documents.

A quality plan may include, or reference, the plans for unit, integration, system and acceptance tests. Guidance on test planning and the test environment is provided as part of inspection and testing .

NOTE: Guidance on quality plans is given in ISO 10005 and on configuration management in ISO 10007. For further information, see ISO/IEC 12207:1995, subclauses 6.2 to 6.5.

## 4.3 Contract Review

### 4.3.1 General

> The supplier shall establish and maintain documented procedures for contract review and for the coordination of these activities.

Software may be developed as part of a contract, as a product available for a market sector, as software embedded in a hardware product, or in support of the business processes of the supplier. Contract review is applicable in all these circumstances.

### 4.3.2 Review

> Before submission of a tender, or the acceptance of a contract or order (statement of requirement), the tender, contract or order shall be reviewed by the supplier to ensure that:
>
> a) the requirements are adequately defined and documented; where no written statement of requirement is available for an order received by verbal means, the supplier shall ensure that the order requirements are agreed before their acceptance;
> b) any differences between the contract or order requirements and those in the tender are resolved
> c) the supplier has the capability to meet the contract or order requirements.

The following concerns may also be relevant during the supplier's review of software tenders, contracts, or orders.

a) Customer-related concerns:
- the terminology to be used, is agreed by the relevant parties;
- the customer has the capability and resources to meet the defined contractual obligations;
- agreed criteria for customer to accept or reject product;
- the customer's responsibilities in the provision of data and related facilities;
- the extent to which the customer is to participate in joint development or in subcontracted work;
- the arrangements for joint reviews to monitor progress of the contract;
- the agreed procedure for handling changes in customer's requirements during the development and/or maintenance;
- life-cycle processes imposed by the customer;
- handling of problems detected after acceptance, including customer complaints and claims;
- the responsibility for removal of nonconformities after any warranty period;
- any obligations for the customer to upgrade to a subsequent version when the supplier dictates, or for the supplier to maintain historic versions;
- deployment and associated user training.
b) Technical concerns:
- the feasibility of meeting the requirements;
- the software development standards and procedures to be used;
- facilities, tools, software items and data, to be provided by the customer, are identified and methods defined and documented to assess their suitability for use;

- the operating system or hardware platform;
- agreement on the control of interfaces with the software product; replication and distribution requirements.

c) Management concerns:
- possible contingencies or risks are identified and the impact of these on subsequent activities are assessed;
- the supplier's responsibility with regard to subcontracted work;
- scheduling of progress, technical reviews and deliverables; installation, maintenance and support requirements;
- the timely availability of technical, human and financial resources

d) Legal, security and confidentiality concerns:
- information handled under the contract may be subject to concerns regarding Intellectual Property Rights, licence agreements, confidentiality and the protection of such information;
- guardianship of the master copy of the product and the rights of the customer to access or verify that master copy;
- the level of information disclosure to the customer needs to be mutually agreed to by the parties;
- definition of warranty terms;
- liabilities/penalties associated with the contract.

NOTE: For further information, see ISO/IEC 12207:1995, subclauses 5.2.1, 5.2.6 and 6.4.2.1.

### 4.3.3 Amendment to a contract

> The supplier shall identify how an amendment to a contract is made and correctly transferred to the functions concerned within the supplier's organization.

No further software-related guidance is provided.
NOTE: For further information, see ISO/IEC 12207:1995, subclauses 5.1.3.5 and 5.2.3.2.

### 4.3.4 Records

> Records of contract reviews shall be maintained (see 4.16).
>
> NOTE 9 Channels for communication and interfaces with the customer's organization in these contract matters should be established.

No further software-related guidance is provided.

## 4.4 Design Control

### 4.4.1 General

> The supplier shall establish and maintain documented procedures to control and verify the design of the product in order to ensure that the specified requirements are met.

This subclause provides guidance on the development activities of requirements analysis, architectural design, detailed design and coding. This subclause also contains guidance on development planning.

A software development project should be organized according to one or more life cycle models. Processes, activities and tasks should be planned and performed according to the nature of the life cycle model used. The life cycle model used may be adapted to suit particular project needs. This part of ISO 9000 is intended for application irrespective of the life cycle model used and is not intended to indicate a specific life cycle model.

A life cycle model identifies a set of processes and specifies when and how the processes are invoked. The sequence in which the processes are described in this part of ISO 9000 does not suggest in any way the sequence in which they are performed.

The development process is that which transforms the requirements into a software product. This process should be carried out in a disciplined manner, in order to prevent the introduction of errors. This approach reduces dependence on the verification and validation processes as the sole methods for identifying problems. The supplier should therefore establish and maintain documented procedures that ensure that the software products are developed in compliance with specified requirements and in accordance with the development plan and/or quality plan.

The following aspects inherent to the design activities should be taken into account:

a) Design method: a design method should be systematically used. Consideration should be given to the suitability of that method for the type of task, product or project and the compatibility of the application, the methods and the tools to be used.

b) Use of past experiences: utilizing lessons learned from past experiences, the supplier should avoid recurrences of the same or similar problems by applying lessons learned from previous projects, analysis of metrics and postproject reviews.

c) Subsequent processes: to the extent practical, the software product should be designed to facilitate testing, installation, maintenance and use.

d) Security and safety: special consideration should be given to the design for testability or validation. For products where failure may cause damage to people, property or the environment, design of such software should ensure definition of specific design requirements that specify desired immunity from, and system response to, potential failure conditions.

For coding, rules for the use of programming languages, consistent naming conventions, coding and adequate commentary rules should be specified and observed. Such rules should be documented and controlled.

The supplier may use tools, facilities and techniques in order to make the quality system guidelines in this part of ISO 9000 effective. These tools, facilities and techniques can be effective for management purposes as well as for product development and/or servicing. Whether these tools and techniques are developed internally, or purchased, the supplier should evaluate whether or not they are fit for purpose. Tools used in the implementation of the product, such as analysis and design tools, compilers, and assemblers should be approved and placed under an appropriate level of configuration management control, prior to use. The scope of use of such tools and techniques should be documented and their use reviewed as appropriate, to determine whether there is a need to improve and/or upgrade them.

Personnel may need training in the use of such tools and techniques, at commencement of usage, or after any improvements/upgrades.

### 4.4.2  Design and development planning

> The supplier shall prepare plans for each design and development activity. The plans shall describe or reference these activities, and define responsibility for their implementation. The design and development activities shall be assigned to qualified personnel equipped with adequate resources. The plans shall be updated as the design evolves.

For software products, development planning should determine the activities of requirements analysis, design, coding, integration, testing, installation and support for acceptance of software products. Development planning should be documented in a development plan.

A development plan should be reviewed and approved. A development plan may have other names such as "software development plan" or "software project plan".

A development plan may define how the project is to be managed, the progress reviews required and the type and frequency of reports to management, the customer, and other relevant parties, taking into account any contractual requirements.

Development planning should address the following items, as appropriate:

a) the definition of the project, including a statement of its objectives and reference to any related customer or supplier projects;
b) the definition of input and output of the project as a whole;
c) the organization of the project resources, including the team structure, responsibilities, use of subcontractors and material resources to be used;
d) organizational and technical interfaces between different individuals or groups, such as
   • sub-project teams,
   • subcontractors,
   • users,
   • customer representatives,
   • quality assurance representative;
e) the identification of, or reference to
   • development activities to be carried out,
   • required inputs to each activity,
   • required outputs from each activity,
   • management and supporting activities to be carried out;
f) the analysis of the possible risks, assumptions, dependencies and problems associated with the development;
g) the schedule identifying
   • the phases of the project,
   • the work to be performed (the inputs to, outputs from and definition of each task),
   • the associated resources and timing,
   • the associated dependencies,
   • the milestones;
h) the identification of
   • standards, rules, practices and conventions,
   • tools and techniques for development, including the qualification of, and configuration controls placed on, such tools and techniques,
   • configuration management practices,
   • method of controlling nonconforming products,
   • methods of control of nondeliverable software used to support development,
   • procedures for archiving, back-up and recovery, including contingency planning,
   • methods of control for virus protection;
i) the identification of related plans (including system level plans) such as
   • quality plan,
   • risk management plan,
   • configuration management plan,

- integration plan,
- test plan,
- installation plan,
- migration plan,
- training plan,
- maintenance plan,
- re-use plan.

The development plan and any of these related plans may be an independent document, a part of another document or composed of several documents.
NOTE: For further information, see ISO/IEC 12207:1995, subclause 5.2.4.

### 4.4.3  Organizational and technical interfaces

> Organizational and technical interfaces between different groups which input into the design process shall be defined and the necessary information documented, transmitted and regularly reviewed.

The boundaries of responsibility for each part of the software product and the way that technical information will be transmitted between all parties should be clearly defined in the development plans of suppliers or subcontractors. The supplier may require submission of a subcontractor's development plan, for review.

In defining interfaces, care should be taken to consider parties, other than the customer and supplier, who need input to the design, installation, maintenance and training activities. These may include subcontractors, regulatory authorities, associated development projects and help-desk staff. In particular, the end-users and any intermediate operations function may need to be involved to ensure that appropriate capacity and training are available to achieve committed service levels.

The customer may have certain responsibilities under the contract. Particular concerns include the need for the customer to cooperate with the supplier, to provide necessary information in a timely manner, and resolve action items. Where a customer representative is appropriate, he/she may represent the eventual user of the product, as well as executive management, and have the authority to deal with contractual matters which include, but are not limited to, the following:

a) defining the customer's requirements to the supplier;
b) answering questions from the supplier;
c) approving the supplier's proposals;
d) concluding agreements with the supplier;
e) ensuring that the customer's organization observes the agreements made with the supplier;
f) defining the acceptance criteria and procedures;
g) dealing with customer-supplied software items, data, facilities and tools that are found unsuitable for use;
h) defining the customer's responsibility.

Where mutually agreed, joint reviews involving the supplier and customer may be scheduled on a regular basis, or at significant project events, to cover the following aspects, as appropriate:

a) the progress of software development work undertaken by the supplier;
b) the progress of agreed activities being undertaken by the customer;

c) conformance of the development products to the customer's agreed requirements specification;

d) the progress of activities concerning the eventual users of the system under development, such as system conversion and training;

e) verification results;

f) acceptance test results.

NOTE: For further information, see ISO/IEC 12207:1995, subclauses 5.2.6.1 and 6.6.2.

### 4.4.4   Design input

> Design input requirements relating to the product, including applicable statutory and regulatory requirements, shall be identified, documented and their selection reviewed by the supplier for adequacy. Incomplete, ambiguous or conflicting requirements shall be resolved with those responsible for imposing these requirements.
>
> Design input shall take into consideration the results of any contract review activities.

The requirements specification should be provided by the customer. However, where mutually agreed, the supplier may provide the specification. In such a case, the supplier should, where appropriate:

a) establish documented procedures for developing the requirements specification, including
  - methods for agreeing on requirements and authorizing changes, especially during iterative development, methods for the evaluation of prototypes or demonstrations, where used,
  - recording and reviewing discussion results on both sides,

b) develop the requirements specification in close cooperation with the customer and make efforts to prevent misunderstandings by, for example, provision of definition of terms, explanation of the background of requirements,

c) obtain the customer's approval of the requirements specification.

Interviews, surveys, studies, prototypes, demonstrations and analysis methods may be used for developing the requirements specification, as appropriate.

The requirements specification may be provided and agreed in the form of a system specification. In this case, procedures should be in place to ensure the correct allocation of system requirements into hardware and software aspects and also the appropriate interface specifications.

The requirements specification may not be fully defined at contract acceptance, and may be developed during a project. The contract may be amended when the requirements specification changes. Changes to the requirements specification should be controlled.

The requirements should include all aspects necessary to satisfy the customer's agreed needs. The requirements specification may need to take the operational environment into account. The requirements may include, but not be limited to the following characteristics: functionality; reliability; usability; efficiency; maintainability and portability (see also ISOIEC 9126). Sub-characteristics may be specified, for example security. Safety considerations and statutory obligations may also be specified.

If the software product needs to interface with other software or hardware products, the interfaces between the software product to be developed and other software or hardware products should be specified, as far as possible, either directly or by reference, in the requirements specification.

The requirements should be expressed in terms which allow validation during product acceptance.

NOTE: For further information, see ISO/IEC 12207:1995, subclauses 5.3.2 to 5.3.4.

## 4.4.5 Design output

> Design output shall be documented and expressed in terms that can be verified and validated against design input requirements.
>
> Design output shall:
>
> a) meet the design input requirements;
> b) contain or make reference to acceptance criteria;
> c) identify those characteristics of the design that are crucial to the safe and proper functioning of the product (e.g., operating, storage, handling, maintenance and disposal requirements).
>
> Design output documents shall be reviewed before release.

The required output from the design activity should be defined and documented in accordance with the chosen method. This documentation should be correct, complete and consistent with the requirements. Design outputs may include:

- architectural design specification;
- detailed design specification;
- source code;
- user guides.

NOTE: For further information, see ISO/IEC 12207:1995, subclauses 5.3.5 to 5.3.7.

## 4.4.6 Design review

> At appropriate stages of design, formal documented reviews of the design results shall be planned and conducted. Participants at each design review shall include representatives of all functions concerned with the design stage being reviewed, as well as other specialist personnel, as required. Records of such reviews shall be maintained (see 4.16).

The supplier should plan and implement review processes for all software development projects. The degree of formality and rigour of the activities associated with the review processes should be appropriate to the complexity of the product and the degree of risk associated with the specified use of the software product. The supplier should establish documented procedures for dealing with process and product deficiencies or nonconformities identified during these activities.

During design reviews, aspects inherent to the design activities should be taken into account, for example feasibility, security and safety, programming rules and testability.

The results of review, and any further activities required to ensure that the specified requirements are met, should be recorded and checked when they are completed.

Most design reviews during the development are scheduled, but there may also be unscheduled design reviews.

A documented design review procedure should address the following:

a) what is to be reviewed, when, and the type of review;
b) what functional groups would be concerned in each type of review and, if there is to be a review meeting, who would chair it;
c) what records have to be produced, for example meeting minutes, issues, problems, actions and action status.

Also the following may be addressed in the design review procedure:

a) the methods for monitoring the application of rules, practices and conventions to ensure compliance, such as peer reviews, walkthroughs, code inspections;
b) what has to be done prior to the conduct of a review, such as establishment of objectives, meeting agenda, documents required and roles of review personnel;
c) what has to be done during the review, including the techniques to be used and guidelines for all participants;
d) the success criteria for the review;
e) what follow-up methods are to be used to ensure that issues identified at the review are resolved.

Where specified in the contract, the supplier should conduct design review meetings in cooperation with the customer. Both parties should agree on the results of such reviews.

It is recommended that further design activities should proceed only when the consequences of all known deficiencies are satisfactorily resolved, or the risk of proceeding otherwise is known.
NOTE: For further information, see ISO/IEC 12207:1995, subclauses 5.3.4.2, 5.3.5.6, 5.3.6.7 and 6.6.3.

### 4.4.7   Design verification

> At appropriate stages of design, design verification shall be performed to ensure that the design stage output meets the design stage input requirements. The design verification measures shall be recorded (see 4.16).
>
> NOTE 10 In addition to conducting design reviews (see 4.4.6), design verification may include activities such as
>
> • performing alternative calculations,
> • comparing the new design with a similar proven design, if available,
> • undertaking tests and demonstrations, and
> • reviewing the design stage documents before release.

Verification of design should be performed as appropriate during the development process. Design verification may comprise reviews of design output, demonstrations including prototypes and simulations, or tests. Verification may be conducted on the output from other development activities. These verification activities should be planned and conducted in accordance with the quality plan or documented procedures to ensure that process outputs meet the process input requirements.

The verification results and any further actions required to ensure that the design stage input requirements are met should be recorded and checked when the actions are completed.

Only verified design outputs should be submitted for acceptance and subsequent use. Any findings should be adequately addressed and resolved.
NOTE: For further information, see ISO/IEC 12207:1995, subclauses 5.3.4.2, 5.3.5.6, 5.3.5.7, 5.3.7.5, 5.3.9 and 6.4.

### 4.4.8   Design validation

> Design validation shall be performed to ensure that product conforms to defined user needs and/or requirements.
>
> NOTES
>
> 11   Design validation follows successful design verification (see 4.4.7).
> 12   Validation is normally performed under defined operating conditions.
> 13   Validation is normally performed on the final product, but may be necessary in earlier stages prior to product completion.
> 14   Multiple validations may be performed if there are different intended uses.

Before offering the product for customer acceptance, the supplier should validate the product in accordance with its specified intended use, for example during final inspection and test.

In software development, it is important that the validation results and any further actions required to ensure that the specified requirements are met should be recorded and checked when the actions are completed. Only validated products should be submitted for acceptance and subsequent use.
NOTE: For further information, see ISO/IEC 12207:1995, subclauses 5.3.1 and 6.5.

## 4.4.9 Design changes

> All design changes and modifications shall be identified, documented, reviewed and approved by authorized personnel before their implementation.

The supplier should establish and maintain procedures for controlling the implementation of any design changes, which may arise at any time during the product life cycle, in order to:

a) document and justify the change;
b) evaluate consequences of the change;
c) approve or disapprove the change;
d) implement and verify the change.

In the software development environment, control of design changes is usually addressed under the discipline of configuration management.
NOTE: For further information, see ISO/IEC 12207:1995, subclauses 5.5.2, 5.5.3 and 6.2.3.

## 4.5 Document and Data Control

### 4.5.1 General

> The supplier shall establish and maintain documented procedures to control all documents and data that relate to the requirements of this International Standard including, to the extent applicable, documents of external origin such as standards and customer drawings.
>
> NOTE 15 Documents and data can be in the form of any type of media, such as hard copy or electronic media.

Configuration management procedures may be used to implement document and data control. In establishing the procedures to control documents and data, the supplier should identify those documents and data, including those of external origin such as standards and customer data, which should be subject to the control procedures.

The document and data control procedures should be applied to relevant documents and data, including the following:

a) contractual documents including specification of requirements;
b) procedural documents describing the quality system to be applied in the software life cycle;
c) planning documents describing the planning and progress of activities of the supplier and the suppliers interactions with the customer;
d) product documents and data describing or associated with a particular software product.

NOTE: For further information, see ISO/IEC 12207, clause 6.1.

## 4.5.2   Document and data approval and issue

> The documents and data shall be reviewed and approved for adequacy by authorized personnel prior to issue. A master list or equivalent document control procedure identifying the current revision status of documents shall be established and be readily available to preclude the use of invalid and/or obsolete documents.
>
> This control shall ensure that:
>
> a) the pertinent issues of appropriate documents are available at all locations where operations essential to the effective functioning of the quality system are performed;
> b) invalid and/or obsolete documents are promptly removed from all points of issue or use, or otherwise assured against unintended use;
> c) any obsolete documents retained for legal and/or knowledge-preservation purposes are suitably identified.

Where document control is achieved by electronic means, special attention should be given to appropriate approval, access, distribution, media and archiving procedures.

## 4.5.3   Document and data changes

> Changes to documents and data shall be reviewed and approved by the same functions/organizations that performed the original review and approval, unless specifically designated otherwise. The designated functions/organizations shall have access to pertinent background information upon which to base their review and approval.
>
> Where practicable, the nature of the change shall be identified in the document or the appropriate attachments.

No further software-related guidance is provided.

## 4.6   Purchasing

### 4.6.1   General

> The supplier shall establish and maintain documented procedures to ensure that purchased product (see 3.1) conforms to specified requirements.

In developing, supplying, installing and maintaining software products, purchased products may include:

- commercial off-the-shelf software;
- subcontracted development;
- computer and communications hardware;
- a tool intended to assist in the development of software;
- contract staff;
- maintenance and customer support services;
- training courses and materials.

NOTE: For further information, see ISO/IEC 12207:1995, subclause 5.1.

## 4.6.2   Evaluation of subcontractors

> The supplier shall:
>
> a) evaluate and select subcontractors on the basis of their ability to meet subcontract requirements including the quality system and any specific quality assurance requirements;
> b) define the type and extent of control exercised by the supplier over subcontractors. This shall be dependent upon the type of product, the impact of subcontracted product on the quality of final product and, where applicable, on the quality audit reports and/or quality records of the previously demonstrated capability and performance of subcontractors;
> c) establish and maintain quality records of acceptable subcontractors (see 4.16).

No further software-related guidance is provided.

## 4.6.3   Purchasing data

> Purchasing documents shall contain data clearly describing the product ordered, including where applicable:
>
> a) the type, class, grade or other precise identification;
> b) the title or other positive identification, and applicable issues of specifications, drawings, process requirements, inspection instructions and other relevant technical data, including requirements for approval or qualification of product, procedures, process equipment and personnel;
> c) the title, number and issue of the quality system standard to be applied.
>
> The supplier shall review and approve purchasing documents for adequacy of specified requirements prior to release.

Purchasing documents for software development should contain data clearly describing the product ordered, including, where applicable:

a) precise identification of the product ordered, such as product name and/or product number;
b) requirements specification, or the identity of it (or the procedure to identify requirements specifications where not fixed at the time ordered);
c) standards to be applied (e.g. communications protocol, architectural specification);
d) procedures and/or work instructions;
e) development environment;
f) requirements on personnel.

The considerations under contract review may also be applied to subcontracts.

## 4.6.4   Verification of purchased product

### 4.6.4.1   Supplier verification at subcontractor's premises

> Where the supplier proposes to verify purchased product at the subcontractor's premises, the supplier shall specify verification arrangements and the method of product release in the purchasing documents.

No further software-related guidance is provided.

### 4.6.4.2   *Customer verification of subcontracted product*

> Where specified in the contract, the supplier's customer or the customer's representative shall be afforded the right to verify at the subcontractor's premises and the supplier's premises that subcontracted product conforms to specified requirements. Such verification shall not be used by the supplier as evidence of effective control of quality by the subcontractor.
>
> Verification by the customer shall not absolve the supplier of the responsibility to provide acceptable product, nor shall it preclude subsequent rejection by the customer.

No further software-related guidance is provided.

## 4.7   Control of Customer-Supplied Product

> The supplier shall establish and maintain documented procedures for the control of verification, storage and maintenance of customer-supplied product provided for incorporation into the supplies or for related activities. Any such product that is lost, damaged or is otherwise unsuitable for use shall be recorded and reported to the customer (see 4.16).
>
> Verification by the supplier does not absolve the customer of the responsibility to provide acceptable product.

The supplier may be required to acquire and include product, including data, supplied by the customer. For example:

a) software products including commercial software product supplied by the customer;
b) development tools;
c) development environment including network services;
d) test and operational data;
e) interface or other specifications;
f) hardware;
g) customer proprietary information, including specifications.

Consideration should be given to the requirements of the contract in addressing required licensing and the support of such software product, in any maintenance agreement related to the product to be delivered.

The means by which updates to customer-supplied items are accepted and integrated should be defined. The supplier may apply the same kinds of verification activities to customer-supplied product as to purchased product.

NOTE: For further information, see ISO/IEC 12207:1995, subclause 6.1.

## 4.8   Product Identification and Traceability

> Where appropriate, the supplier shall establish and maintain documented procedures for identifying the product by suitable means from receipt and during all stages of production, delivery and installation.
>
> Where and to the extent that traceability is a specified requirement, the supplier shall establish and maintain documented procedures for unique identification of individual product or batches. This identification shall be recorded (see 4.16).

The supplier should establish and maintain procedures for identifying software items during all phases, starting from specification through development, replication and delivery. Where required by contract, these procedures may also apply after delivery of the product.

Throughout the product life cycle, there should be procedures to trace the components of the software item or product. Such tracing may vary in scope according to the requirements of the contract or marketplace, from being able to place a certain change request in a specific release, to recording the destination and usage of each variant of the product.

In software, a means by which identification and traceability may be achieved, is configuration management. Configuration management is a management discipline that applies technical and administrative direction to the development and support life cycle of configuration items, including software items. This discipline is also applicable to related documentation and hardware. Use of configuration management is dependent on the project size and complexity, and the risk level.

One objective of configuration management is to document and provide full visibility of the product's present configuration and on the status of achievement of its requirements. Another objective is that everyone working on the product at any time in its life cycle uses correct and accurate information.

A configuration management system may provide the capability to:

a) identify uniquely the versions of each software item;
b) identify the versions of each software item which together constitute a specific version of a complete product;
c) identify the build status of software products under development, delivered or installed;
d) control simultaneous updating of a given software item by two or more people working independently;
e) provide coordination for the updating of multiple products in one or more locations as required;
f) identify and track all actions and changes resulting from a change request, or problem, from initiation through to release.

The supplier should identify the configuration with the following:

a) product structure and selection of configuration items;
b) documentation and computer files;
c) naming conventions;
d) establishment of configuration baselines.

The products that may be managed by a configuration management system include:

a) documents and data pertaining to contract, process, planning and product;
b) source, object and executable code;
c) incorporated products including
   • software tools,
   • re-used software including libraries,
   • purchased software,
   • customer supplied software.

Procedures should be applied to ensure that the following can be identified for each software item:

a) the documentation;
b) any associated development tools;

c) interfaces to other software items and to hardware;
d) the hardware and software environment.

The supplier should establish and maintain configuration status accounting procedures to record, manage and report on the status of software items, of change requests and of the implementation of approved changes.

The supplier should develop and implement a configuration management plan which includes the following:

a) organizations involved in configuration management and responsibilities assigned to each of them;
b) configuration management activities to be carried out;
c) configuration management tools, techniques and methodologies to be used;
d) the point at which items should be brought under configuration control.

NOTE: For further information on configuration management, refer to ISO 10007, and ISO/IEC 12207:1995, subclauses 6.1 and 6.2.

## 4.9  Process Control

> The supplier shall identify and plan the production, installation and servicing processes which directly affect quality and shall ensure that these processes are carried out under controlled conditions. Controlled conditions shall include the following:
>
> a) documented procedures defining the manner of production, installation and servicing, where the absence of such procedures could adversely affect quality;
> b) use of suitable production, installation and servicing equipment, and a suitable working environment;
> c) compliance with reference standards/codes, quality plans and/or documented procedures;
> d) monitoring and control of suitable process parameters and product characteristics;
> e) the approval of processes and equipment, as appropriate;
> f) criteria for workmanship, which shall be stipulated, in the clearest practicable manner (e.g., written standards, representative samples or illustrations);
> g) suitable maintenance of equipment to ensure continuing process capability.
>
> Where the results of processes cannot be fully verified by subsequent inspection and testing of the product and where, for example, processing deficiencies may become apparent only after the product is in use, the processes shall be carried out by qualified operators and/or shall require continuous monitoring and control of process parameters to ensure that the specified requirements are met.
>
> The requirements for any qualification of process operations, including associated equipment and personnel (see 4.18), shall be specified.
>
> NOTE 16 Such processes requiring pre-qualification of their process capability are frequently referred to as special processes.
>
> Records shall be maintained for qualified processes, equipment and personnel, as appropriate (see 4.16).

As stated in guidance for the design control element of ISO 9001, a software development project should be organized according to a set of processes which transform the requirements into a software product. The "process control" element, as applied to software development, is applicable to the replication, delivery and installation of software items or products.

Where required by contract, the supplier should establish and perform the replication procedures, considering the following, to ensure that replication is conducted correctly:

a) identification of the master and the copies including format, variant and version;
b) the number of copies of each software item to be delivered;
c) disaster recovery plans including custody of master and back-up copies where applicable;
d) the period of obligation of the supplier to supply copies and the capability of reading master copies;
e) the type of media for each software item and associated labelling;
f) checks against the possibility of software viruses;
g) the stipulation of required documentation such as manuals and user guides, including identification and packaging;
h) copyright and licensing concerns addressed and agreed;
i) controlling the environment under which the replication is effected to ensure repeatability.

For software product releases, the supplier and customer should agree and document procedures for initial subsequent releases.

It is recommended that the release of software should establish a baseline that records the tests completed and the resolution of identified deficiencies. Quantitative analysis, to predict system reliability, may be carried out for software with safety and/or security requirements.

These procedures should include the following:

a) descriptions of the types (or classes) of release, depending on the frequency and/or impact on the customer's operations and ability to implement changes at any point in time;
b) methods by which the customer will be advised of current or planned future changes;
c) methods to confirm that changes implemented will not introduce other problems; such methods should include the determination of the level of regression testing to be applied to each release;
d) ground rules to determine where localized temporary fixes may be incorporated or release of a complete updated copy of the software product is necessary;
e) requirements for records indicating which changes have been implemented and at what locations, for multiple products and sites.

When installation of the software product is contractually required, the supplier and customer should agree on their respective roles, responsibilities and obligations, and such agreements should be documented. In providing for installation, consideration should be given to the following:

a) the need for validation at each installation required by contract;
b) an installation procedure;
c) a procedure for approval of each installation upon completion;
d) a schedule;
e) access to customer's facilities (e.g. security badges, passwords, escorts);
f) availability of skilled personnel;
g) availability of and access to customer's systems and equipment;
h) identification of what the customer is to provide at the site;
i) training for the use of new facilities.

NOTE: For further information, see ISO/IEC 12207:1995, subclauses 5.3.12 and 6.3.3.

## 4.10  Inspection and Testing

### 4.10.1  General

> The supplier shall establish and maintain documented procedures for inspection and testing activities in order to verify that the specified requirements for the product are met. The required inspection and testing, and the records to be established, shall be detailed in the quality plan or documented procedures.

Testing may be required at several levels, from the individual software item to the complete software product. There are several different approaches to testing, and the extent of testing, the degree of controls on the test environment, test inputs and test outputs, may vary with the approach, complexity of the product, and the risks. Software testing is also performed during software integration. The techniques described under Design review may also be relevant in inspection and testing activities.

The supplier should establish, document and review plans for unit, integration, system and acceptance tests, in accordance with the quality plan or documented procedures, covering, as appropriate:

a) test objectives;

b) configurations to be tested;

c) types of tests to be performed (e.g. functional tests, boundary tests, performance tests, usability tests);

d) sequence of tests, test cases, test procedures, test data and expected results;

e) scope of tests to be performed, in terms of coverage and volumes;

f) relevancy of the tests to the test objectives and to operational use;

g) special concerns such as security and safety;

h) test environment, tools and test software, including any associated qualification and controls;

i) testing of end-user documentation;

j) personnel required and associated training requirements, including training material;

k) the degree of independence between the personnel developing the software and the personnel performing the tests;

l) the responsibilities for specification and performance of the tests;

m) the test completion criteria;

n) the method of recording results;

o) procedures for analysing and approving results;

p) procedures for handling problems found during test execution, including suspension criteria and resumption requirements;

q) the need for, and extent of, regression tests;

r) the repeatability of the tests.

NOTE: For further information, see ISO/IEC 12207:1995, subclauses 5.1.5, 5.3.5.5, 5.3.6.5, 5.3.6.6, 5.3.7, 5.3.11 and 5.3.13.

### 4.10.2  Receiving inspection and testing

> **4.10.2.1** The supplier shall ensure that incoming product is not used or processed (except in the circumstances described in 4.10.2.3) until it has been inspected or otherwise verified as conforming to specified requirements. Verification of conformance to the specified requirements shall be in accordance with the quality plan and/or documented procedures.
>
> **4.10.2.2** In determining the amount and nature of receiving inspection, consideration shall be given to the amount of control exercised at the subcontractor's premises and recorded evidence of conformance provided.

> **4.10.2.3** Where incoming product is released for urgent production purposes prior to verification, it shall be positively identified and recorded (see 4.16) in order to permit immediate recall and replacement in the event of nonconformity to specified requirements.

The supplier may be required to acquire and include software product, including data, supplied by a third party. The supplier should establish and maintain documented procedures for verification (upon receipt) of such product, taking into account the requirements of the contract.

The supplier may apply the same kinds of verification activities to customer-supplied product as to purchased product.

### 4.10.3 In-process inspection and testing

> The supplier shall:
>
> a) inspect and test product as required by the quality plan and/or documented procedures;
> b) hold product until the required inspection and tests have been completed or necessary reports have been received and verified, except when product is released under positive-recall procedures (see 4.10.2.3). Release under positive-recall procedures shall not preclude the activities outlined in 4.10.3a).

The general considerations for testing apply.

### 4.10.4 Final inspection and testing

> The supplier shall carry out all final inspection and testing in accordance with the quality plan and/or documented procedures to complete the evidence of conformance of the finished product to the specified requirements.
>
> The quality plan and/or documented procedures for final inspection and testing shall require that all specified inspection and tests, including those specified either on receipt of product or in-process, have been carried out and that the results meet specified requirements.
>
> No product shall be dispatched until all the activities specified in the quality plan and/or documented procedures have been satisfactorily completed and the associated data and documentation are available and authorized.

Before offering the product for customer acceptance, the supplier should validate the operation of the product in accordance with its specified intended use, under conditions similar to the application environment, as specified in the contract. Any differences between the validation environment and the actual application environment, and the risks associated with such differences, should be identified and justified as early in the life cycle as possible, and recorded.

In the course of validation, configuration audits or evaluations may be performed, where appropriate, before release of a configuration baseline to confirm, by examination of the review, inspection and test records, that the software product complies with its contractual or specified requirements.

When considering the test environment, the following concerns should be addressed:

a) the features to be tested;
b) the controls to be placed on the test environment, including the test tools;
c) any limitations placed on the tests, by the environment.

Where testing in the target environment is required, the following concerns should be addressed:

a) the specific responsibilities of the supplier and customer for carrying out and evaluating the test;
b) restoration of the user environment (after test).

Acceptance test support may be required when the supplier is ready to deliver the validated product. The customer should judge whether or not the product is acceptable according to previously agreed criteria and in a manner specified in the contract. Acceptance tests should be performed by the customer or may be performed on behalf of the customer by the supplier or a third party. The supplier should cooperate in acceptance activities with the customer as stipulated in the contract.

When acceptance testing, if required by the contract, is performed by the supplier, the acceptance test activities may be recognized as relating to final inspection and test, and validation. In some instances, validation, field testing and acceptance testing may be one and the same activity.

Before carrying out acceptance activities, the supplier should assist the customer to identify the following:

a) time schedule;
b) procedures for evaluation, including acceptance criteria;
c) software/hardware environments, including the controls on them;
d) human resources required and associated training.

The method of handling problems detected during the acceptance procedure and their disposition should be agreed between the customer and supplier and documented.

### 4.10.5  Inspection and test records

> The supplier shall establish and maintain records which provide evidence that the product has been inspected and/or tested. These records shall show clearly whether to product has passed or failed the inspections and/or tests according to defined acceptance criteria. Where the product fails to pass any inspection and/or test, the procedures for control of nonconforming product shall apply (see 4.13).
>
> Records shall identify the inspection authority responsible for the release of product (see 4.16).

The supplier should ensure that the test results are recorded as defined in the relevant specification.

## 4.11  Control of Inspection, Measuring and Test Equipment

### 4.11.1  General

> The supplier shall establish and maintain documented procedures to control, calibrate and maintain inspection, measuring and test equipment (including test software), used by the supplier to demonstrate the conformance of product to the specified requirements. Inspection, measuring and test equipment shall be used in a manner which ensures that the measurement uncertainty is known and is consistent with the required measurement capability.
>
> Where test software or comparative references such as test hardware are used as suitable forms of inspection, they shall be checked to prove that they are capable of verifying the acceptability of product, prior to release for use during production, installation or servicing and shall be rechecked at prescribed intervals. The supplier shall establish the extent and frequency of such checks and shall maintain records as evidence of control (see 4.16).
>
> Where the availability of technical data pertaining to the inspection, measuring and test equipment is a specified requirement, such data shall be made available, when required by the customer or customer's representative, for verification that the inspection, measuring and test equipment is functionally adequate.
>
> NOTE 17 For the purposes of this International Standard, the term "measuring equipment" includes measurement devices.

Where the supplier uses tools, facilities and techniques, in the conduct of any tests verifying conformance of the software product to specified requirements, the supplier should consider the effect of such tools on the quality of the software product, when approving them. In addition, such tools may be placed under configuration management prior to use.

The scope of use of test tools and techniques should be documented and their use reviewed at defined intervals, to determine if there is a need to improve and/or upgrade them.

NOTE: For further information, see ISO/IEC 12207:1995, subclause 7.2.

### 4.11.2   Control procedure

The supplier shall:

a) determine the measurements to be made and the accuracy required, and select the appropriate inspection, measuring and test equipment that is capable of the necessary accuracy and precision;

b) identify all inspection, measuring and test equipment that can affect product quality, and calibrate and adjust them at prescribed intervals, or prior to use, against certified equipment having a known valid relationship to internationally or nationally recognized standards. Where no such standards exist, the basis used for calibration shall be documented;

c) define the process employed for the calibration of inspection, measuring and test equipment, including details of equipment type, unique identification, location, frequency of checks, check method, acceptance criteria and the action to be taken when results are unsatisfactory;

d) identify inspection, measuring and test equipment with a suitable indicator or approved identification record to show the calibration status;

e) maintain calibration records for inspection, measuring and test equipment (see 4.16);

f) assess and document the validity of previous inspection and test results when inspection, measuring and test equipment is found to be out of calibration;

g) ensure that the environmental conditions are suitable for the calibrations, inspections, measurements and tests being carried out;

h) ensure that the handling, preservation and storage of inspection, measuring and test equipment is such that the accuracy and fitness for use are maintained;

i) safeguard inspection, measuring and test facilities, including both test hardware and test software, from adjustments which would invalidate the calibration setting.

NOTE 18 The metrological confirmation system for measuring equipment given in ISO 10012 may be used for guidance.

Calibration is a verification technique that is not directly applicable to software. However, it may be applicable to hardware and tools used to test and validate the software. Consequently, items b) to f) above are not applicable to the software itself but may be applicable to the environment used when testing the software.

## 4.12   Inspection and Test Status

The inspection and test status of product shall be identified by suitable means, which indicate the conformance or nonconformance of product with regard to inspection and tests performed. The identification of inspection and test status shall be maintained, as defined in the quality plan and/or documented procedures, throughout production, installation and servicing of the product to ensure that only product that has passed the required inspections and tests [or released under an authorized concession (see 4.13.2)] is dispatched, used or installed.

The supplier should have a means of identifying the stage of development of the components of the product and the test status. For example: untested; tested with error; tested successfully or approved for release to any further development activity. The creation of builds or the movement of software items between development, test and operational environments may serve to indicate this status. Inspection and test records may be used to identify inspection and test status.

NOTE: For further information, see ISO/IEC 12207:1995, subclause 6.2.

## 4.13    Control of Nonconforming Product

### 4.13.1    General

> The supplier shall establish and maintain documented procedures to ensure that product that does not conform to specified requirements is prevented from unintended use or installation. This control shall provide for identification, documentation, evaluation, segregation (when practical), disposition of nonconforming product, and for notification to the functions concerned.

In software development, segregation of nonconforming items may be effected by transferring the item out of a production or a testing environment, into a separate environment. In the case of embedded software it may become necessary to segregate the nonconforming item (hardware) which contains the nonconforming software.

The supplier should identify at what points control and recording of nonconforming product is required. Where a software item manifests a defect during the development or maintenance process, the investigation and resolution of such defects should be controlled and recorded.

A configuration management process may be invoked to implement part or the whole of this requirement.

NOTE: For further information, see ISO/IEC 12207:1995, subclauses 6.2 and 6.8.

### 4.13.2    Review and disposition of nonconforming product

> The responsibility for review and authority for the disposition of nonconforming product shall be defined.
>
> Nonconforming product shall be reviewed in accordance with documented procedures. It may be:
>
> a) reworked to meet the specified requirements;
> b) accepted with or without repair by concession;
> c) regraded for alternative applications; or
> d) rejected or scrapped.
>
> Where required by the contract, the proposed use or repair of product [see 4.13.2b)] which does not conform to specified requirements shall be reported for concession to the customer or customer's representative. The description of the nonconformity that has been accepted, and of repairs, shall be recorded to denote the actual condition (see 4.16).
>
> Repaired and/or reworked product shall be re-inspected in accordance with the quality plan and/or documented procedures.

Attention should be paid to the following aspects, in the disposition of nonconformities:

a) any discovered problems and their possible impacts to any other parts of the software should be noted and those responsible notified so the problems can be tracked until they are resolved;
b) areas impacted by any modifications should be identified and re-tested, and the method for determining the scope of re-testing should be identified in a documented procedure;
c) the priority of the non conformities.

With software, repair or rework to achieve fulfilment of specified requirements creates a new software version. In software development, disposition of nonconforming product may be achieved by:

a) repair or rework (i.e. to fix defects) to meet the requirement;
b) acceptance with or without repair by concession;

c) treatment as a conforming product after the amendment of requirements;

d) rejection.

## 4.14 Corrective and Preventive Action

### 4.14.1 General

> The supplier shall establish and maintain documented procedures for implementing corrective and preventive action.
>
> Any corrective or preventive action taken to eliminate the causes of actual or potential nonconformities shall be to a degree appropriate to the magnitude of problems and commensurate with the risks encountered.
>
> The supplier shall implement and record any changes to the documented procedures resulting from corrective and preventive action.

Where corrective action directly affects the software product, the configuration management process may be invoked to manage the changes. Corrective actions that involve changes to the software life cycle processes should be reviewed by management and implemented by means of document and data control procedures.

NOTE: For further information, see ISO/IEC 12207:1995, subclauses 6.2, 6.8 and 7.3.

### 4.14.2 Corrective action

> The procedures for corrective action shall include:
>
> a) the effective handling of customer complaints and reports of product nonconformities;
> b) investigation of the cause of nonconformities relating to product, process and quality system, and recording the results of the investigation (see 4.16);
> c) determination of the corrective action needed to eliminate the cause of nonconformities;
> d) application of controls to ensure that corrective action is taken and that it is effective.

No further software-related guidance is provided.

### 4.14.3 Preventive action

> The procedures for preventive action shall include:
>
> a) the use of appropriate sources of information such as processes and work operations which affect product quality, concessions, audit results, quality records, service reports and customer complaints to detect, analyze and eliminate potential causes of non conformities;
> b) determination of the steps needed to deal with any problems requiring preventive action;
> c) initiation of preventive action and application of controls to ensure that it is effective;
> d) ensuring that relevant information on actions taken is submitted for management review (see 4.1.3).

Analysis of the root causes of nonconformities may provide input to preventive action. The measures taken to reverse unfavorable trends in metric levels may be considered as preventive actions.

## 4.15 Handling, Storage, Packaging, Preservation and Delivery

### 4.15.1 General

> The supplier shall establish and maintain documented procedures for handling, storage, packaging, preservation and delivery of product.

No further software-related guidance is provided.

NOTE: For further information, see ISO/IEC 12207:1995, subclauses 5.2.7.1, 5.3.13.2 and 6.2.6.

### 4.15.2    Handling

> The supplier shall provide methods of handling product that prevent damage or deterioration.

Damage to software means alteration of contents. Software infection by computer virus should be treated as damaged software.

Software information does not deteriorate; however, the media on which it is stored may be subject to deterioration, and appropriate precautions should be taken by the supplier.

Virus protection requirements, as applicable to software products designated for delivery, are described as part of the guidance on replication.

### 4.15.3    Storage

> The supplier shall use designated storage areas or stock rooms to prevent damage or deterioration of product, pending use or delivery. Appropriate methods for authorizing receipt to and dispatch from such areas shall be stipulated.
>
> In order to detect deterioration, the condition of product in stock shall be assessed at appropriate intervals.

A system should be established for:

  a)  storing software items;
  b)  controlling access to software items;
  c)  maintaining versions of products in established baselines.

To protect the integrity of the product and provide a basis for the control of change, it is essential that software items be held in an environment which:

  a)  protects them from unauthorized change or corruption;
  b)  permits the controlled access to and retrieval of the master and any copies.

Consideration should be given to the storage of computer media, particularly with respect to the electromagnetic and electrostatic environment.

### 4.15.4    Packaging

> The supplier shall control packing, packaging and marking processes (including materials used) to the extent necessary to ensure conformance to specified requirements.

Packaging requirements, as applicable to software products, designated for delivery, are described as part of the guidance on replication. In cases where electronic storage is used there may be no physical activity related to this clause. During packaging, software may be compressed and/or encrypted.

### 4.15.5    Preservation

> The supplier shall apply appropriate methods for preservation and segregation of product when the product is under the supplier's control.

A system should be established to provide the capability to preserve software items, for example:

a) regular back-up of software;
b) ensuring the timely copying of software to replacement media;
c) the storage of software media in a protected environment;
d) the storage of software media in redundant environments to ensure disaster recovery.

## *4.15.6   Delivery*

> The supplier shall arrange for the protection of the quality of product after final inspection and test. Where contractually specified, this protection shall be extended to include delivery to destination.

Delivery may be achieved by physical movement of media containing software, or by electronic transmission. In cases where electronic transmission is used, consideration should be given to protection against damage by viruses.

Documented procedures should be established and maintained for verifying the correctness and completeness of copies of the software product delivered. These procedures should provide for appropriate preventive action to protect the software product from damage during delivery. In addition, documented procedures should exist to verify that an appropriate level of software virus checking has been performed and that appropriate measures have been taken to protect product integrity.

## 4.16   Control of Quality Records

> The supplier shall establish and maintain documented procedures for identification, collection, indexing, access, filing, storage, maintenance and disposition of quality records.
>
> Quality records shall be maintained to demonstrate conformance to specified requirements and the effective operation of the quality system. Pertinent quality records from the subcontractor shall be an element of these data.
>
> All quality records shall be legible and shall be stored and retained in such a way that they are readily retrievable in facilities that provide a suitable environment to prevent damage or deterioration and to prevent loss. Retention times of quality records shall be established and recorded. Where agreed contractually, quality records shall be made available for evaluation by the customer or the customer's representative for an agreed period.
>
> NOTE 19 Records may be in the form of any type of media, such as hard copy or electronic media.

Examples of quality records are:

- documented test results;
- problem reports;
- change requests;
- annotated documents;
- review records;
- minutes;
- audit reports.

Where records are held on electronic media, consideration of the retention times and accessibility of the records should take into account the rate of degradation of the electronic images and the availability of the devices and software needed to access the records.

NOTE: For further information, see ISO/IEC 12207:1995, subclause 6.1.6.2.

## 4.17  Internal Quality Audits

> The supplier shall establish and maintain documented procedures for planning and implementing internal quality audits to verify whether quality activities and related results comply with planned arrangements and to determine the effectiveness of the quality system.
>
> Internal quality audits shall be scheduled on the basis of the status and importance of the activity to be audited and shall be carried out by personnel independent of those having direct responsibility for the activity being audited.
>
> The results of the audits shall be recorded (see 4.16) and brought to the attention of the personnel having responsibility in the area audited. The management personnel responsible for the area shall take timely corrective action on deficiencies found during the audit.
>
> Follow-up audit activities shall verify and record the implementation and effectiveness of the corrective action taken (see 4.16).
>
> NOTES
>
> 20  The results of internal quality audits form an integral part of the input to management review activities (see 4.1.3).
> 21  Guidance on quality system audits is given in ISO 10011.

When software suppliers organize their work into projects, the audit plan should define a selection of projects. Consideration should be given to covering progressively the whole quality system of the supplier. This may be achieved by auditing several projects at different stages throughout the life cycle. Where a single project is consuming most of the resources of the organization, audits of that project may be scheduled as it progresses. Where the intended project changes its timescales, the internal audit schedule should be reviewed, either to change the timing of the audit, or to consider a different project.

The supplier's internal auditor should consider the consistency between the project quality plans with the organization's quality system.
NOTE: For further information, see ISO/IEC 12207:1995, subclauses 6.7, 6.8 and 7.3.2.

## 4.18  Training

> The supplier shall establish and maintain documented procedures for identifying training needs and provide for the training of all personnel performing activities affecting quality. Personnel performing specific assigned tasks shall be qualified on the basis of appropriate education, training and/or experience, as required. Appropriate records of training shall be maintained (see 4.16).

The training needs to be addressed should be determined considering the specific tools, techniques, methods and computer resources to be used in the development and management of the software product. It might also be required to include training in the skills and knowledge of the specific field with which the software is to deal. Qualification and training requirements should be documented.
NOTE: For further information, see ISO/IEC 12207:1995, subclause 7.4.

## 4.19  Servicing

> Where servicing is a specified requirement, the supplier shall establish and maintain documented procedures for performing, verifying and reporting that the servicing meets the specified requirements.

For the purposes of this part of ISO 9000, servicing is recognized as being related to the software terms: maintenance and customer support. Customer support is described in ISO 9000-2.

Maintenance activities for software products are typically classified into the following.

a) Problem resolution: this involves the detection and analysis of software nonconformities causing operational problems, and correction of the underlying software faults. When resolving problems, temporary fixes may be used to minimize downtime and permanent modifications carried out later.
b) Interface modifications: these may be required when additions or changes are made to the hardware system, or components, controlled by the software.
c) Functional expansion or performance improvement.

For interface modifications and functional expansion, depending upon the scale of work, change control procedures should apply, or a new and separate development project should be initiated, and the whole of this part of ISO 9000 becomes relevant. The maintenance activities described in this clause are therefore limited to problem resolution (often called corrective maintenance).

When maintenance of the software product is requested by the customer after initial delivery and installation, this should be stipulated in the contract. The supplier should establish and maintain documented procedures for performing maintenance activities and verifying that such activities meet the specified requirements for maintenance. Maintenance activities may also be performed on the development environment, tools and documentation.

The items to be maintained, and the period of time for which they should be maintained, should be specified in the contract. The following are examples of such items:

a) program(s);
b) data and their structures;
c) specifications;
d) documents for customer and/or user;
e) documents for supplier's use;
f) test plans.

All maintenance activities should be carried out and managed in accordance with a maintenance plan and/or procedures defined and agreed beforehand by the supplier and customer. The plan should include the following, as appropriate:

a) scope of maintenance;
b) identification of the initial status of the product;
c) support organization(s);
d) maintenance activities;
e) maintenance records and reports;
f) configuration management activities;
g) proposed release schedule.

Where appropriate, maintenance activities should be recorded and the records retained. Rules for the submission of maintenance reports should be established and agreed upon by the supplier and customer.

The maintenance records should include the following items, as appropriate, for each software product being maintained:

a) problem reports that have been received and the current status of each;
b) organization responsible for responding to requests for assistance or implementing the appropriate corrective actions;
c) priorities that have been assigned to the corrective actions;
d) results of the corrective actions;
e) statistical data on failure occurrences and maintenance activities.

The record of the maintenance activities may be used for evaluation and enhancement of the software product and for improvement of the quality system itself.
NOTE: For further information, see ISO/IEC 12207:1995, subclauses 5.4.4, 5.5. and 6.8.

## 4.20  Statistical Techniques

> **4.20.1** Identification of need
>
> The supplier shall identify the need for statistical techniques required for establishing, controlling and verifying process capability and product characteristics.
>
> **4.20.2** Procedures
>
> The supplier shall establish and maintain documented procedures to implement and control the application of the statistical techniques identified in 4.20.1.

Statistical techniques may be used to analyse measurements of process capability and product characteristics, with the objective of producing data that can be used to evaluate product quality and process capability. The data may be used to assess conformance with quality requirements, when they are expressed quantitatively.
Examples of product characteristics to which statistical techniques may be applied are:

• testability;
• usability;
• reliability;
• maintainability;
• availability.

Examples of software process capability characteristics to which statistical techniques may be applied are:

• process maturity;
• number and type of defects in process outputs;
• defect removal efficiency;
• milestone slippage.

The term "metric" means a measurable characteristic.
Metrics should comply with the following principles:

a) the metric should bring value to the process or product;
b) the metric is clearly defined;
c) the meaning of the metric in relation to software product quality or development process quality is understood;
d) the way in which the metric may be influenced (such as by changes in design and development techniques) is identified;
e) the direction of change in the metric, indicative of improved quality, is understood.

Independent of the metrics used, it is important that levels are known and used for process control and improvement rather than which specific metrics are used.

Different process metrics may be appropriate for different software products produced by the same supplier.

NOTE: Further guidance may be found in ISO/IEC 9126.

# ANNEX A
## (NORMATIVE)

## BIBLIOGRAPHY

[1] ISO 9000-2:1997, *Quality management and quality assurance standards — Part 2: Generic guidelines for the application of ISO 9001, ISO 9002 and ISO 9003.*
[2] ISO 10005:1995, *Quality management — Guidelines for quality plans.*
[3] ISO 10006:-l), *Quality management — Guidelines to quality in project management.*
[4] ISO 10007:1995, *Quality management — Guidelines for configuration management.*
[5] ISO 10011-1:1990, *Guidelines for auditing quality systems — Part 1: Auditing.*
[6] ISO 10011-2:1991, *Guidelines for auditing quality systems — Part 2: Qualification criteria for quality systems auditors.*
[7] ISO 10011-3:1991, *Guidelines for auditing quality systems — Part 3: Management of audit programmes.*
[8] ISO 10012-1:1992, *Quality assurance requirements for measuring equipment — Part 1: Metrological confirmation system for measuring equipment.*
[9] ISO 10013:1995, *Guidelines for developing quality manuals.*
[10] ISOIIEC 9126:1991, *Information technology — Software product evaluation — Quality characteristics and guidelines for their use.*
[11] ISOII EC 12207:1995, *Information technology — Software life cycle processes.*

# ANNEX B
## (INFORMATIVE)

## CROSS-REFERENCES TO ISO/1EC 12207

This cross-reference table:

- identifies ISO/IEC 12207 subclauses that may assist in the construction of a quality system that conforms to ISO 9001;
- summarizes the notes to be found within the body of this part of ISO 9000.
- is not intended to be a complete description of the relation between ISO/IEC 12207 and ISO 9001, in particular how ISO/IEC 12207 covers the requirements of ISO 9001 and vice versa.

| ISO 9000-3 | ISO/IEC 12207 |
|---|---|
| 4.1.2 | 7.2, 6.3.1.6 |
| 4.1.3 | 7.1.4 |
| 4.2.3 | 6.2, 6.3, 6.4, 6.5 |
| 4.3.2 | 5.2.1, 5.2.6, 6.4.2.1 |
| 4.3.3 | 5.1.3.5, 5.2.3.2 |
| 4.4.2 | 5.2.4 |
| 4.4.3 | 5.2.6.1, 6.6.2 |
| 4.4.4 | 5.3.2, 5.3.3, 5.3.4 |
| 4.4.5 | 5.3.5, 5.3.6, 5.3.7 |
| 4.4.6 | 5.3.4.2, 5.3.5.6, 5.3.6.7, 6.6.3 |
| 4.4.7 | 5.3.4.2,5.3.5.6,5.3.5.7,5.3.7.5,5.3.9,6.4 |
| 4.4.8 | 5.3.1, 6.5 |
| 4.4.9 | 5.5.2, 5.5.3, 6.2.3 |
| 4.5.1 | 6.1 |
| 4.6.1 | 5.1 |
| 4.7 | 6.1 |
| 4.8 | 6.1, 6.2 |
| 4.9 | 5.3.12, 6.3.3 |
| 4.10.1 | 5.1.5,5.3.5.5,5.3.6.5,5.3.6.6,5.3.7,5.3.8,5.3.9,5.3.10,5.3.11,5.3.13 |
| 4.11.1 | 7.2 |
| 4.12 | 6.2 |
| 4.13.1 | 6.2, 6.8 |
| 4.14.1 | 6.2, 6.8, 7.3 |
| 4.15.1 | 5.2.7.1, 5.3.13.2, 6.2.6 |
| 4.16 | 6.1.6.2 |
| 4.17 | 6.7, 6.8, 7.3.2 |
| 4.18 | 7.4 |
| 4.19 | 5.4.4, 5.5, 6.8 |

# INTERNATIONAL STANDARD ISO 9001*

## QUALITY SYSTEMS — MODEL FOR QUALITY ASSURANCE IN DESIGN, DEVELOPMENT, PRODUCTION, INSTALLATION AND SERVICING**

## CONTENTS

---

\* Second edition 1994-07-01
\*\* Reference number ISO 9001:1994(E)

## FOREWORD

ISO (the International Organization for Standardization) is a worldwide federation of national standards bodies (ISO member bodies). The work of preparing International Standards is normally carried out through SO technical committees. Each member body interested in a subject for which a technical committee has been established has the right to be represented on that committee. International organizations, governmental and non-governmental, in liaison with ISO, also take part in the work. ISO collaborates closely with the International Electrotechnical Commission (IEC) on all matters of electrotechnical standardization.

Draft International Standards adopted by the technical committees are circulated to the member bodies for voting. Publication as an International Standard requires approval by at least 75% of the member bodies casting a vote.

International Standard ISO 9001 was prepared by Technical Committee ISO/7C 176, Quality management and quality assurance, Subcommittee SC 2, Quality systems.

This second edition cancels and replaces the first edition (ISO 9001:1987), which has been technically revised.

Annex A of this International Standard is for information only.

## INTRODUCTION

This International Standard is one of three International Standards dealing with quality system requirements that can be used for external quality assurance purposes. The quality assurance modes, set out in the three International Standards listed below, represent three distinct forms of quality system requirements suitable for the purpose of a supplier demonstrating its capability, and for the assessment of the capability of a supplier by external parties.

a) *ISO 9001, Quality systems — Model for quality assurance in design, development, production, installation and servicing*
   - for use when conformance to specified requirements is to be assured by the supplier during design, development, production, installation and servicing.
b) *ISO 9002, Quality systems — Model for quality assurance in production, installation and servicing*
   - for use when conformance to specified requirements is to be assured by the supplier during production, installation and servicing.
c) *ISO 9003, Quality systems — Model for quality assurance in final inspection and test*
   - for use when conformance to specified requirements is to be assured by the supplier solely at final inspection and test.

It is emphasized that the quality system requirements specified in this International Standard, ISO 9002 and ISO 9003 are complementary (not alternative) to the technical (product) specified requirements. They specify requirements which determine what elements quality systems have to encompass, but it is not the purpose of these International Standards to enforce uniformity of quality systems. They are generic and independent of any specific industry or economic sector. The design and implementation of a quality system will be influenced by the varying needs of an organization, its particular objectives, the products and services supplied, and the processes and specific practices employed.

It is intended that these International Standards will be adopted in their present form, but on occasions they may need to be tailored by adding or deleting certain quality system requirements for specific contractual situations. ISO 9000-1 provides guidance on such tailoring as well as on selection of the appropriate quality assurance model, viz. ISO 9001, ISO 9002 or ISO 9003.

## QUALITY SYSTEMS — MODEL FOR QUALITY ASSURANCE IN DESIGN, DEVELOPMENT, PRODUCTION, INSTALLATION AND SERVICING

### 1 SCOPE

This International Standard specifies quality system requirements for use where a supplier's capability to design and supply conforming product needs to be demonstrated.

The requirements specified are aimed primarily at achieving customer satisfaction by preventing nonconformity at all stages from design through to servicing.

This International Standard is applicable in situations when

a) design is required and the product requirements are stated principally in performance terms, or they need to be established, and

b) confidence in product conformance can be attained by adequate demonstration of a supplier's capabilities in design, development, production, installation and servicing.

NOTE 1   For informative references, see annex A.

### 2 NORMATIVE REFERENCE

The following standard contains provisions which, through reference in this text, constitute provisions of this International Standard. At the time of publication, the edition indicated was valid. All standards are subject to revision, and parties to agreements based on this International Standard are encouraged to investigate the possibility of applying the most recent edition of the standard indicated below. Members of IEC and ISO maintain registers of currently valid International Standards. *ISO 8402:1994, Quality management and quality assurance — Vocabulary.*

### 3 DEFINITIONS

For the purposes of this International Standard, the definitions given in ISO 8402 and the following definitions apply.

### 3.1   Product: Result of activities or processes.

NOTES
2   A product may include service, hardware, processed materials, software or a combination thereof.
3   A product can be tangible (e.g. assemblies or processed materials) or intangible (e.g. knowledge or concepts), or a combination thereof.
4   For the purposes of this International Standard, the term "product" applies to the intended product offering only and not to unintended "by-products" affecting the environment. This differs from the definition given in ISO 8402.

**3.2 Tender: Offer made by a supplier in response to an invitation to satisfy a contract award to provide product.**

**3.3 Contract: Agreed requirements between a supplier and customer transmitted by any means.**

## 4 QUALITY SYSTEM REQUIREMENTS

### 4.1 Management responsibility

#### 4.1.1 Quality policy

The supplier's management with executive responsibility shall define and document its policy for quality, including objectives for quality and its commitment to quality. The quality policy shall be relevant to the supplier's organizational goals and the expectations and needs of its customers. The supplier shall ensure that this policy is understood, implemented and maintained at all levels of the organization.

#### 4.1.2 Organization

##### 4.1.2.1 Responsibility and authority

The responsibility, authority and the interrelation of personnel who manage, perform and verify work affecting quality shall be defined and documented, particularly for personnel who need the organizational freedom and authority to:

a) initiate action to prevent the occurrence of any nonconformities relating to the product, process and quality system;
b) identify and record any problems relating to the product, process and quality system;
c) initiate, recommend or provide solutions through designated channels;
d) verify the implementation of solutions;
e) control further processing, delivery or installation of nonconforming product until the deficiency or unsatisfactory condition has been corrected.

##### 4.1.2.2 Resources

The supplier shall identify resource requirements and provide adequate resources, including the assignment of trained personnel (see 4.18), for management, performance of work and verification activities including internal quality audits.

##### 4.1.2.3 Management representative

The supplier's management with executive responsibility shall appoint a member of the supplier's own management who, irrespective of other responsibilities, shall have defined authority for

a) ensuring that a quality system is established, implemented and maintained in accordance with this International Standard, and
b) reporting on the performance of the quality system to the supplier's management for review and as a basis for improvement of the quality system.

NOTE 5    The responsibility of a management representative may also include liaison with external parties on matters relating to the supplier's quality system.

### 4.1.3   Management review

The supplier's management with executive responsibility shall review the quality system at defined intervals sufficient to ensure its continuing suitability and effectiveness in satisfying the requirements of this International Standard and the supplier's stated quality policy and objectives (see 4.1.1). Records of such reviews shall be maintained (see 4.16).

## 4.2   Quality system

### 4.2.1   General

The supplier shall establish, document and maintain a quality system as a means of ensuring that product conforms to specified requirements. The supplier shall prepare a quality manual covering the requirements of this International Standard. The quality manual shall include or make reference to the quality system procedures and outline the structure of the documentation used in the quality system.

NOTE 6   Guidance on quality manuals is given in ISO 10013.

### 4.2.2   Quality system procedures

The supplier shall

 a) prepare documented procedures consistent with the requirements of this International Standard and the supplier's stated quality policy, and
 b) effectively implement the quality system and its documented procedures.

For the purposes of this International Standard, the range and detail of the procedures that form part of the quality system shall be dependent upon the complexity of the work, the methods used, and the skills and training needed by personnel involved in carrying out the activity.

NOTE 7   Documented procedures may make reference to work instructions that define how an activity is performed.

### 4.2.3   Quality planning

The supplier shall define and document how the requirements for quality will be met. Quality planning shall be consistent with all other requirements of a supplier's quality system and shall be documented in a format to suit the supplier's method of operation.

The supplier shall give consideration to the following activities, as appropriate, in meeting the specified requirements for products, projects or contracts:

 a) the preparation of quality plans;
 b) the identification and acquisition of any controls, processes, equipment (including inspection and test equipment), fixtures, resources and skills that may be needed to achieve the required quality;
 c) ensuring the compatibility of the design, the production process, installation, servicing, inspection and test procedures and the applicable documentation;
 d) the updating, as necessary, of quality control, inspection and testing techniques, including the development of new instrumentation;
 e) the identification of any measurement requirement involving capability that exceeds the known state of the art, in sufficient time for the needed capability to be developed;
 f) the identification of suitable verification at appropriate stages in the realization of product;

g) the clarification of standards of acceptability for all features and requirements, including those which contain a subjective element;

h) the identification and preparation of quality records (see 4.16).

NOTE 8　The quality plans referred to [see 4.2.3a)] may be in the form of a reference to the appropriate documented procedures that form an integral part of the supplier's quality system.

## 4.3　Contract review

### 4.3.1　General

The supplier shall establish and maintain documented procedures for contract review and for the coordination of these activities.

### 4.3.2　Review

Before submission of a tender, or the acceptance of a contract or order (statement of requirement), the tender, contract or order shall be reviewed by the supplier to ensure that:

a) the requirements are adequately defined and documented; where no written statement of requirement is available for an order received by verbal means, the supplier shall ensure that the order requirements are agreed before their acceptance;

b) any differences between the contract or order requirements and those in the tender are resolved;

c) the supplier has the capability to meet the contract or order requirements.

### 4.3.3　Amendment to a contract

The supplier shall identify how an amendment to a contract is made and correctly transferred to the functions concerned within the supplier's organization.

### 4.3.4　Records

Records of contract reviews shall be maintained (see 4.16).

NOTE 9　Channels for communication and interfaces with the customer's organization in these contract matters should be established.

## 4.4　Design control

### 4.4.1　General

The supplier shall establish and maintain documented procedures to control and verify the design of the product in order to ensure that the specified requirements are met.

### 4.4.2　Design and development planning

The supplier shall prepare plans for each design and development activity. The pans shall describe or reference these activities, and define responsibility for their implementation. The design and development activities shall be assigned to qualified personnel equipped with adequate resources. The plans shall be updated as the design evolves.

### 4.4.3　Organizational and technical interfaces

Organizational and technical interfaces between different groups which input into the design process shall be defined and the necessary information documented, transmitted and regularly reviewed.

### 4.4.4  Design input

Design input requirements relating to the product, including applicable statutory and regulatory requirements, shall be identified, documented and their selection reviewed by the supplier for adequacy. Incomplete, ambiguous or conflicting requirements shall be resolved with those responsible for imposing these requirements.

Design input shall take into consideration the results of any contract review activities.

### 4.4.5  Design output

Design output shall be documented and expressed in terms that can be verified and validated against design input requirements.

Design output shall:

a) meet the design input requirements;
b) contain or make reference to acceptance criteria;
c) identify those characteristics of the design that are crucial to the safe and proper functioning of the product (e.g. operating, storage, handing, maintenance and disposal requirements).

Design output documents shall be reviewed before release.

### 4.4.6  Design review

At appropriate stages of design, formal documented reviews of the design results shall be planned and conducted. Participants at each design review shall include representatives of al functions concerned with the design stage being reviewed, as well as other specialist personnel, as required. Records of such reviews shall be maintained (see 4.16).

### 4.4.7  Design verification

At appropriate stages of design, design verification shall be performed to ensure that the design stage output meets the design stage input requirements. The design verification measures shall be recorded (see 4.16).

NOTE 10   In addition to conducting design reviews (see 4.4.6), design verification may include activities such as

- performing alternative calculations,
- comparing the new design with a similar proven design, if available,
- undertaking tests and demonstrations, and
- reviewing the design stage documents before release.

### 4.4.8  Design validation

Design validation shall be performed to ensure that product conforms to defined user needs and/or requirements.

NOTES
11   Design validation follows successful design verification (see 4.4.7).
12   Validation is normally performed under defined operating conditions.
13   Validation is normally performed on the final product, but may be necessary in earlier stages prior to product completion.
14   Multiple validations may be performed if there are different intended uses.

### 4.4.9   Design changes

All design changes and modifications shall be identified, documented, reviewed and approved by authorized personnel before their implementation.

## 4.5   Document and data control

### 4.5.1   General

The supplier shall establish and maintain documented procedures to control all documents and data that relate to the requirements of this International Standard including, to the extent applicable, documents of external origin such as standards and customer drawings.

NOTE 15   Documents and data can be in the form of any type of media, such as hard copy or electronic media.

### 4.5.2   Document and data approval and issue

The documents and data shall be reviewed and approved for adequacy by authorized personnel prior to issue. A master list or equivalent document control procedure identifying the current revision status of documents shall be established and be readily available to preclude the use of invalid and/or obsolete documents.

   This control shall ensure that:

a) the pertinent issues of appropriate documents are available at all locations where operations essential to the effective functioning of the quality system are performed;
b) invalid and/or obsolete documents are promptly removed from all points of issue or use, or otherwise assured against unintended use;
c) any obsolete documents retained for legal and/or knowledge-preservation purposes are suitably identified.

### 4.5.3   Document and data changes

Changes to documents and data shall be reviewed and approved by the same functions organizations that performed the original review and approval, unless specifically designated otherwise. The designated functions organizations shall have access to pertinent background information upon which to base their review and approval.

   Where practicable, the nature of the change shall be identified in the document or the appropriate attachments.

## 4.6   Purchasing

### 4.6.1   General

The supplier shall establish and maintain documented procedures to ensure that purchased product (see 3.1) conforms to specified requirements.

### 4.6.2   Evaluation of subcontractors

The supplier shall:

a) evaluate and select subcontractors on the basis of their ability to meet subcontract require-ments including the quality system and any specific quality assurance requirements;

b) define the type and extent of control exercised by the supplier over subcontractors. This shall be dependent upon the type of product, the impact of subcontracted product on the quality of final product and, where applicable, on the quality audit reports and/or quality records of the previously demonstrated capability and performance of subcontractors;

c) establish and maintain quality records of acceptable subcontractors (see 4.16).

### 4.6.3 Purchasing data

Purchasing documents shall contain data clearly describing the product ordered, including where applicable:

a) the type, class, grade or other precise identification;

b) the title or other positive identification, and applicable issues of specifications, drawings, process requirements, inspection instructions and other relevant technical data, including requirements for approval or qualification of product, procedures, process equipment and personnel;

c) the title, number and issue of the quality system standard to be applied.

The supplier shall review and approve purchasing documents for adequacy of the specified requirements prior to release.

### 4.6.4 Verification of purchased product

#### 4.6.4.1 Supplier verification at subcontractor's premises

Where the supplier proposes to verify purchased product at the subcontractor's premises, the supplier shall specify verification arrangements and the method of product release in the purchasing documents.

#### 4.6.4.2 Customer verification of subcontracted product

Where specified in the contract, the supplier's customer or the customer's representative shall be afforded the right to verify at the subcontractor's premises and the supplier's premises that subcontracted product conforms to specified requirements. Such verification shall not be used by the supplier as evidence of effective control of quality by the subcontractor.

Verification by the customer shall not absolve the supplier of the responsibility to provide acceptable product, nor shall it preclude subsequent rejection by the customer.

## 4.7 Control of customer-supplied product

The supplier shall establish and maintain documented procedures for the control of verification, storage and maintenance of customer-supplied product provided for incorporation into the supplies or for related activities. Any such product that is lost, damaged or is otherwise unsuitable for use shall be recorded and reported to the customer (see 4.16).

Verification by the supplier does not absolve the customer of the responsibility to provide acceptable product.

## 4.8 Product identification and traceability

Where appropriate, the supplier shall establish and maintain documented procedures for identifying the product by suitable means from receipt and during a stages of production, delivery and installation.

Where and to the extent that traceability is a specified requirement, the supplier shall establish and maintain documented procedures for unique identification of individual product or batches. This identification shall be recorded (see 4.16).

## 4.9  Process control

The supplier shall identify and plan the production, installation and servicing processes which directly affect quality and shall ensure that these processes are carried out under controlled conditions. Controlled conditions shall include the following:

a) documented procedures defining the manner of production, installation and servicing, where the absence of such procedures could adversely affect quality;
b) use of suitable production, installation and servicing equipment, and a suitable working environment;
c) compliance with reference standards codes, quality plans and/or documented procedures;
d) monitoring and control of suitable process parameters and product characteristics;
e) the approval of processes and equipment, as appropriate;
f) criteria for workmanship, which shall be stipulated in the clearest practical manner (e.g. written standards, representative samples or illustrations);
g) suitable maintenance of equipment to ensure continuing process capability.

Where the results of processes cannot be fully verified by subsequent inspection and testing of the product and where, for example, processing deficiencies may become apparent only after the product is in use, the processes shall be carried out by qualified operators and/or shall require continuous monitoring and control of process parameters to ensure that the specified requirements are met.

The requirements for any qualification of process operations, including associated equipment and personnel (see 4.18), shall be specified.

NOTE 16 Such processes requiring pre-qualification of their process capability are frequently referred to as special processes.

Records shall be maintained for qualified processes, equipment and personnel, as appropriate (see 4.16).

## 4.10  Inspection and testing

### 4.10.1  General

The supplier shall establish and maintain documented procedures for inspection and testing activities in order to verify that the specified requirements for the product are met. The required inspection and testing, and the records to be established, shall be detailed in the quality plan or documented procedures.

### 4.10.2  Receiving inspection and testing

4.10.2.1 The supplier shall ensure that incoming product is not used or processed (except in the circumstances described in 4.10.2.3) until it has been inspected or otherwise verified as conforming to specified requirements. Verification of conformance to the specified requirements shall be in accordance with the quality plan and/or documented procedures.

*4.10.2.2* In determining the amount and nature of receiving inspection, consideration shall be given to the amount of control exercised at the subcontractor's premises and the recorded evidence of conformance provided.

*4.10.2.3* Where incoming product is released for urgent production purposes prior to verification, it shall be positively identified and recorded (see 4.16) in order to permit immediate recall and replacement in the event of nonconformity to specified requirements.

### 4.10.3  In-process inspection and testing

The supplier shall:

a) inspect and test the product as required by the quality plan and/or documented procedures;
b) hold product until the required inspection and tests have been completed or necessary reports have been received and verified, except when product is released under positive-recall procedures (see 4.10.2.3). Release under positive-recall procedures shall not preclude the activities outlined in 4.10.3(a).

### 4.10.4  Final inspection and testing

The supplier shall carry out all final inspection and testing in accordance with the quality plan and/or documented procedures to complete the evidence of conformance of the finished product to the specified requirements.

The quality plan and/or documented procedures for final inspection and testing shall require that a specified inspection and tests, including those specified either on receipt of product or in-process, have been carried out and that the results meet specified requirements.

No product shall be dispatched until all the activities specified in the quality plan and/or documented procedures have been satisfactorily completed and the associated data and documentation are available and authorized.

### 4.10.5  Inspection and test records

The supplier shall establish and maintain records which provide evidence that the product has been inspected and/or tested. These records shall show clearly whether the product has passed or failed the inspections and/or tests according to defined acceptance criteria. Where the product fails to pass any inspection and/or test, the procedures for control of nonconforming product shall apply (see 4.13).

Records shall identify the inspection authority responsible for the release of product (see 4.16).

## 4.11  Control of inspection, measuring and test equipment

### 4.11.1  General

The supplier shall establish and maintain documented procedures to control, calibrate and maintain inspection, measuring and test equipment (including test software) used by the supplier to demonstrate the conformance of product to the specified requirements. Inspection, measuring and test equipment shall be used in a manner which ensures that the measurement uncertainty is known and is consistent with the required measurement capability.

Where test software or comparative references such as test hardware are used as suitable forms of inspection, they shall be checked to prove that they are capable of verifying the acceptability of product, prior to release for use during production, installation or servicing, and shall be rechecked at prescribed intervals. The supplier shall establish the extent and frequency of such checks and shall maintain records as evidence of control (see 4.16).

Where the availability of technical data pertaining to the inspection, measuring and test equipment is a specified requirement, such data shall be made available, when required by the customer or customer's representative, for verification that the inspection, measuring and test equipment is functionally adequate.

NOTE 17    For the purposes of this International Standard, the term "measuring equipment" includes measurement devices.

### 4.11.2    Control procedure

The supplier shall:

a) determine the measurements to be made and the accuracy required, and select the appropriate inspection, measuring and test equipment that is capable of the necessary accuracy and precision;

b) identify al inspection, measuring and test equipment that can affect product quality, and calibrate and adjust them at prescribed intervals, or prior to use, against certified equipment having a known valid relationship to internationally or nationally recognized standards. Where no such standards exist, the basis used for calibration shall be documented;

c) define the process employed for the calibration of inspection, measuring and test equipment, including details of equipment type, unique identification, location, frequency of checks, check method, acceptance criteria and the action to be taken when results are unsatisfactory;

d) identify inspection, measuring and test equipment with a suitable indicator or approved identification record to show the calibration status;

e) maintain calibration records for inspection, measuring and test equipment (see 4.16);

f) assess and document the validity of previous inspection and test results when inspection, measuring or test equipment is found to be out of calibration;

g) ensure that the environmental conditions are suitable for the calibrations, inspections, measurements and tests being carried out;

h) ensure that the handing, preservation and storage of inspection, measuring and test equipment is such that the accuracy and fitness for use are maintained;

i) safeguard inspection, measuring and test facilities, including both test hardware and test software, from adjustments which would invalidate the calibration setting.

NOTE 18    The metrological confirmation system for measuring equipment given in ISO 10012 may be used for guidance.

### 4.12    Inspection and test status

The inspection and test status of product shall be identified by suitable means, which indicate the conformance or nonconformance of product with regard to inspection and tests performed. The identification of inspection and test status shall be maintained, as defined in the quality plan and/or documented procedures, throughout production, installation and servicing of the product to ensure that only product that has passed the required inspections and tests [or released under an authorized concession (see 4.13.2)] is dispatched, used or installed.

### 4.13    Control of nonconforming product

#### 4.13.1    General

The supplier shall establish and maintain documented procedures to ensure that product that does not conform to specified requirements is prevented from unintended use or installation. This control shall provide for identification, documentation, evaluation, segregation (when practical), disposition of nonconforming product, and for notification to the functions concerned.

### 4.13.2  Review and disposition of nonconforming product

The responsibility for review and authority for the disposition of nonconforming product shall be defined.

Nonconforming product shall be reviewed in accordance with documented procedures. It may be

a) reworked to meet the specified requirements,
b) accepted with or without repair by concession,
c) regraded for alternative applications, or
d) rejected or scrapped.

Where required by the contract, the proposed use or repair of product [see 4.13.2(b)] which does not conform to specified requirements shall be reported for concession to the customer or customer's representative. The description of the nonconformity that has been accepted, and of repairs, shall be recorded to denote the actual condition (see 4.16).

Repaired and/or reworked product shall be re-inspected in accordance with the quality plan and/or documented procedures.

## 4.14  Corrective and preventive action

### 4.14.1  General

The supplier shall establish and maintain documented procedures for implementing corrective and preventive action.

Any corrective or preventive action taken to eliminate the causes of actual or potential nonconformities shall be to a degree appropriate to the magnitude of problems and commensurate with the risks encountered.

The supplier shall implement and record any changes to the documented procedures resulting from corrective and preventive action.

### 4.14.2  Corrective action

The procedures for corrective action shall include:

a) the effective handing of customer complaints and reports of product nonconformities;
b) investigation of the cause of nonconformities relating to product, process and quality system, and recording the results of the investigation (see 4.16);
c) determination of the corrective action needed to eliminate the cause of nonconformities;
d) application of controls to ensure that corrective action is taken and that it is effective.

### 4.14.3  Preventive action

The procedures for preventive action shall include:

a) the use of appropriate sources of information such as processes and work operations which affect product quality, concessions, audit results, quality records, service reports and customer complaints to detect, analyze and eliminate potential causes of nonconformities;
b) determination of the steps needed to deal with any problems requiring preventive action;
c) initiation of preventive action and application of controls to ensure that it is effective;

d) ensuring that relevant information on actions taken is submitted for management review (see 4.1.3).

## 4.15 Handling, storage, packaging, preservation and delivery

### 4.15.1 General

The supplier shall establish and maintain documented procedures for handling, storage, packaging, preservation and delivery of product.

### 4.15.2 Handling

The supplier shall provide methods of handling product that prevent damage or deterioration.

### 4.15.3 Storage

The supplier shall use designated storage areas or stock rooms to prevent damage or deterioration of product, pending use or delivery. Appropriate methods for authorizing receipt to and dispatch from such areas shall be stipulated.

In order to detect deterioration, the condition of product in stock shall be assessed at appropriate intervals.

### 4.15.4 Packaging

The supplier shall control packing, packaging and marking processes (including materials used) to the extent necessary to ensure conformance to specified requirements.

### 4.15.5 Preservation

The supplier shall apply appropriate methods for preservation and segregation of product when the product is under the supplier's control.

### 4.15.6 Delivery

The supplier shall arrange for the protection of the quality of product after final inspection and test. Where contractually specified, this protection shall be extended to include delivery to destination.

## 4.16 Control of quality records

The supplier shall establish and maintain documented procedures for identification, collection, indexing, access, filing, storage, maintenance and disposition of quality records.

Quality records shall be maintained to demonstrate conformance to specified requirements and the effective operation of the quality system. Pertinent quality records from the subcontractor shall be an element of these data.

All quality records shall be legible and shall be stored and retained in such a way that they are readily retrievable in facilities that provide a suitable environment to prevent damage or deterioration and to prevent loss. Retention times of quality records shall be established and recorded. Where agreed contractually, quality records shall be made available for evaluation by the customer or the customer's representative for an agreed period.

NOTE 19     Records may be in the form of any type of media, such as hard copy or electronic media.

## 4.17 Internal quality audits

The supplier shall establish and maintain documented procedures for planning and implementing internal quality audits to verify whether quality activities and related results comply with panned arrangements and to determine the effectiveness of the quality system.

Internal quality audits shall be scheduled on the basis of the status and importance of the activity to be audited and shall be carried out by personnel independent of those having direct responsibility for the activity being audited.

The results of the audits shall be recorded (see 4.16) and brought to the attention of the personnel having responsibility in the area audited. The management personnel responsible for the area shall take timely corrective action on deficiencies found during the audit.

Follow-up audit activities shall verify and record the implementation and effectiveness of the corrective action taken (see 4.16).

NOTES
20   The results of internal quality audits form an integral part of the input to management review activities (see 4.1.3).
21   Guidance on quality system audits is given in ISO 10011.

## 4.18   Training

The supplier shall establish and maintain documented procedures for identifying training needs and provide for the training of all personnel performing activities affecting quality. Personnel performing specific assigned tasks shall be qualified on the basis of appropriate education, training and/or experience, as required. Appropriate records of training shall be maintained (see 4.16).

## 4.19   Servicing

Where servicing is a specified requirement, the supplier shall establish and maintain documented procedures for performing, verifying and reporting that the servicing meets the specified requirements.

## 4.20   Statistical techniques

### 4.20.1   Identification of need

The supplier shall identify the need for statistical techniques required for establishing, controlling and verifying process capability and product characteristics.

### 4.20.2 Procedures

The supplier shall establish and maintain documented procedures to implement and control the application of the statistical techniques identified in 4.20.1.

## ANNEX A
(INFORMATIVE)

## BIBLIOGRAPHY

[1] ISO 9000-1:1994, *Quality management and quality assurance standards — Part 1: Guidelines for selection and use.*
[2] ISO 9000-2:1993, *Quality management and quality assurance standards — Part 2: Generic guidelines for the application of ISO 9001, ISO 9002 and ISO 9003.*
[3] ISO 9000-3:1991, *Quality management and quality assurance standards — Part 3: Guidelines for the application of ISO 9001 to the development, supply and maintenance of software.*
[4] ISO 9002:1994, *Quality systems — Model for quality assurance in production, installation and servicing.*
[5] ISO 9003:1994, *Quality systems — Model for quality assurance in final inspection and test.*
[6] ISO 10011-1:1990, *Guidelines for auditing quality systems — Part 1: Auditing.*

[7] ISO 10011-2:1991, *Guidelines for auditing quality systems — Part 2: Qualification criteria for quality systems auditors.*

[8] ISO 10011-3:1991, *Guidelines for auditing quality systems — Part 3: Management of audit programmes.*

[9] ISO 10012-1:1992, *Quality assurance requirements for measuring equipment — Part 1: Metrological confirmation system for measuring equipment.*

[10] ISO 10013:—[1], *Guidelines for developing quality manuals.*

[11] ISO/TR 13425:—[1] *Guidelines for the selection of statistical methods in standardization and specification.*

---

1) To be published.

# INTERNATIONAL STANDARD ISO 9001*

## QUALITY MANAGEMENT SYSTEMS — REQUIREMENTS**

## CONTENTS

---

\*    Third edition 2000-12-15
\*\*  Reference number ISO 9001:2000(E)

## FOREWORD

ISO (the International Organization for Standardization) is a worldwide federation of national standards bodies (ISO member bodies). The work of preparing International Standards is normally carried out through ISO technical committees. Each member body interested in a subject for which a technical committee has been established has the right to be represented on that committee. International organizations, governmental and non-governmental, in liaison with ISO, also take part in the work. ISO collaborates closely with the International Electrotechnical Commission (IEC) on all matters of electrotechnical standardization.

International Standards are drafted in accordance with the rules given in the ISOIIEC Directives, Part 3.

Draft International Standards adopted by the technical committees are circulated to the member bodies for voting. Publication as an International Standard requires approval by at least 75% of the member bodies casting a vote.

Attention is drawn to the possibility that some of the elements of this International Standard may be the subject of patent rights. ISO shall not be held responsible for identifying any or all such patent rights.

International Standard ISO 9001 was prepared by Technical Committee ISOITC 176, *Quality management and quality assurance*, Subcommittee SC 2, *Quality systems*.

This third edition of ISO 9001 cancels and replaces the second edition (ISO 9001:1994) together with ISO 9002:1994 and ISO 9003:1994. It constitutes a technical revision of these documents. Those organizations which have used ISO 9002:1994 and ISO 9003:1994 in the past may use this International Standard by excluding certain requirements in accordance with 1.2.

The title of ISO 9001 has been revised in this edition and no longer includes the term "Quality assurance". This reflects the fact that the quality management system requirements specified in this edition of ISO 9001, in addition to quality assurance of product, also aim to enhance customer satisfaction.

Annexes A and B of this International Standard are for information only.

## INTRODUCTION

### 0.1   GENERAL

The adoption of a quality management system should be a strategic decision of an organization. The design and implementation of an organization's quality management system is influenced by varying needs, particular objectives, the products provided, the processes employed and the size and structure of the organization. It is not the intent of this International Standard to imply uniformity in the structure of quality management systems or uniformity of documentation.

The quality management system requirements specified in this International Standard are complementary to requirements for products. Information marked "NOTE" is for guidance in understanding or clarifying the associated requirement.

This International Standard can be used by internal and external parties, including certification bodies, to assess the organization's ability to meet customer, regulatory and the organization's own requirements.

The quality management principles stated in ISO 9000 and ISO 9004 have been taken into consideration during the development of this International Standard.

## 0.2 Process Approach

This International Standard promotes the adoption of a process approach when developing, implementing and improving the effectiveness of a quality management system, to enhance customer satisfaction by meeting customer requirements.

For an organization to function effectively, it has to identify and manage numerous linked activities. An activity using resources, and managed in order to enable the transformation of inputs into outputs, can be considered as a process. Often the output from one process directly forms the input to the next.

The application of a system of processes within an organization, together with the identification and interactions of these processes, and their management, can be referred to as the "process approach".

An advantage of the process approach is the ongoing control that it provides over the linkage between the individual processes within the system of processes, as well as over their combination and interaction.

When used within a quality management system, such an approach emphasizes the importance of

a) understanding and meeting requirements,
b) the need to consider processes in terms of added value,
c) obtaining results of process performance and effectiveness, and
d) continual improvement of processes based on objective measurement.

The model of a process-based quality management system shown in Figure 1 illustrates the process linkages presented in clauses 4 to 8. This illustration shows that customers play a significant role in defining requirements as inputs. Monitoring of customer satisfaction requires the evaluation of information relating to customer perception as to whether the organization has met the customer requirements. The model shown in Figure 1 covers all the requirements of this International Standard, but does not show processes at a detailed level.

NOTE   In addition, the methodology known as "Plan-Do-Check-Act" (PDCA) can be applied to all processes. PDCA can be briefly described as follows.

Plan:   establish the objectives and processes necessary to deliver results in accordance with customer requirements and the organization's policies.

Do:   implement the processes.

Check: monitor and measure processes and product against policies, objectives and requirements for the product and report the results.

Act:   take actions to continually improve process performance.

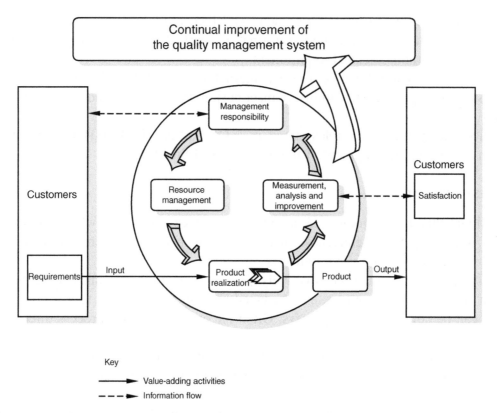

**FIGURE 1** Model of a process-based quality management system.

## 0.3   Relationship with ISO 9004

The present editions of ISO 9001 and ISO 9004 have been developed as a consistent pair of quality management system standards which have been designed to complement each other, but can also be used independently. Although the two International Standards have different scopes, they have similar structures in order to assist their application as a consistent pair.

ISO 9001 specifies requirements for a quality management system that can be used for internal application by organizations, or for certification, or for contractual purposes. It focuses on the effectiveness of the quality management system in meeting customer requirements.

ISO 9004 gives guidance on a wider range of objectives of a quality management system than does ISO 9001, particularly for the continual improvement of an organization's overall performance and efficiency, as well as its effectiveness. ISO 9004 is recommended as a guide for organizations whose top management wishes to move beyond the requirements of ISO 9001, in pursuit of continual improvement of performance. However, it is not intended for certification or for contractual purposes.

## 0.4   Compatibility with Other Management Systems

This International Standard has been aligned with ISO 14001:1996 in order to enhance the compatibility of the two standards for the benefit of the user community.

This International Standard does not include requirements specific to other management systems, such as those particular to environmental management, occupational health and safety management, financial management or risk management. However, this International Standard enables an organization to align or integrate its own quality management system with related management system requirements. It is possible for an organization to adapt its existing management system(s) in order to establish a quality management system that complies with the requirements of this International Standard.

## 1 SCOPE

### 1.1 GENERAL

This International Standard specifies requirements for a quality management system where an organization

a) needs to demonstrate its ability to consistently provide product that meets customer and applicable regulatory requirements, and

b) aims to enhance customer satisfaction through the effective application of the system, including processes for continual improvement of the system and the assurance of conformity to customer and applicable regulatory requirements.

NOTE   In this International Standard, the term "product" applies only to the product intended for, or required by, a customer.

### 1.2 APPLICATION

All requirements of this International Standard are generic and are intended to be applicable to all organizations, regardless of type, size and product provided.

Where any requirement(s) of this International Standard cannot be applied due to the nature of an organization and its product, this can be considered for exclusion.

Where exclusions are made, claims of conformity to this International Standard are not acceptable unless these exclusions are limited to requirements within clause 7, and such exclusions do not affect the organization's ability, or responsibility, to provide product that meets customer and applicable regulatory requirements.

## 2 NORMATIVE REFERENCE

The following normative document contains provisions which, through reference in this text, constitute provisions of this International Standard. For dated references, subsequent amendments to, or revisions of, any of these publications do not apply. However, parties to agreements based on this International Standard are encouraged to investigate the possibility of applying the most recent edition of the normative document indicated below. For undated references, the latest edition of the normative document referred to applies. Members of ISO and IEC maintain registers of currently valid International Standards.

ISO 9000:2000, *Quality management systems — Fundamentals and vocabulary.*

## 3 TERMS AND DEFINITIONS

For the purposes of this International Standard, the terms and definitions given in ISO 9000 apply.

The following terms, used in this edition of ISO 9001 to describe the supply chain, have been changed to reflect the vocabulary currently used:

Supplier ⟶ Organization ⟶ Customer

The term "organization" replaces the term "supplier" used in ISO 9001:1994, and refers to the unit to which this International Standard applies. Also, the term "supplier" now replaces the term "subcontractor".

Throughout the text of this International Standard, wherever the term "product" occurs, it can also mean "service".

# 4    QUALITY MANAGEMENT SYSTEM

## 4.1    General Requirements

The organization shall establish, document, implement and maintain a quality management system and continually improve its effectiveness in accordance with the requirements of this International Standard.

The organization shall

a) identify the processes needed for the quality management system and their application throughout the organization (see 1.2),
b) determine the sequence and interaction of these processes,
c) determine criteria and methods needed to ensure that both the operation and control of these processes are effective,
d) ensure the availability of resources and information necessary to support the operation and monitoring of these processes,
e) monitor, measure and analyze these processes, and
f) implement actions necessary to achieve planned results and continual improvement of these processes.

These processes shall be managed by the organization in accordance with the requirements of this International Standard.

Where an organization chooses to outsource any process that affects product conformity with requirements, the organization shall ensure control over such processes. Control of such outsourced processes shall be identified within the quality management system.

NOTE    Processes needed for the quality management system referred to above should include processes for management activities, provision of resources, product realization and measurement.

## 4.2    Documentation Requirements

### 4.2.1    General

The quality management system documentation shall include

a) documented statements of a quality policy and quality objectives,
b) a quality manual,
c) documented procedures required by this International Standard,
d) documents needed by the organization to ensure the effective planning, operation and control of its processes, and
e) records required by this International Standard (see 4.2.4).

NOTE 1   Where the term "documented procedure" appears within this International Standard, this means that the procedure is established, documented, implemented and maintained.

NOTE 2   The extent of the quality management system documentation can differ from one organization to another due to

- a) the size of organization and type of activities,
- b) the complexity of processes and their interactions, and
- c) the competence of personnel.

NOTE 3   The documentation can be in any form or type of medium.

### 4.2.2   Quality Manual

The organization shall establish and maintain a quality manual that includes

- a) the scope of the quality management system, including details of and justification for any exclusions (see 1.2),
- b) the documented procedures established for the quality management system, or reference to them, and
- c) a description of the interaction between the processes of the quality management system.

### 4.2.3   Control of Documents

Documents required by the quality management system shall be controlled. Records are a special type of document and shall be controlled according to the requirements given in 4.2.4.

A documented procedure shall be established to define the controls needed

- a) to approve documents for adequacy prior to issue,
- b) to review and update as necessary and re-approve documents,
- c) to ensure that changes and the current revision status of documents are identified,
- d) to ensure that relevant versions of applicable documents are available at points of use,
- e) to ensure that documents remain legible and readily identifiable,
- f) to ensure that documents of external origin are identified and their distribution controlled, and
- g) to prevent the unintended use of obsolete documents, and to apply suitable identification to them if they are retained for any purpose.

### 4.2.4   Control of Records

Records shall be established and maintained to provide evidence of conformity to requirements and of the effective operation of the quality management system. Records shall remain legible, readily identifiable and retrievable. A documented procedure shall be established to define the controls needed for the identification, storage, protection, retrieval, retention time and disposition of records.

## 5   MANAGEMENT RESPONSIBILITY

### 5.1   MANAGEMENT COMMITMENT

Top management shall provide evidence of its commitment to the development and implementation of the quality management system and continually improving its effectiveness by

a) communicating to the organization the importance of meeting customer as well as statutory and regulatory requirements,
b) establishing the quality policy,
c) ensuring that quality objectives are established,
d) conducting management reviews, and
e) ensuring the availability of resources.

## 5.2 CUSTOMER FOCUS

Top management shall ensure that customer requirements are determined and are met with the aim of enhancing customer satisfaction (see 7.2.1 and 8.2.1).

## 5.3 QUALITY POLICY

Top management shall ensure that the quality policy

a) is appropriate to the purpose of the organization,
b) includes a commitment to comply with requirements and continually improve the effectiveness of the quality management system,
c) provides a framework for establishing and reviewing quality objectives,
d) is communicated and understood within the organization, and
e) is reviewed for continuing suitability.

## 5.4 PLANNING

### 5.4.1 Quality Objectives

Top management shall ensure that quality objectives, including those needed to meet requirements for product [see 7.1(a)], are established at relevant functions and levels within the organization. The quality objectives shall be measurable and consistent with the quality policy.

### 5.4.2 Quality Management System Planning

Top management shall ensure that

a) the planning of the quality management system is carried out in order to meet the requirements given in 4.1, as well as the quality objectives, and
b) the integrity of the quality management system is maintained when changes to the quality management system are planned and implemented.

## 5.5 RESPONSIBILITY, AUTHORITY AND COMMUNICATION

### 5.5.1 Responsibility and Authority

Top management shall ensure that responsibilities and authorities are defined and communicated within the organization.

### 5.5.2 Management Representative

Top management shall appoint a member of management who, irrespective of other responsibilities, shall have responsibility and authority that includes

a) ensuring that processes needed for the quality management system are established, implemented and maintained,

b) reporting to top management on the performance of the quality management system and any need for improvement, and

c) ensuring the promotion of awareness of customer requirements throughout the organization.

NOTE The responsibility of a management representative can include liaison with external parties on matters relating to the quality management system.

### 5.5.3 Internal Communication

Top management shall ensure that appropriate communication processes are established within the organization and that communication takes place regarding the effectiveness of the quality management system.

## 5.6 MANAGEMENT REVIEW

### 5.6.1 General

Top management shall review the organization's quality management system, at planned intervals, to ensure its continuing suitability, adequacy and effectiveness. This review shall include assessing opportunities for improvement and the need for changes to the quality management system, including the quality policy and quality objectives.

Records from management reviews shall be maintained (see 4.2.4).

### 5.6.2 Review Input

The input to management review shall include information on

a) results of audits,

b) customer feedback,

c) process performance and product conformity,

d) status of preventive and corrective actions,

e) follow-up actions from previous management reviews,

f) changes that could affect the quality management system, and

g) recommendations for improvement.

### 5.6.3 Review Output

The output from the management review shall include any decisions and actions related to

a) improvement of the effectiveness of the quality management system and its processes,

b) improvement of product related to customer requirements, and

c) resource needs.

## 6 RESOURCE MANAGEMENT

## 6.1 PROVISION OF RESOURCES

The organization shall determine and provide the resources needed

a) to implement and maintain the quality management system and continually improve its effectiveness, and

b) to enhance customer satisfaction by meeting customer requirements.

## 6.2   HUMAN RESOURCES

### 6.2.1   General

Personnel performing work affecting product quality shall be competent on the basis of appropriate education, training, skills and experience.

### 6.2.2   Competence, Awareness and Training

The organization shall

a) determine the necessary competence for personnel performing work affecting product quality,

b) provide training or take other actions to satisfy these needs,

c) evaluate the effectiveness of the actions taken,

d) ensure that its personnel are aware of the relevance and importance of their activities and how they contribute to the achievement of the quality objectives, and

e) maintain appropriate records of education, training, skills and experience (see 4.2.4).

## 6.3   INFRASTRUCTURE

The organization shall determine, provide and maintain the infrastructure needed to achieve conformity to product requirements. Infrastructure includes, as applicable

a) buildings, workspace and associated utilities,

b) process equipment (both hardware and software), and

c) supporting services (such as transport or communication).

## 6.4   WORK ENVIRONMENT

The organization shall determine and manage the work environment needed to achieve conformity to product requirements.

## 7   PRODUCT REALIZATION

### 7.1   PLANNING OF PRODUCT REALIZATION

The organization shall plan and develop the processes needed for product realization. Planning of product realization shall be consistent with the requirements of the other processes of the quality management system (see 4.1).

In planning product realization, the organization shall determine the following, as appropriate:

a) quality objectives and requirements for the product;

b) the need to establish processes, documents, and provide resources specific to the product;

c) required verification, validation, monitoring, inspection and test activities specific to the product and the criteria for product acceptance;

d) records needed to provide evidence that the realization processes and resulting product meet requirements (see 4.2.4).

The output of this planning shall be in a form suitable for the organization's method of operations.

NOTE 1    A document specifying the processes of the quality management system (including the product realization processes) and the resources to be applied to a specific product, project or contract, can be referred to as a quality plan.

NOTE 2    The organization may also apply the requirements given in 7.3 to the development of product realization processes.

## 7.2    CUSTOMER-RELATED PROCESSES

### 7.2.1    Determination of Requirements Related to the Product

The organization shall determine

a) requirements specified by the customer, including the requirements for delivery and post-delivery activities,

b) requirements not stated by the customer but necessary for specified or intended use, where known,

c) statutory and regulatory requirements related to the product, and

d) any additional requirements determined by the organization.

### 7.2.2    Review of Requirements Related to the Product

The organization shall review the requirements related to the product. This review shall be conducted prior to the organization's commitment to supply a product to the customer (e.g. submission of tenders, acceptance of contracts or orders, acceptance of changes to contracts or orders) and shall ensure that

a) product requirements are defined,

b) contract or order requirements differing from those previously expressed are resolved, and

c) the organization has the ability to meet the defined requirements.

Records of the results of the review and actions arising from the review shall be maintained (see 4.2.4).

Where the customer provides no documented statement of requirement, the customer requirements shall be confirmed by the organization before acceptance.

Where product requirements are changed, the organization shall ensure that relevant documents are amended and that relevant personnel are made aware of the changed requirements.

NOTE    In some situations, such as internet sales, a formal review is impractical for each order. Instead the review can cover relevant product information such as catalogues or advertising material.

### 7.2.3  Customer Communication

The organization shall determine and implement effective arrangements for communicating with customers in relation to

a) product information,
b) enquiries, contracts or order handling, including amendments, and
c) customer feedback, including customer complaints.

### 7.3  Design and Development

### 7.3.1  Design and Development Planning

The organization shall plan and control the design and development of product.
  During the design and development planning, the organization shall determine

a) the design and development stages,
b) the review, verification and validation that are appropriate to each design and development stage, and
c) the responsibilities and authorities for design and development.

  The organization shall manage the interfaces between different groups involved in design and development to ensure effective communication and clear assignment of responsibility.
  Planning output shall be updated, as appropriate, as the design and development progresses.

### 7.3.2  Design and Development Inputs

Inputs relating to product requirements shall be determined and records maintained (see 4.2.4). These inputs shall include

a) functional and performance requirements,
b) applicable statutory and regulatory requirements,
c) where applicable, information derived from previous similar designs, and
d) other requirements essential for design and development.

  These inputs shall be reviewed for adequacy. Requirements shall be complete, unambiguous and not in conflict with each other.

### 7.3.3  Design and Development Outputs

The outputs of design and development shall be provided in a form that enables verification against the design and development input and shall be approved prior to release.
  Design and development outputs shall

a) meet the input requirements for design and development,
b) provide appropriate information for purchasing, production and for service provision,
c) contain or reference product acceptance criteria, and
d) specify the characteristics of the product that are essential for its safe and proper use.

### 7.3.4   Design and Development Review

At suitable stages, systematic reviews of design and development shall be performed in accordance with planned arrangements (see 7.3.1)

a) to evaluate the ability of the results of design and development to meet requirements, and
b) to identify any problems and propose necessary actions.

Participants in such reviews shall include representatives of functions concerned with the design and development stage(s) being reviewed. Records of the results of the reviews and any necessary actions shall be maintained (see 4.2.4).

### 7.3.5   Design and Development Verification

Verification shall be performed in accordance with planned arrangements (see 7.3.1) to ensure that the design and development outputs have met the design and development input requirements. Records of the results of the verification and any necessary actions shall be maintained (see 4.2.4).

### 7.3.6   Design and Development Validation

Design and development validation shall be performed in accordance with planned arrangements (see 7.3.1) to ensure that the resulting product is capable of meeting the requirements for the specified application or intended use, where known. Wherever practicable, validation shall be completed prior to the delivery or implementation of the product. Records of the results of validation and any necessary actions shall be maintained (see 4.2.4).

### 7.3.7   Control of Design and Development Changes

Design and development changes shall be identified and records maintained. The changes shall be reviewed, verified and validated, as appropriate, and approved before implementation. The review of design and development changes shall include evaluation of the effect of the changes on constituent parts and product already delivered.

Records of the results of the review of changes and any necessary actions shall be maintained (see 4.2.4).

### 7.4   PURCHASING

### 7.4.1   Purchasing Process

The organization shall ensure that purchased product conforms to specified purchase requirements. The type and extent of control applied to the supplier and the purchased product shall be dependent upon the effect of the purchased product on subsequent product realization or the final product.

The organization shall evaluate and select suppliers based on their ability to supply product in accordance with the organization's requirements. Criteria for selection, evaluation and re-evaluation shall be established. Records of the results of evaluations and any necessary actions arising from the evaluation shall be maintained (see 4.2.4).

### 7.4.2   Purchasing Information

Purchasing information shall describe the product to be purchased, including where appropriate

a) requirements for approval of product, procedures, processes and equipment,
b) requirements for qualification of personnel, and
c) quality management system requirements.

The organization shall ensure the adequacy of specified purchase requirements prior to their communication to the supplier.

### 7.4.3    Verification of Purchased Product

The organization shall establish and implement the inspection or other activities necessary for ensuring that purchased product meets specified purchase requirements.

Where the organization or its customer intends to perform verification at the supplier's premises, the organization shall state the intended verification arrangements and method of product release in the purchasing information.

## 7.5    PRODUCTION AND SERVICE PROVISION

### 7.5.1    Control of Production and Service Provision

The organization shall plan and carry out production and service provision under controlled conditions. Controlled conditions shall include, as applicable

a) the availability of information that describes the characteristics of the product,
b) the availability of work instructions, as necessary,
c) the use of suitable equipment,
d) the availability and use of monitoring and measuring devices,
e) the implementation of monitoring and measurement, and
f) the implementation of release, delivery and post-delivery activities.

### 7.5.2    Validation of Processes for Production and Service Provision

The organization shall validate any processes for production and service provision where the resulting output cannot be verified by subsequent monitoring or measurement. This includes any processes where deficiencies become apparent only after the product is in use or the service has been delivered.

Validation shall demonstrate the ability of these processes to achieve planned results.

The organization shall establish arrangements for these processes including, as applicable

a) defined criteria for review and approval of the processes,
b) approval of equipment and qualification of personnel,
c) use of specific methods and procedures,
d) requirements for records (see 4.2.4), and
e) revalidation.

### 7.5.3    Identification and Traceability

Where appropriate, the organization shall identify the product by suitable means throughout product realization.

The organization shall identify the product status with respect to monitoring and measurement requirements.

Where traceability is a requirement, the organization shall control and record the unique identification of the product (see 4.2.4).

NOTE   In some industry sectors, configuration management is a means by which identification and traceability are maintained.

### 7.5.4   Customer Property

The organization shall exercise care with customer property while it is under the organization's control or being used by the organization. The organization shall identify, verify, protect and safeguard customer property provided for use or incorporation into the product. If any customer property is lost, damaged or otherwise found to be unsuitable for use, this shall be reported to the customer and records maintained (see 4.2.4).

NOTE   Customer property can include intellectual property.

### 7.5.5   Preservation of Product

The organization shall preserve the conformity of product during internal processing and delivery to the intended destination. This preservation shall include identification, handling, packaging, storage and protection. Preservation shall also apply to the constituent parts of a product.

### 7.6   Control of Monitoring and Measuring Devices

The organization shall determine the monitoring and measurement to be undertaken and the monitoring and measuring devices needed to provide evidence of conformity of product to determined requirements (see 7.2.1).

The organization shall establish processes to ensure that monitoring and measurement can be carried out and are carried out in a manner that is consistent with the monitoring and measurement requirements.

Where necessary to ensure valid results, measuring equipment shall

a) be calibrated or verified at specified intervals, or prior to use, against measurement standards traceable to international or national measurement standards; where no such standards exist, the basis used for calibration or verification shall be recorded;
b) be adjusted or re-adjusted as necessary;
c) be identified to enable the calibration status to be determined;
d) be safeguarded from adjustments that would invalidate the measurement result;
e) be protected from damage and deterioration during handling, maintenance and storage.

In addition, the organization shall assess and record the validity of the previous measuring results when the equipment is found not to conform to requirements. The organization shall take appropriate action on the equipment and any product affected. Records of the results of calibration and verification shall be maintained (see 4.2.4).

When used in the monitoring and measurement of specified requirements, the ability of computer software to satisfy the intended application shall be confirmed. This shall be undertaken prior to initial use and reconfirmed as necessary.

NOTE   See ISO 10012-1 and ISO 10012-2 for guidance.

# 8   MEASUREMENT, ANALYSIS AND IMPROVEMENT

## 8.1   GENERAL

The organization shall plan and implement the monitoring, measurement, analysis and improvement processes needed

a) to demonstrate conformity of the product,
b) to ensure conformity of the quality management system, and
c) to continually improve the effectiveness of the quality management system.

This shall include determination of applicable methods, including statistical techniques, and the extent of their use.

## 8.2   MONITORING AND MEASUREMENT

### 8.2.1   Customer Satisfaction

As one of the measurements of the performance of the quality management system, the organization shall monitor information relating to customer perception as to whether the organization has met customer requirements. The methods for obtaining and using this information shall be determined.

### 8.2.2   Internal Audit

The organization shall conduct internal audits at planned intervals to determine whether the quality management system

a) conforms to the planned arrangements (see 7.1), to the requirements of this International Standard and to the quality management system requirements established by the organization, and
b) is effectively implemented and maintained.

An audit programme shall be planned, taking into consideration the status and importance of the processes and areas to be audited, as well as the results of previous audits. The audit criteria, scope, frequency and methods shall be defined. Selection of auditors and conduct of audits shall ensure objectivity and impartiality of the audit process. Auditors shall not audit their own work.

The responsibilities and requirements for planning and conducting audits, and for reporting results and maintaining records (see 4.2.4) shall be defined in a documented procedure.

The management responsible for the area being audited shall ensure that actions are taken without undue delay to eliminate detected nonconformities and their causes. Follow-up activities shall include the verification of the actions taken and the reporting of verification results (see 8.5.2).

NOTE   See ISO 10011-1, ISO 10011-2 and ISO 10011-3 for guidance.

### 8.2.3   Monitoring and Measurement of Processes

The organization shall apply suitable methods for monitoring and, where applicable, measurement of the quality management system processes. These methods shall demonstrate the ability of the processes to achieve planned results. When planned results are not achieved, correction and corrective action shall be taken, as appropriate, to ensure conformity of the product.

### 8.2.4   Monitoring and Measurement of Product

The organization shall monitor and measure the characteristics of the product to verify that product requirements have been met. This shall be carried out at appropriate stages of the product realization process in accordance with the planned arrangements (see 7.1).

Evidence of conformity with the acceptance criteria shall be maintained. Records shall indicate the person(s) authorizing release of product (see 4.2.4).

Product release and service delivery shall not proceed until the planned arrangements (see 7.1) have been satisfactorily completed, unless otherwise approved by a relevant authority and, where applicable, by the customer.

### 8.3   CONTROL OF NONCONFORMING PRODUCT

The organization shall ensure that product which does not conform to product requirements is identified and controlled to prevent its unintended use or delivery. The controls and related responsibilities and authorities for dealing with nonconforming product shall be defined in a documented procedure.

The organization shall deal with nonconforming product by one or more of the following ways:

a) by taking action to eliminate the detected nonconformity;
b) by authorizing its use, release or acceptance under concession by a relevant authority and, where applicable, by the customer;
c) by taking action to preclude its original intended use or application.

Records of the nature of nonconformities and any subsequent actions taken, including concessions obtained, shall be maintained (see 4.2.4).

When nonconforming product is corrected it shall be subject to re-verification to demonstrate conformity to the requirements.

When nonconforming product is detected after delivery or use has started, the organization shall take action appropriate to the effects, or potential effects, of the nonconformity.

### 8.4   ANALYSIS OF DATA

The organization shall determine, collect and analyze appropriate data to demonstrate the suitability and effectiveness of the quality management system and to evaluate where continual improvement of the effectiveness of the quality management system can be made. This shall include data generated as a result of monitoring and measurement and from other relevant sources.

The analysis of data shall provide information relating to

a) customer satisfaction (see 8.2.1),
b) conformity to product requirements (see 7.2.1),
c) characteristics and trends of processes and products including opportunities for preventive action, and
d) suppliers.

## 8.5    Improvement

### 8.5.1    Continual Improvement

The organization shall continually improve the effectiveness of the quality management system through the use of the quality policy, quality objectives, audit results, analysis of data, corrective and preventive actions and management review.

### 8.5.2    Corrective Action

The organization shall take action to eliminate the cause of nonconformities in order to prevent recurrence. Corrective actions shall be appropriate to the effects of the nonconformities encountered.

A documented procedure shall be established to define requirements for

a) reviewing nonconformities (including customer complaints),
b) determining the causes of nonconformities,
c) evaluating the need for action to ensure that nonconformities do not recur,
d) determining and implementing action needed,
e) records of the results of action taken (see 4.2.4), and
f) reviewing corrective action taken.

### 8.5.3    Preventive Action

The organization shall determine action to eliminate the causes of potential nonconformities in order to prevent their occurrence. Preventive actions shall be appropriate to the effects of the potential problems.

A documented procedure shall be established to define requirements for

a) determining potential nonconformities and their causes,
b) evaluating the need for action to prevent occurrence of nonconformities,
c) determining and implementing action needed,
d) records of results of action taken (see 4.2.4), and
e) reviewing preventive action taken.

# ANNEX A

(informative)

## Correspondence between ISO 9001:2000 and ISO 14001:1996

**TABLE A.1**
**Correspondence between ISO 9001:2000 and ISO 14001:1996**

| ISO 9001:2000 | | | ISO 14001:1996 |
|---|---|---|---|
| **Introduction** | | | **Introduction** |
| General | 0.1 | | |
| Process approach | 0.2 | | |
| Relationship with ISO 9004 | 0.3 | | |
| Compatibility with other management systems | 0.4 | | |
| **Scope** | **1** | **1** | **Scope** |
| General | 1.1 | | |
| Application | 1.2 | | |
| **Normative reference** | **2** | **2** | **Normative references** |
| **Terms and definitions** | **3** | **3** | **Definitions** |
| **Quality management system** | **4** | **4** | **Environmental management system requirements** |
| General requirements | 4.1 | 4.1 | General requirements |
| Documentation requirements | 4.2 | | |
| General | 4.2.1 | 4.4.4 | Environmental management system documentation |
| Quality manual | 4.2.2 | 4.4.4 | Environmental management system documentation |
| Control of documents | 4.2.3 | 4.4.5 | Document control |
| Control of records | 4.2.4 | 4.5.3 | Records |
| **Management responsibility** | **5** | 4.4.1 | Structure and responsibility |
| Management commitment | 5.1 | 4.2 | Environmental policy |
| | | 4.4.1 | Structure and responsibility |
| Customer focus | 5.2 | 4.3.1 | Environmental aspects |
| | | 4.3.2 | Legal and other requirements |
| Quality policy | 5.3 | 4.2 | Environmental policy |
| Planning | 5.4 | 4.3 | Planning |
| Quality objectives | 5.4.1 | 4.3.3 | Objectives and targets |
| Quality management system planning | 5.4.2 | 4.3.4 | Environmental management programme(s) |
| Responsibility, authority and communication | 5.5 | 4.1 | General requirements |
| Responsibility and authority | 5.5.1 | 4.4.1 | Structure and responsibility |
| Management representative | 5.5.2 | | |
| Internal communication | 5.5.3 | 4.4.3 | Communication |
| Management review | 5.6 | 4.6 | Management review |
| General | 5.6.1 | | |
| Review input | 5.6.2 | | |
| Review output | 5.6.3 | | |
| **Resource management** | **6** | 4.4.1 | Structure and responsibility |
| Provision of resources | 6.1 | | |
| Human resources | 6.2 | | |
| General | 6.2.1 | | |

**TABLE A.1**
**Correspondence between ISO 9001:2000 and ISO 14001:1996 (continued)**

| ISO 9001:2000 | | ISO 14001:1996 | |
|---|---|---|---|
| Competence, awareness and training | 6.2.2 | 4.4.2 | Training, awareness and competence |
| Infrastructure | 6.3 | 4.4.1 | Structure and responsibility |
| Work environment | 6.4 | | |
| **Product realization** | **7** | 4.4 | Implementation and operation |
| | | 4.4.6 | Operational control |
| Planning of product realization | 7.1 | 4.4.6 | Operational control |
| Customer-related processes | 7.2 | | |
| Determination of requirements related to the product | 7.2.1 | 4.3.1 | Environmental aspects |
| | | 4.3.2 | Legal and other requirements |
| | | 4.4.6 | Operational control |
| Review of requirements related to the product | 7.2.2 | 4.4.6 | Operational control |
| | | 4.3.1 | Environmental aspects |
| Customer communication | 7.2.3 | 4.4.3 | Communications |
| Design and development | 7.3 | | |
| Design and development planning | 7.3.1 | 4.4.6 | Operational control |
| Design and development inputs | 7.3.2 | | |
| Design and development outputs | 7.3.3 | | |
| Design and development review | 7.3.4 | | |
| Design and development verification | 7.3.5 | | |
| Design and development validation | 7.3.6 | | |
| Control of design and development changes | 7.3.7 | | |
| Purchasing | 7.4 | 4.4.6 | Operational control |
| Purchasing process | 7.4.1 | | |
| Purchasing information | 7.4.2 | | |
| Verification of purchased product | 7.4.3 | | |
| Production and service provision | 7.5 | 4.4.6 | Operational control |
| Control of production and service provision | 7.5.1 | | |
| Validation of processes for production and service provision | 7.5.2 | | |
| Identification and traceability | 7.5.3 | | |
| Customer property | 7.5.4 | | |
| Preservation of product | 7.5.5 | | |
| Control of monitoring and measuring devices | 7.6 | 4.5.1 | Monitoring and measurement |
| **Measurement, analysis and improvement** | **8** | 4.5 | Checking and corrective action |
| General | 8.1 | 4.5.1 | Monitoring and measurement |
| Monitoring and measurement | 8.2 | | |
| Customer satisfaction | 8.2.1 | | |
| Internal audit | 8.2.2 | 4.5.4 | Environmental management system audit |
| Monitoring and measurement of processes | 8.2.3 | 4.5.1 | Monitoring and measurement |
| Monitoring and measurement of product | 8.2.4 | | |
| Control of nonconforming product | 8.3 | 4.5.2 | Nonconformance and corrective and preventive action |
| | | 4.4.7 | Emergency preparedness and response |
| Analysis of data | 8.4 | 4.5.1 | Monitoring and measurement |
| Improvement | 8.5 | 4.2 | Environmental policy |
| Continual improvement | 8.5.1 | 4.3.4 | Environmental management programme(s) |
| Corrective action | 8.5.2 | 4.5.2 | Nonconformance and corrective and preventive action |
| Preventive action | 8.5.3 | | |

**TABLE A.2**
**Correspondence between ISO 14001:1996 and ISO 9001:2000**

| ISO 14001:1996 | | ISO 9001:2000 | |
|---|---|---|---|
| **Introduction** | - | **0** | **Introduction** |
| | | 0.1 | General |
| | | 0.2 | Process approach |
| | | 0.3 | Relationship with ISO 9004 |
| | | 0.4 | Compatibility with other management systems |
| **Scope** | 1 | **1** | **Scope** |
| | | 1.1 | General |
| | | 1.2 | Application |
| **Normative references** | 2 | **2** | **Normative reference** |
| **Definitions** | 3 | **3** | **Terms and definitions** |
| **Environmental management system requirements** | 4 | **4** | **Quality management system** |
| General requirements | 4.1 | 4.1 | General requirements |
| | | 5.5 | Responsibility, authority and communication |
| | | 5.5.1 | Responsibility and authority |
| Environmental policy | 4.2 | 5.1 | Management commitment |
| | | 5.3 | Quality policy |
| | | 8.5 | Improvement |
| Planning | 4.3 | 5.4 | Planning |
| Environmental aspects | 4.3.1 | 5.2 | Customer focus |
| | | 7.2.1 | Determination of requirements related to the product |
| | | 7.2.2 | Review of requirements related to the product |
| Legal and other requirements | 4.3.2 | 5.2 | Customer focus |
| | | 7.2.1 | Determination of requirements related to the product |
| Objectives and targets | 4.3.3 | 5.4.1 | Quality objectives |
| Environmental management programme(s) | 4.3.4 | 5.4.2 | Quality management system planning |
| | | 8.5.1 | Continual improvement |
| Implementation and operation | 4.4 | **7** | **Product realization** |
| | | 7.1 | Planning of product realization |
| Structure and responsibility | 4.4.1 | **5** | **Management responsibility** |
| | | 5.1 | Management commitment |
| | | 5.5.1 | Responsibility and authority |
| | | 5.5.2 | Management representative |
| | | **6** | **Resource management** |
| | | 6.1 | Provision of resources |
| | | 6.2 | Human resources |
| | | 6.2.1 | General |
| | | 6.3 | Infrastructure |
| | | 6.4 | Work environment |
| Training, awareness and competence | 4.4.2 | 6.2.2 | Competence, awareness and training |
| Communication | 4.4.3 | 5.5.3 | Internal communication |
| | | 7.2.3 | Customer communication |
| Environmental management system documentation | 4.4.4 | 4.2 | Documentation requirements |
| | | 4.2.1 | General |
| | | 4.2.2 | Quality manual |
| Document control | 4.4.5 | 4.2.3 | Control of documents |

**TABLE A.2**
**Correspondence between ISO 14001:1996 and ISO 9001:2000 (continued)**

| ISO 14001:1996 | | ISO 9001:2000 | |
|---|---|---|---|
| Operational control | 4.4.6 | **7** | **Product realization** |
| | | 7.1 | Planning of product realization |
| | | 7.2 | Customer-related processes |
| | | 7.2.1 | Determination of requirements related to the product |
| | | 7.2.2 | Review of requirements related to the product |
| | | 7.3 | Design and development |
| | | 7.3.1 | Design and development planning |
| | | 7.3.2 | Design and development inputs |
| | | 7.3.3 | Design and development outputs |
| | | 7.3.4 | Design and development review |
| | | 7.3.5 | Design and development verification |
| | | 7.3.6 | Design and development validation |
| | | 7.3.7 | Control of design and development changes |
| | | 7.4 | Purchasing |
| | | 7.4.1 | Purchasing process |
| | | 7.4.2 | Purchasing information |
| | | 7.4.3 | Verification of purchased product |
| | | 7.5 | Production and service provision |
| | | 7.5.1 | Control of production and service provision |
| | | 7.5.3 | Identification and traceability |
| | | 7.5.4 | Customer property |
| | | 7.5.5 | Preservation of product |
| | | 7.5.2 | Validation of processes for production and service provision |
| Emergency preparedness and response | 4.4.7 | 8.3 | Control of nonconforming product |
| Checking and corrective action | 4.5 | **8** | **Measurement, analysis and improvement** |
| Monitoring and measurement | 4.5.1 | 7.6 | Control of monitoring and measuring devices |
| | | 8.1 | General |
| | | 8.2 | Monitoring and measurement |
| | | 8.2.1 | Customer satisfaction |
| | | 8.2.3 | Monitoring and measurement of processes |
| | | 8.2.4 | Monitoring and measurement of product |
| | | 8.4 | Analysis of data |
| Nonconformance and corrective and preventive action | 4.5.2 | 8.3 | Control of nonconforming product |
| | | 8.5.2 | Corrective action |
| | | 8.5.3 | Preventive action |
| Records | 4.5.3 | 4.2.4 | Control of records |
| Environmental management system audit | 4.5.4 | 8.2.2 | Internal audit |
| Management review | 4.6 | 5.6 | Management review |
| | | 5.6.1 | General |
| | | 5.6.2 | Review input |
| | | 5.6.3 | Review output |

# ANNEX B

(informative)

## Correspondence between ISO 9001:2000 and ISO 9001:1994

**TABLE B.1**
**Correspondence between ISO 9001:1994 and ISO 9001:2000**

| ISO 9001:1994 | ISO 9001:2000 |
|---|---|
| **1 Scope** | **1** |
| **2 Normative reference** | **2** |
| **3 Definitions** | **3** |
| **4 Quality system requirements** [title only] | |
| 4.1 Management responsibility [title only]<br>4.1.1 Quality policy<br>4.1.2 Organization [title only]<br>4.1.2.1 Responsibility and authority<br>4.1.2.2 Resources<br>4.1.2.3 Management representative<br>4.1.3 Management review | <br>5.1 + 5.3 + 5.4.1<br><br>5.5.1<br>6.1 + 6.2.1<br>5.5.2<br>5.6.1 + 8.5.1 |
| 4.2 Quality system [title only]<br>4.2.1 General<br>4.2.2 Quality system procedures<br>4.2.3 Quality planning | <br>4.1 + 4.2.2<br>4.2.1<br>5.4.2 + 7.1 |
| 4.3 Contract review [title only]<br>4.3.1 General<br>4.3.2 Review<br>4.3.3 Amendment to a contract<br>4.3.4 Records | <br><br>5.2 + 7.2.1 + 7.2.2 + 7.2.3<br>7.2.2<br>7.2.2 |
| 4.4 Design control [title only]<br>4.4.1 General<br>4.4.2 Design and development planning<br>4.4.3 Organizational and technical interfaces<br>4.4.4 Design input<br>4.4.5 Design output<br>4.4.6 Design review<br>4.4.7 Design verification<br>4.4.8 Design validation<br>4.4.9 Design changes | <br><br>7.3.1<br>7.3.1<br>7.2.1 + 7.3.2<br>7.3.3<br>7.3.4<br>7.3.5<br>7.3.6<br>7.3.7 |
| 4.5 Document and data control [title only]<br>4.5.1 General<br>4.5.2 Document and data approval and issue<br>4.5.3 Document and data changes | <br>4.2.3<br>4.2.3<br>4.2.3 |
| 4.6 Purchasing [title only]<br>4.6.1 General<br>4.6.2 Evaluation of subcontractors<br>4.6.3 Purchasing data<br>4.6.4 Verification of purchased product | <br><br>7.4.1<br>7.4.2<br>7.4.3 |
| 4.7 Control of customer-supplied product | 7.5.4 |
| 4.8 Product identification and traceability | 7.5.3 |

**TABLE B.1**
**Correspondence between ISO 9001:1994 and ISO 9001:2000** (continued)

| ISO 9001:1994 | ISO 9001:2000 |
|---|---|
| 4.9 Process control | 6.3 + 6.4 + 7.5.1 + 7.5.2 |
| 4.10 Inspection and testing [title only]<br>4.10.1 General<br>4.10.2 Receiving inspection and testing<br>4.10.3 In-process inspection and testing<br>4.10.4 Final inspection and testing<br>4.10.5 Inspection and test records | <br>7.1 +8.1<br>7.4.3 + 8.2.4<br>8.2.4<br>8.2.4<br>7.5.3 + 8.2.4 |
| 4.11 Control of inspection, measuring and test equipment [title only]<br>4.11.1 General<br>4.11.2 Control procedure | <br><br>7.6<br>7.6 |
| 4.12 Inspection and test status | 7.5.3 |
| 4.13 Control of nonconforming product [title only]<br>4.13.1 General<br>4.13.2 Review and disposition of nonconforming product | <br>8.3<br>8.3 |
| 4.14 Corrective and preventive action [title only]<br>4.14.1 General<br>4.14.2 Corrective action<br>4.14.3 Preventive action | <br>8.5.2 + 8.5.3<br>8.5.2<br>8.5.3 |
| 4.15 Handling, storage, packaging, preservation & delivery [title only]<br>4.15.1 General<br>4.15.2 Handling<br>4.15.3 Storage<br>4.15.4 Packaging<br>4.15.5 Preservation<br>4.15.6 Delivery | <br><br>7.5.5<br>7.5.5<br>7.5.5<br>7.5.5<br>7.5.1 |
| 4.16 Control of quality records | 4.2.4 |
| 4.17 Internal quality audits | 8.2.2 + 8.2.3 |
| 4.18 Training | 6.2.2 |
| 4.19 Servicing | 7.5.1 |
| 4.20 Statistical techniques [title only]<br>4.20.1 Identification of need<br>4.20.2 Procedures | <br>8.1 + 8.2.3 + 8.2.4 + 8.4<br>8.1 + 8.2.3 + 8.2.4 + 8.4 |

**TABLE B.2**
**Correspondence between ISO 9001:2000 and ISO 9001:1994**

| ISO 9001:2000 | ISO 9001:1994 |
|---|---|
| **1 Scope** | 1 |
| 1.1 General | |
| 1.2 Application | |
| **2 Normative reference** | 2 |
| **3 Terms and definitions** | 3 |
| **4 Quality management system** [title only] | |
| 4.1 General requirements | 4.2.1 |
| 4.2 Documentation requirements [title only]<br>4.2.1 General<br>4.2.2 Quality manual<br>4.2.3 Control of documents<br>4.2.4 Control of records | <br>4.2.2<br>4.2.1<br>4.5.1 + 4.5.2 + 4.5.3<br>4.16 |
| **5 Management responsibility** [title only] | |
| 5.1 Management commitment | 4.1.1 |
| 5.2 Customer focus | 4.3.2 |
| 5.3 Quality policy | 4.1.1 |
| 5.4 Planning [title only]<br>5.4.1 Quality objectives<br>5.4.2 Quality management system planning | <br>4.1.1<br>4.2.3 |
| 5.5 Responsibility, authority and communication [title only]<br>5.5.1 Responsibility and authority<br>5.5.2 Management representative<br>5.5.3 Internal communication | <br>4.1.2.1<br>4.1.2.3<br> |
| 5.6 Management review [title only]<br>5.6.1 General<br>5.6.2 Review input<br>5.6.3 Review output | <br>4.1.3<br><br> |
| **6 Resource management** [title only] | |
| 6.1 Provision of resources | 4.1.2.2 |
| 6.2 Human resources [title only]<br>6.2.1 General<br>6.2.2 Competence, awareness and training | <br>4.1.2.2<br>4.18 |
| 6.3 Infrastructure | 4.9 |
| 6.4 Work environment | 4.9 |
| **7 Product realization** [title only] | |
| 7.1 Planning of product realization | 4.2.3 + 4.10.1 |
| 7.2 Customer-related processes [title only]<br>7.2.1 Determination of requirements related to the product<br>7.2.2 Review of requirements related to the product<br>7.2.3 Customer communication | <br>4.3.2 + 4.4.4<br>4.3.2 + 4.3.3 + 4.3.4<br>4.3.2 |

**TABLE B.2**
**Correspondence between ISO 9001:2000 and ISO 9001:1994 (continued)**

| ISO 9001:2000 | ISO 9001:1994 |
|---|---|
| 7.3 Design and development [title only]<br>7.3.1 Design and development planning<br>7.3.2 Design and development inputs<br>7.3.3 Design and development outputs<br>7.3.4 Design and development review<br>7.3.5 Design and development verification<br>7.3.6 Design and development validation<br>7.3.7 Control of design and development changes | <br>4.4.2 + 4.4.3<br>4.4.4<br>4.4.5<br>4.4.6<br>4.4.7<br>4.4.8<br>4.4.9 |
| 7.4 Purchasing [title only]<br>7.4.1 Purchasing process<br>7.4.2 Purchasing information<br>7.4.3 Verification of purchased product | <br>4.6.2<br>4.6.3<br>4.6.4 + 4.10.2 |
| 7.5 Production and service provision [title only]<br>7.5.1 Control of production and service provision<br>7.5.2 Validation of processes for production and service provision<br>7.5.3 Identification and traceability<br>7.5.4 Customer property<br>7.5.5 Preservation of product | <br>4.9 + 4.15.6 + 4.19<br>4.9<br>4.8 + 4.10.5 + 4.12<br>4.7<br>4.15.2 + 4.15.3 + 4.15.4 +.4.15.5 |
| 7.6 Control of monitoring and measuring devices | 4.11.1 + 4.11.2 |
| **8 Measurement, analysis and improvement** [title only] | |
| 8.1 General | 4.10.1 + 4.20.1 +4.20.2 |
| 8.2 Monitoring and measurement [title only]<br>8.2.1 Customer satisfaction<br>8.2.2 Internal audit<br>8.2.3 Monitoring and measurement of processes<br>8.2.4 Monitoring and measurement of product | <br><br>4.17<br>4.17 + 4.20.1 + 4.20.2<br>4.10.2 + 4.10.3 + 4.10.4 + 4.10.5 + 4.20.1 + 4.20.2 |
| 8.3 Control of nonconforming product | 4.13.1 + 4.13.2 |
| 8.4 Analysis of data | 4.20.1 + 4.20.2 |
| 8.5 Improvement [title only]<br>8.5.1 Continual improvement<br>8.5.2 Corrective action<br>8.5.3 Preventive action | <br>4.1.3<br>4.14.1 + 4.14.2<br>4.14.1 + 4.14.3 |

## BIBLIOGRAPHY

[1] ISO 9000-3:1997, Quality management and quality assurance standards — Part 3: Guidelines for the application of ISO 9001:1994 to the development, supply, installation and maintenance of computer software.

[2] ISO 9004:2000, Quality management systems — Guidelines for performance improvements.

[3] ISO 10005:1995, Quality management — Guidelines for quality plans.

[4] ISO 10006:1997, Quality management — Guidelines to quality in project management.

[5] ISO 10007:1995, Quality management — Guidelines for configuration management.

[6] ISO 10011-1:1990, Guidelines for auditing quality systems — Part 1: Auditing[1].

[7] ISO 10011-2:1991, Guidelines for auditing quality systems — Part 2: Qualification criteria for quality systems auditors[1].

[8] ISO 10011-3:1991, Guidelines for auditing quality systems — Part 3: Management of audit programmes[1].

[9] ISO 10012-1:1992, Quality assurance requirements for measuring equipment — Part 1: Metrological confirmation system for measuring equipment.

[10] ISO 10012-2:1997, Quality assurance for measuring equipment — Part 2: Guidelines for control of measurement processes.

[11] ISO 10013:1995, Guidelines for developing quality manuals.

[12] ISO/TR 10014:1998, Guidelines for managing the economics of quality.

[13] ISO 10015:1999, Quality management — Guidelines for training.

[14] ISO/TR 10017:1999, Guidance on statistical techniques for ISO 9001:1994.

[15] ISO 14001:1996, Environmental management systems — Specification with guidance for use.

[16] I EC 60300-1:—[2], Dependability management — Part 1: Dependability programme management.

[17] Quality Management Principles Brochure[3].

[18] ISO 9000 + ISO 14000 News (a bimonthly publication which provides comprehensive coverage of international developments relating to ISO's management system standards, including news of their implementation by diverse organizations around the world )[4].

[19] Reference websites: http://www.iso.ch http://www.bsi.org.uk/iso-tcl 76-sc2

1) To be revised as ISO 19011, *Guidelines on quality and/or environmental management systems auditing.*
2) To be published. (Revision of ISO 9000-4:1993)
3) Available from website: http://www.iso.ch
4) Available from ISO Central Secretariat (sales@iso.ch).

# INTERNATIONAL STANDARD ISO 13485

First edition 1996-12-15

# Quality Systems — Medical Devices — Particular Requirements for the Application of ISO 9001

## CONTENT

## FOREWORD

ISO (the International Organization for Standardization) is a worldwide federation of national standards bodies (ISO member bodies). The work of preparing International Standards is normally carried out through ISO technical committees. Each member body interested in a subject for which a technical committee has been established has the right to be represented on that committee. International organizations, governmental and nongovernmental, in liaison with ISO, also take part in the work. ISO collaborates closely with the International Electrotechnical Commission (IEC) on all matters of electrotechnical standardization.

Draft International Standards adopted by the technical committees are circulated to the member bodies for voting. Publication as an International Standard requires approval by at least 75% of the member bodies casting a vote.

International Standard ISO 13485 was prepared by Technical Committee ISOITC 210, *Quality management and corresponding general aspects for medical devices.*

## INTRODUCTION

ISO 9001 is a general standard defining quality system requirements. ISO 13485 provides particular requirements for suppliers of medical devices that are more specific than the general requirements specified in ISO 9001.

In conjunction with ISO 9001, this International Standard defines requirements for quality systems relating to the design/development, production, installation and servicing of medical devices. It embraces all the principles of good manufacturing practice (GMP) widely used in the manufacture of medical devices. It can only be used in combination with ISO 9001 and is not an independent standard.

There are a wide variety of medical devices and some of the particular requirements of this International Standard only apply to named groups of medical devices. These groups are described in clause 3.

Other International Standards specify more detailed particular requirements that are additional to those specified here. Suppliers should review the requirements and consider using the relevant International Standards in these areas.

To assist in the understanding of the requirements of this International Standard, an international guidance standard is being prepared.

## 1 SCOPE

This International Standard specifies, in conjunction with ISO 9001, the quality system requirements for the design/development, production and, when relevant, installation and servicing of medical devices.

This International Standard, in conjunction with ISO 9001, is applicable when there is a need to assess a medical device supplier's quality system.

As part of an assessment by a third party for the purpose of regulatory requirements, the supplier may be required to provide access to confidential data in order to demonstrate compliance with this International Standard. The supplier may be required to exhibit these data but is not obliged to provide copies for retention.

NOTE In this International Standard the term "if appropriate" is used several times. When a requirement is qualified by this phrase, it is deemed to be "appropriate" unless the supplier can document a justification otherwise. A requirement is considered "appropriate" if its non-implementation could result in

- the product not meeting its specified requirements, and/or
- the supplier being unable to carry out corrective action.

## 2 NORMATIVE REFERENCES

The following standards contain provisions which, through reference in this text, constitute provisions of this International Standard. At the time of publication, the editions indicated were valid. All standards are subject to revision, and parties to agreements based on this International Standard are encouraged to investigate the possibility of applying the most recent editions of the standards indicated below. Members of IEC and ISO maintain registers of currently valid International Standards.

ISO 8402:1994, Quality management and quality assurance — Vocabulary.

ISO 9001:1994, Quality systems — Model for quality assurance in design, development, production, installation and servicing.

ISO 11137:1995, Sterilization of healthcare products — Requirements for validation and routine control — Radiation sterilization.

# 3 DEFINITIONS

For the purposes of this International Standard, the definitions given in ISO 8402 apply, with the exception that the definition of "product" as given in ISO 9001 applies. In addition, the following definitions apply.

NOTE   These definitions should be regarded as generic, as definitions provided in national regulations may differ slightly.

**3.1 medical device:** Any instrument, apparatus, appliance, material or other article, whether used alone or in combination, including the software necessary for its proper application, intended by the manufacturer to be used for human beings for the purpose of

- diagnosis, prevention, monitoring, treatment or alleviation of disease,
- diagnosis, monitoring, treatment, alleviation of, or compensation for, an injury or handicap,
- investigation, replacement or modification of the anatomy or of a physiological process,
- control of conception,

and which does not achieve its principal intended action in or on the human body by pharmacological, immunological or metabolic means, but which may be assisted in its function by such means.

NOTE   In addition to the medical device categories defined hereinafter, the term "medical device" also includes non-active medical devices and *in vitro* diagnostic devices.

**3.2 active medical device**: Any medical device (see 3.1) relying for its functioning on a source of electrical energy or any source of power other than that directly generated by the human body or gravity.

**3.3 active implantable medical device**: Any active medical device (see 3.1 and 3.2) which is intended to be totally or partially introduced, surgically or medically, into the human body or by medical intervention into a natural orifice, and which is intended to remain after the procedure.

**3.4 implantable medical device**: Any medical device (see 3.1) intended

- to be totally or partially introduced into the human body or a natural orifice, or
- to replace an epithelial surface or the surface of the eye,

by surgical intervention, and which is intended to remain after the procedure for at least 30 days, and which can only be removed by medical or surgical intervention.

NOTE   This definition applies to implantable medical devices other than active implantable medical devices.

**3.5 sterile medical device**: Any medical device labelled as sterile. (See 3.6.1 of ISO 11137:1995.)

NOTE   Requirements for labelling a medical device as sterile may be subject to national or regional regulations or standards.

**3.6 labelling**: Written, printed or graphic matter

- affixed to a medical device or any of its containers or wrappers, or
- accompanying a medical device, related to identification, technical description and use of the medical device, but excluding shipping documents.

NOTE   For the purposes of this International Standard, the term "marking" as used in ISO 9001 is interpreted to mean "labelling."

**3.7 customer complaint**: Any written, electronic or oral communication that alleges deficiencies related to the identity, quality, durability, reliability, safety or performance of a medical device (see 3.1) that has been placed on the market.

**3.8 advisory notice**: Notice issued by the supplier, subsequent to delivery of the medical device, to provide supplementary information and/or to advise what action should be taken in

- the use of a medical device,
- the modification of a medical device,
- the return to the supplier of a medical device,
- the destruction of a medical device,

for the purpose of corrective or preventive action and in compliance with national and regional regulatory requirements.

# 4 QUALITY SYSTEM REQUIREMENTS

## 4.1 MANAGEMENT RESPONSIBILITY

The requirements given in 4.1 of ISO 9001:1994 apply.

## 4.2 QUALITY SYSTEM

### 4.2.1 General

The requirements given in 4.2.1 of ISO 9001:1994 apply.

*Particular requirement for all medical devices*:

The supplier shall establish and document the specified requirements.

NOTE   If this International Standard is used for compliance with regulatory requirements, the relevant requirements of the regulations should be included in the specified requirements.

### 4.2.2 Quality System Procedures

The requirements given in 4.2.2 of ISO 9001:1994 apply.

### 4.2.3 Quality Planning

The requirements given in 4.2.3 of ISO 9001:1994 apply.

*Particular requirement for all medical devices:*

The supplier shall establish and maintain a file containing documents defining product specifications and quality system requirements (process and quality assurance) for

- complete manufacturing, and
- installation and servicing, if appropriate,

for each type/model of medical device, or referring to the location(s) of this information (see 4.5.2 and 4.16).

### 4.3 CONTRACT REVIEW

The requirements given in 4.3 of ISO 9001:1994 apply.

### 4.4 DESIGN CONTROL

### 4.4.1 General

The requirements given in 4.4.1 of ISO 9001:1994 apply.

*Particular requirement for all medical devices:*

Throughout the design process, the supplier shall evaluate the need for risk analysis and maintain records of any risk analyses performed.

### 4.4.2 Design and Development Planning

The requirements given in 4.4.2 of ISO 9001:1994 apply.

### 4.4.3 Organizational and Technical Interfaces

The requirements given in 4.4.3 of ISO 9001:1994 apply.

### 4.4.4 Design Input

The requirements given in 4.4.4 of ISO 9001:1994 apply.

### 4.4.5 Design Output

The requirements given in 4.4.5 of ISO 9001:1994 apply.

### 4.4.6 Design Review

The requirements given in 4.4.6 of ISO 9001:1994 apply.

### 4.4.7 Design Verification

The requirements given in 4.4.7 of ISO 9001:1994 apply.

### 4.4.8 Design Validation

The requirements given in 4.4.8 of ISO 9001:1994 apply.

*Particular requirement for all medical devices:*

As part of design validation, the supplier shall perform and maintain records of clinical evaluations.

NOTE   Clinical evaluation may include a compilation of relevant scientific literature, historical evidence that similar designs and/or materials are clinically safe, or a clinical investigation or trial, to ensure that the device performs as intended. National or regional regulations may require actual clinical investigations or trials.

### 4.4.9 Design Changes

The requirements given in 4.4.9 of ISO 9001:1994 apply.

### 4.5 DOCUMENT AND DATA CONTROL

### 4.5.1 General

The requirements given in 4.5.1 of ISO 9001:1994 apply.

### 4.5.2 Document and Data Approval and Issue

The requirements given in 4.5.2 of ISO 9001:1994 apply.

*Particular requirement for all medical devices:*

The supplier shall define the period for which at least one copy of obsolete controlled documents shall be retained. This period shall ensure that specifications to which medical devices have been manufactured are available for at least the lifetime of the medical device as defined by the supplier.

### 4.5.3 Document and Data Changes

The requirements given in 4.5.3 of ISO 9001:1994 apply.

### 4.6 PURCHASING

### 4.6.1 General

The requirements given in 4.6.1 of ISO 9001:1994 apply.

### 4.6.2 Evaluation of Subcontractors

The requirements given in 4.6.2 of ISO 9001:1994 apply.

### 4.6.3 Purchasing Data

The requirements given in 4.6.3 of ISO 9001:1994 apply.

*Particular requirement for all medical devices:*

To the extent required by the particular requirements for traceability given in 4.8, the supplier shall retain copies (see 4.16) of relevant purchasing documents.

### 4.6.4 Verification of Purchased Product

The requirements given in 4.6.4 of ISO 9001:1994 apply.

## 4.7 CONTROL OF CUSTOMER-SUPPLIED PRODUCT

The requirements given in 4.7 of ISO 9001:1994 apply.

## 4.8 PRODUCT IDENTIFICATION AND TRACEABILITY

The requirements given in 4.8 of ISO 9001:1994 apply.

*Particular requirements for all medical devices:*

a) Identification

The supplier shall establish and maintain procedures to ensure that medical devices returned to the supplier for reprocessing to specified requirements are identified and distinguished at all times from normal production (see 4.15.1).

b) Traceability

The supplier shall establish, document and maintain procedures for traceability. The procedures shall define the extent of traceability and shall facilitate corrective and preventive action (see 4.14).

*Additional requirements for active implantable medical devices and implantable medical devices:*

When defining the extent of traceability, the supplier shall include all components and materials used and records of the environmental conditions [see 4.9 b) 4)] when these could cause the medical device not to satisfy its specified requirements.

The supplier shall require that its agents or distributors maintain records of the distribution of medical devices with regard to traceability and that such records are available for inspection.

## 4.9 PROCESS CONTROL

The requirements given in 4.9 of ISO 9001:1994 apply.

*Particular requirements for all medical devices:*

a) Personnel

The supplier shall establish, document and maintain requirements for health, cleanliness and clothing of personnel if contact between such personnel and the product or the environment could adversely affect the quality of the product.

The supplier shall ensure that all personnel who are required to work temporarily under special environmental conditions are appropriately trained or supervised by a trained person (see 4.18).

b) Environmental control in manufacture

For medical devices

1) that are supplied sterile, or
2) that are supplied non-sterile and intended for sterilization before use, or
3) if the microbiological and/or particulate cleanliness or other environmental conditions are of significance in their use, or
4) if the environmental conditions are of significance in their manufacture,

the supplier shall establish and document requirements for the environment to which the product is exposed.

If appropriate, the environmental conditions shall be controlled and/or monitored.

c) Cleanliness of product

The supplier shall establish, document and maintain requirements for the cleanliness of the product if

1) product is cleaned by the supplier prior to sterilization and/or its use, or
2) product is supplied non-sterile to be subjected to a cleaning process prior to sterilization and/or its use, or
3) product is supplied to be used non-sterile and its cleanliness is of significance in use, or
4) process agents are to be removed from product during manufacture.

If appropriate, product cleaned in accordance with 1) or 2) above need not be subject to the preceding particular requirements [i.e., a) Personnel, and b) Environmental control in manufacture] prior to the cleaning procedure.

d) Maintenance

The supplier shall establish and document requirements for maintenance activities when such activities may affect product quality.

Records of such maintenance shall be kept (see 4.16).

e) Installation

If appropriate, the supplier shall establish and document both instructions and acceptance criteria for installing and checking the medical device.

Records of installation and checking performed by the supplier or its authorized representative shall be retained (see 4.16).

If the contract (see 4.3) allows installation other than by the supplier or its authorized representative, the supplier shall provide the purchaser with written instructions for installation and checking.

f) Computer software used in process control

The supplier shall establish and maintain documented procedures for the validation of the application of computer software which is used for process control. The results of the validation shall be recorded (see 4.16).

*Additional requirement for sterile medical devices:*

The supplier shall subject the medical device to a validated sterilization process and record (see 4.16) all the control parameters of the sterilization process.

## 4.10 INSPECTION AND TESTING

### 4.10.1 General

The requirements given in 4.10.1 of ISO 9001:1994 apply.

## 4.10.2 Receiving Inspection and Testing

The requirements given in 4.10.2 of ISO 9001:1994 apply.

## 4.10.3 In-Process Inspection and Testing

The requirements given in 4.10.3 of ISO 9001:1994 apply.

## 4.10.4 Final Inspection and Testing

The requirements given in 4.10.4 of ISO 9001:1994 apply.

## 4.10.5 Inspection and Test Records

The requirements given in 4.10.5 of ISO 9001:1994 apply.

*Particular requirement for active implantable medical devices and implantable medical devices:*

The supplier shall record (see 4.16) the identity of personnel performing any inspection or testing.

## 4.11 CONTROL OF INSPECTION, MEASURING AND TEST EQUIPMENT

The requirements given in 4.11 of ISO 9001:1994 apply.

## 4.12 INSPECTION AND TEST STATUS

The requirements given in 4.12 of ISO 9001:1994 apply.

## 4.13 CONTROL OF NONCONFORMING PRODUCT

## 4.13.1 General

The requirements given in 4.13.1 of ISO 9001:1994 apply.

## 4.13.2 Review and Disposition of Nonconforming Product

The requirements given in 4.13.2 of ISO 9001:1994 apply.

*Particular requirements for all medical devices:*

The supplier shall ensure that nonconforming product is accepted by concession only if regulatory requirements are met. The identity of the person(s) authorizing the concession shall be recorded (see 4.16).

If product needs to be reworked (one or more times), the supplier shall document the rework in a work instruction that has undergone the same authorization and approval procedure as the original work instruction. Prior to authorization and approval, a determination of any adverse effect of the rework upon product shall be made and documented.

## 4.14 CORRECTIVE AND PREVENTIVE ACTION

## 4.14.1 General

The requirements given in 4.14.1 of ISO 9001:1994 apply.

*Particular requirements for all medical devices:*

The supplier shall establish and maintain a documented feedback system to provide early warning of quality problems and for input into the corrective and/or preventive action system.

If this International Standard is used for compliance with regulatory requirements which require the supplier to gain experience from the post-production phase, the review of this experience shall form part of the feedback system.

The supplier shall maintain records (see 4.16) of all customer complaint investigations. When the investigation determines that the activities at remote premises contributed to the customer complaint, relevant information shall be communicated between the supplier and the remote premises.

If any customer complaint is not followed by corrective and/or preventive action, the reason shall be recorded.

If this International Standard is used for compliance with regulatory requirements, the supplier shall establish, document and maintain procedures to notify the regulatory authority of those incidents which meet the reporting criteria.

The supplier shall establish, document and maintain procedures for the issue of advisory notices for medical devices. These procedures shall be capable of being implemented at any time.

### 4.14.2 Corrective Action

The requirements given in 4.14.2 of ISO 9001:1994 apply.

### 4.14.3 Preventive Action

The requirements given in 4.14.3 of ISO 9001:1994 apply.

### 4.15 HANDLING, STORAGE, PACKAGING, PRESERVATION AND DELIVERY

### 4.15.1 General

The requirements given in 4.15.1 of ISO 9001:1994 apply.

*Particular requirements for all medical devices:*

The supplier shall establish and maintain documented procedures for the control of product with a limited shelflife or requiring special storage conditions. Such special storage conditions shall be controlled and recorded (see 4.16).

If appropriate, special arrangements shall be established, documented and maintained for the control of used product in order to prevent contamination of other product, the manufacturing environment or personnel.

### 4.15.2 Handling

The requirements given in 4.15.2 of ISO 9001:1994 apply.

### 4.15.3 Storage

The requirements given in 4.15.3 of ISO 9001:1994 apply.

### 4.15.4 Packaging

The requirements given in 4.15.4 of ISO 9001:1994 apply.

*Particular requirement for active implantable medical devices and implantable medical devices:*

The supplier shall record the identity of persons who perform the final labelling operation (see 4.16).

### 4.15.5 Preservation

The requirements given in 4.15.5 of ISO 9001:1994 apply.

### 4.15.6 Delivery

The requirements given in 4.15.6 of ISO 9001:1994 apply.

*Particular requirement for active implantable medical devices and implantable medical devices:*

The supplier shall ensure that the name and address of the shipping package consignee is included in the quality records (see 4.16).

### 4.16 CONTROL OF QUALITY RECORDS

The requirements given in 4.16 of ISO 9001:1994 apply.

*Particular requirements for all medical devices:*

The supplier shall retain the quality records for a period of time at least equivalent to the lifetime of the medical device as defined by the supplier, but not less than 2 years from the date of despatch from the supplier.

NOTE 1    National or regional regulations may require a period longer than 2 years.

The supplier shall establish and maintain a record for each batch of medical devices that provides traceability to the extent specified in 4.8 and identifies the quantity manufactured and quantity approved for distribution. The batch record shall be verified and authorized.

NOTE 2    A batch may be a single medical device.

### 4.17 INTERNAL QUALITY AUDITS

The requirements given in 4.17 of ISO 9001:1994 apply.

### 4.18 TRAINING

The requirements given in 4.18 of ISO 9001:1994 apply.

### 4.19 SERVICING

The requirements given in 4.19 of ISO 9001:1994 apply.

### 4.20 STATISTICAL TECHNIQUES

The requirements given in 4.20 of ISO 9001:1994 apply.

# International Standard ISO 13488

First edition 1996-12-15

# Quality Systems — Medical Devices — Particular Requirements for the Application of ISO 9002

Reference number
ISO 13488:1996(E)

## CONTENT

## FOREWORD

ISO (the International Organization for Standardization) is a worldwide federation of national standards bodies (ISO member bodies). The work of preparing International Standards is normally carried out through ISO technical committees. Each member body interested in a subject for which a technical committee has been established has the right to be represented on that committee. International organizations, governmental and nongovernmental, in liaison with ISO, also take part in the work. ISO collaborates closely with the International Electrotechnical Commission (IEC) on all matters of electrotechnical standardization.

Draft International Standards adopted by the technical committees are circulated to the member bodies for voting. Publication as an International Standard requires approval by at least 75% of the member bodies casting a vote.

International Standard ISO 13488 was prepared by Technical Committee ISOlTC 210, *Quality management and corresponding general aspects for medical devices.*

## INTRODUCTION

ISO 9002 is a general standard defining quality system requirements. ISO 13488 provides particular requirements for suppliers of medical devices that are more specific than the general requirements specified in ISO 9002.

In conjunction with ISO 9002, this International Standard defines requirements for quality systems relating to the production, installation and servicing of medical devices. It embraces all the principles of good manufacturing practice (GMP) widely used in the manufacture of medical devices. It can only be used in combination with ISO 9002 and is not an independent standard.

There are a wide variety of medical devices and some of the particular requirements of this International Standard only apply to named groups of medical devices. These groups are described in clause 3.

Other International Standards specify more detailed particular requirements that are additional to those specified here. Suppliers should review the requirements and consider using the relevant International Standards in these areas.

To assist in the understanding of the requirements of this International Standard, an international guidance standard is being prepared.

# 1 SCOPE

This International Standard specifies, in conjunction with ISO 9002, the quality system requirements for the production and, when relevant, installation and servicing of medical devices.

This International Standard, in conjunction with ISO 9002, is applicable when there is a need to assess a medical device supplier's quality system.

As part of an assessment by a third party for the purpose of regulatory requirements, the supplier may be required to provide access to confidential data in order to demonstrate compliance with this International Standard. The supplier may be required to exhibit these data but is not obliged to provide copies for retention.

NOTE — In this International Standard the term "if appropriate" is used several times. When a requirement is qualified by this phrase, it is deemed to be "appropriate" unless the supplier can document a justification otherwise. A requirement is considered "appropriate" if its non-implementation could result in

- the product not meeting its specified requirements, and/or
- the supplier being unable to carry out corrective action.

# 2 NORMATIVE REFERENCES

The following standards contain provisions which, through reference in this text, constitute provisions of this International Standard. At the time of publication, the editions indicated were valid. All standards are subject to revision, and parties to agreements based on this International Standard are encouraged to investigate the possibility of applying the most recent editions of the standards indicated below. Members of IEC and ISO maintain registers of currently valid International Standards.

ISO 8402:1994, Quality management and quality assurance — Vocabulary.

ISO 9002:1994, Quality systems — Model for quality assurance in production, installation and servicing.

ISO 11137:1995, Sterilization of healthcare products — Requirements for validation and routine control — Radiation sterilization.

## 3 DEFINITIONS

For the purposes of this International Standard, the definitions given in ISO 8402 apply, with the exception that the definition of "product" as given in ISO 9002 applies. In addition, the following definitions apply.

NOTE — These definitions should be regarded as generic, as definitions provided in national regulations may differ slightly.

**3.1 medical device**: Any instrument, apparatus, appliance, material or other article, whether used alone or in combination, including the software necessary for its proper application, intended by the manufacturer to be used for human beings for the purpose of

- diagnosis, prevention, monitoring, treatment or alleviation of disease,
- diagnosis, monitoring, treatment, alleviation of, or compensation for, an injury or handicap,
- investigation, replacement or modification of the anatomy or of a physiological process,
- control of conception,

and which does not achieve its principal intended action in or on the human body by pharmacological, immunological or metabolic means, but which may be assisted in its function by such means.

NOTE — In addition to the medical device categories defined hereinafter, the term "medical device" also includes non-active medical devices and *in vitro* diagnostic devices.

**3.2 active medical device**: Any medical device (see 3.1) relying for its functioning on a source of electrical energy or any source of power other than that directly generated by the human body or gravity.

**3.3 active implantable medical device**: Any active medical device (see 3.1 and 3.2) which is intended to be totally or partially introduced, surgically or medically, into the human body or by medical intervention into a natural orifice, and which is intended to remain after the procedure.

**3.4 implantable medical device**: Any medical device (see 3.1) intended

- to be totally or partially introduced into the human body or a natural orifice, or
- to replace an epithelial surface or the surface of the eye,

by surgical intervention, and which is intended to remain after the procedure for at least 30 days, and which can only be removed by medical or surgical intervention.

NOTE — This definition applies to implantable medical devices other than active implantable medical devices.

**3.5 sterile medical device**: Any medical device labelled as sterile. (See 3.6.1 of ISO 11137:1995.)

NOTE — Requirements for labelling a medical device as sterile may be subject to national or regional regulations or standards.

**3.6 labelling**: Written, printed or graphic matter

- affixed to a medical device or any of its containers or wrappers, or
- accompanying a medical device,

related to identification, technical description and use of the medical device, but excluding shipping documents.

NOTE — For the purposes of this International Standard, the term "marking" as used in ISO 9002 is interpreted to mean "labelling."

**3.7 customer complaint**: Any written, electronic or oral communication that alleges deficiencies related to the identity, quality, durability, reliability, safety or performance of a medical device (see 3.1) that has been placed on the market.

**3.8 advisory notice**: Notice issued by the supplier, subsequent to delivery of the medical device, to provide supplementary information and/or to advise what action should be taken in

- the use of a medical device,
- the modification of a medical device,
- the return to the supplier of a medical device,
- the destruction of a medical device,

for the purpose of corrective or preventive action and in compliance with national and regional regulatory requirements.

# 4 QUALITY SYSTEM REQUIREMENTS

## 4.1 MANAGEMENT RESPONSIBILITY

The requirements given in 4.1 of ISO 9002:1994 apply.

## 4.2 QUALITY SYSTEM

### 4.2.1 General

The requirements given in 4.2.1 of ISO 9002:1994 apply.

*Particular requirement for all medical devices*:

The supplier shall establish and document the specified requirements.

NOTE — If this International Standard is used for compliance with regulatory requirements, the relevant requirements of the regulations should be included in the specified requirements.

### 4.2.2 Quality System Procedures

The requirements given in 4.2.2 of ISO 9002:1994 apply.

### 4.2.3 Quality Planning

The requirements given in 4.2.3 of ISO 9002:1994 apply.

*Particular requirement for all medical devices:*

The supplier shall establish and maintain a file containing documents defining product specifications and quality system requirements (process and quality assurance) for

- complete manufacturing, and
- installation and servicing, if appropriate,

for each type/model of medical device, or referring to the location(s) of this information (see 4.5.2 and 4.16).

## 4.3 CONTRACT REVIEW

The requirements given in 4.3 of ISO 9002:1994 apply.

## 4.4   DESIGN CONTROL

See 4.4 of ISO 9002:1994.

## 4.5 DOCUMENT AND DATA CONTROL

### 4.5.1 General

The requirements given in 4.5.1 of ISO 9002:1994 apply.

### 4.5.2 Document and Data Approval and Issue

The requirements given in 4.5.2 of ISO 9002:1994 apply.

*Particular requirement for all medical devices:*

The supplier shall define the period for which at least one copy of obsolete controlled documents shall be retained. This period shall ensure that specifications to which medical devices have been manufactured are available for at least the lifetime of the medical device as defined by the supplier.

### 4.5.3 Document and Data Changes

The requirements given in 4.5.3 of ISO 9002:1994 apply.

## 4.6 PURCHASING

### 4.6.1 General

The requirements given in 4.6.1 of ISO 9002:1994 apply.

### 4.6.2 Evaluation of Subcontractors

The requirements given in 4.6.2 of ISO 9002:1994 apply.

### 4.6.3 PURCHASING DATA

The requirements given in 4.6.3 of ISO 9002:1994 apply.

*Particular requirement for all medical devices:*

To the extent required by the particular requirements for traceability given in 4.8, the supplier shall retain copies (see 4.16) of relevant purchasing documents.

### 4.6.4 Verification of Purchased Product

The requirements given in 4.6.4 of ISO 9002:1994 apply.

### 4.7 CONTROL OF CUSTOMER-SUPPLIED PRODUCT

The requirements given in 4.7 of ISO 9002:1994 apply.

### 4.8 PRODUCT IDENTIFICATION AND TRACEABILITY

The requirements given in 4.8 of ISO 9002:1994 apply.

*Particular requirements for all medical devices:*

a) Identification

The supplier shall establish and maintain procedures to ensure that medical devices returned to the supplier for reprocessing to specified requirements are identified and distinguished at all times from normal production (see 4.15.1).

b) Traceability

The supplier shall establish, document and maintain procedures for traceability. The procedures shall define the extent of traceability and shall facilitate corrective and preventive action (see 4.14). *Additional requirements for active implantable medical devices and implantable medical devices:*
When defining the extent of traceability, the supplier shall include all components and materials used and records of the environmental conditions [see 4.9 b) 4)] when these could cause the medical device not to satisfy its specified requirements.

The supplier shall require that its agents or distributors maintain records of the distribution of medical devices with regard to traceability and that such records are available for inspection.

### 4.9 PROCESS CONTROL

The requirements given in 4.9 of ISO 9002:1994 apply.

*Particular requirements for all medical devices:*

a) Personnel

The supplier shall establish, document and maintain requirements for health, cleanliness and clothing of personnel if contact between such personnel and the product or the environment could adversely affect the quality of the product.

The supplier shall ensure that all personnel who are required to work temporarily under special environmental conditions are appropriately trained or supervised by a trained person (see 4.18).

b) Environmental control in manufacture

For medical devices

1) that are supplied sterile, or
2) that are supplied non-sterile and intended for sterilization before use, or
3) if the microbiological and/or particulate cleanliness or other environmental conditions are of significance in their use, or
4) if the environmental conditions are of significance in their manufacture,

the supplier shall establish and document requirements for the environment to which the product is exposed.

If appropriate, the environmental conditions shall be controlled and/or monitored.

c) Cleanliness of product

The supplier shall establish, document and maintain requirements for the cleanliness of the product if

1) product is cleaned by the supplier prior to sterilization and/or its use, or
2) product is supplied non-sterile to be subjected to a cleaning process prior to sterilization and/or its use, or
3) product is supplied to be used non-sterile and its cleanliness is of significance in use, or
4) process agents are to be removed from product during manufacture.

If appropriate, product cleaned in accordance with 1) or 2) above need not be subject to the preceding particular requirements [i.e., a) Personnel, and b) Environmental control in manufacture] prior to the cleaning procedure.

d) Maintenance

The supplier shall establish and document requirements for maintenance activities when such activities may affect product quality.

Records of such maintenance shall be kept (see 4.16).

e) Installation

If appropriate, the supplier shall establish and document both instructions and acceptance criteria for installing and checking the, medical device.

Records of installation and, checking performed by the supplier or its authorized representative shall be retained (see 4.16).

If the contract (see 4.3) allows installation other than by the supplier or its authorized representative, the supplier shall provide the purchaser with written instructions for installation and checking.

f) Computer software used in process control

The supplier shall establish and maintain documented procedures for the validation of the application of computer software which is used for process control. The results of the validation shall be recorded (see 4.16).

*Additional requirement for sterile medical devices:*

The supplier shall subject the medical device to a validated sterilization process and record (see 4.16) all the control parameters of the sterilization process.

## 4.10 INSPECTION AND TESTING

### 4.10.1 General

The requirements given in 4.10.1 of ISO 9002:1994 apply.

### 4.10.2 Receiving Inspection and Testing

The requirements given in 4.10.2 of ISO 9002:1994 apply.

### 4.10.3 In-Process Inspection and Testing

The requirements given in 4.10.3 of ISO 9002:1994 apply.

### 4.10.4 Final Inspection and Testing

The requirements given in 4.10.4 of ISO 9002:1994 apply.

### 4.10.5 Inspection and Test Records

The requirements given in 4.10.5 of ISO 9002:1994 apply.

*Particular requirement for active implantable medical devices and implantable medical devices:*

The supplier shall record (see 4.16) the identity of personnel performing any inspection or testing.

## 4.11 CONTROL OF INSPECTION, MEASURING AND TEST EQUIPMENT

The requirements given in 4.11 of ISO 9002:1994 apply.

## 4.12 INSPECTION AND TEST STATUS

The requirements given in 4.12 of ISO 9002:1994 apply.

## 4.13 CONTROL OF NONCONFORMING PRODUCT

### 4.13.1 General

The requirements given in 4.13.1 of ISO 9002:1994 apply.

### 4.13.2 Review and Disposition of Nonconforming Product

The requirements given in 4.13.2 of ISO 9002:1994 apply.

*Particular requirements for all medical devices:*

The supplier shall ensure that nonconforming product is accepted by concession only if regulatory requirements are met. The identity of the person(s) authorizing the concession shall be recorded (see 4.16).

If product needs to be reworked (one or more times), the supplier shall document the rework in a work instruction that has undergone the same authorization and approval procedure as the original work instruction. Prior to authorization and approval, a determination of any adverse effect of the rework upon product shall be made and documented.

## 4.14 CORRECTIVE AND PREVENTIVE ACTION

### 4.14.1 General

The requirements given in 4.14.1 of ISO 9002:1994 apply.

*Particular requirements for all medical devices:*

The supplier shall establish and maintain a documented feedback system to provide early warning of quality problems and for input into the corrective and/or preventive action system.

If this International Standard is used for compliance with regulatory requirements which require the supplier to gain experience from the post-production phase, the review of this experience shall form part of the feedback system.

The supplier shall maintain records (see 4.16) of all customer complaint investigations. When the investigation determines that the activities at remote premises contributed to the customer complaint, relevant information shall be communicated between the supplier and the remote premises.

If any customer complaint is not followed by corrective and/or preventive action, the reason shall be recorded.

If this International Standard is used for compliance with regulatory requirements, the supplier shall establish, document and maintain procedures to notify the regulatory authority of those incidents which meet the reporting criteria.

The supplier shall establish, document and maintain procedures for the issue of advisory notices for medical devices. These procedures shall be capable of being implemented at any time.

### 4.14.2 Corrective Action

The requirements given in 4.14.2 of ISO 9002:1994 apply.

### 4.14.3 Preventive Action

The requirements given in 4.14.3 of ISO 9002:1994 apply.

## 4.15 HANDLING, STORAGE, PACKAGING, PRESERVATION AND DELIVERY

### 4.15.1 General

The requirements given in 4.15.1 of ISO 9002:1994 apply.

*Particular requirements for all medical devices:*

The supplier shall establish and maintain documented procedures for the control of product with a limited shelflife or requiring special storage conditions. Such special storage conditions shall be controlled and recorded (see 4.16).

If appropriate, special arrangements shall be established, documented and maintained for the control of used product in order to prevent contamination of other product, the manufacturing environment or personnel.

### 4.15.2 Handling

The requirements given in 4.15.2 of ISO 9002:1994 apply.

### 4.15.3 Storage

The requirements given in 4.15.3 of ISO 9002:1994 apply.

### 4.15.4 Packaging

The requirements given in 4.15.4 of ISO 9002:1994 apply.

*Particular requirement for active implantable medical devices and implantable medical devices:*

The supplier shall record the identity of persons who perform the final labelling operation (see 4.16).

### 4.15.5 Preservation

The requirements given in 4.15.5 of ISO 9002:1994 apply.

### 4.15.6 Delivery

The requirements given in 4.15.6 of ISO 9002:1994 apply.

*Particular requirement for active implantable medical devices and impllantable medical devices:*

The supplier shall ensure that the name and address of the shipping package consignee is included in the quality records (see 4.16).

### 4.16 Control of Quality Records

The requirements given in 4.16 of ISO 9002:1994 apply.

*Particular requirements for all medical devices:*

The supplier shall retain the quality records for a period of time at least equivalent to the lifetime of the medical device as defined by the supplier, but not less than 2 years from the date of despatch from the supplier.

NOTE 1 — National or regional regulations may require a period longer than 2 years.

The supplier shall establish and maintain a record for each batch of medical devices that provides traceability to the extent specified in 4.8 and identifies the quantity manufactured and quantity approved for distribution. The batch record shall be verified and authorized.

NOTE 2 — A batch may be a single medical device.

### 4.17 Internal Quality Audits

The requirements given in 4.17 of ISO 9002:1994 apply.

## 4.18 TRAINING

The requirements given in 4.18 of ISO 9002:1994 apply.

## 4.19 SERVICING

The requirements given in 4.19 of ISO 9002:1994 apply.

## 4.20 STATISTICAL TECHNIQUES

The requirements given in 4.20 of ISO 9002:1994 apply.

_____

# International Standard ISO/IEC 17025

First edition 1999-12-15

# General Requirements for the Competence of Testing and Calibration Laboratories

Reference number
ISO/IEC 17025:1999(E)

## CONTENTS

## FOREWORD

ISO (the International Organization for Standardization) and IEC (the International Electrotechnical Commission) form the specialized system for worldwide standardization. National bodies that are members of ISO or IEC participate in the development of International Standards through technical committees established by the respective organization to deal with particular fields of technical activity. ISO and IEC technical committees collaborate in fields of mutual interest. Other international organizations, governmental and non-governmental, in liaison with ISO and IEC, also take part in the work.

International Standards are drafted in accordance with the rules given in the ISO/IEC Directives, Part 3.

Draft International Standards adopted by the technical committees are circulated to member bodies for voting. Publication as an International Standard requires approval by at least 75% of the member bodies casting a vote.

Attention is drawn to the possibility that some of the elements of this International Standard may be the subject of patent rights. ISO and IEC shall not be held responsible for identifying any or all such patent rights.

International Standard ISO/IEC 17025 was prepared by ISO/CASCO, *Committee on Conformity Assessment*.

This first edition of ISO/IEC 17025 cancels and replaces ISO/IEC Guide 25:1990.

Annexes A and B of this International Standard are for information only.

## INTRODUCTION

This International Standard has been produced as the result of extensive experience in the implementation of ISO/IEC Guide 25 and EN 45001, both of which it now replaces. It contains all of the requirements that testing and calibration laboratories have to meet if they wish to demonstrate that they operate a quality system, are technically competent, and are able to generate technically valid results.

Accreditation bodies that recognize the competence of testing and calibration laboratories should use this International Standard as the basis for their accreditation. Clause 4 specifies the requirements for sound management. Clause 5 specifies the requirements for technical competence for the type of tests and/or calibrations the laboratory undertakes.

The growth in use of quality systems generally has increased the need to ensure that laboratories which form part of larger organizations or offer other services can operate to a quality system that is seen as compliant with ISO 9001 or ISO 9002 as well as with this International Standard. Care has been taken, therefore, to incorporate all those requirements of ISO 9001 and ISO 9002 that are relevant to the scope of testing and calibration services that are covered by the laboratory's quality system.

Testing and calibration laboratories that comply with this International Standard will therefore also operate in accordance with ISO 9001 or ISO 9002.

Certification against ISO 9001 and ISO 9002 does not of itself demonstrate the competence of the laboratory to produce technically valid data and results.

The acceptance of testing and calibration results between countries should be facilitated if laboratories comply with this International Standard and if they obtain accreditation from bodies which have entered into mutual recognition agreements with equivalent bodies in other countries using this International Standard.

The use of this International Standard will facilitate cooperation between laboratories and other bodies, and assist in the exchange of information and experience, and in the harmonization of standards and procedures.

## 1 SCOPE

**1.1** This International Standard specifies the general requirements for the competence to carry out tests and/or calibrations, including sampling. It covers testing and calibration performed using standard methods, non-standard methods, and laboratory-developed methods.

**1.2** This International Standard is applicable to all organizations performing tests and/or calibrations. These include, for example, first-, second- and third-party laboratories, and laboratories where testing and/or calibration forms part of inspection and product certification.

This International Standard is applicable to all laboratories regardless of the number of personnel or the extent of the scope of testing and/or calibration activities. When a laboratory does not undertake one or more of the activities covered by this International Standard, such as sampling and the design/development of new methods, the requirements of those clauses do not apply.

**1.3**   The notes given provide clarification of the text, examples and guidance. They do not contain requirements and do not form an integral part of this International Standard.

**1.4**   This International Standard is for use by laboratories in developing their quality, administrative and technical systems that govern their operations. Laboratory clients, regulatory authorities and accreditation bodies may also use it in confirming or recognizing the competence of laboratories.

**1.5**   Compliance with regulatory and safety requirements on the operation of laboratories is not covered by this International Standard.

**1.6**   If testing and calibration laboratories comply with the requirements of this International Standard they will operate a quality system for their testing and calibration activities that also meets the requirements of ISO 9001 when they engage in the design/development of new methods, and/or develop test programmes combining standard and non-standard test and calibration methods, and ISO 9002 when they only use standard methods. Annex A provides nominal cross-references between this International Standard and ISO 9001 and ISO 9002. ISO/IEC 17025 covers several technical competence requirements that are not covered by ISO 9001 and ISO 9002.

NOTE 1   It might be necessary to explain or interpret certain requirements in this International Standard to ensure that the requirements are applied in a consistent manner. Guidance for establishing applications for specific fields, especially for accreditation bodies (see ISO/IEC Guide 58:1993, 4.1.3) is given in annex B.

NOTE 2   If a laboratory wishes accreditation for part or all of its testing and calibration activities, it should select an accreditation body that operates in accordance with ISO/IEC Guide 58.

## 2 NORMATIVE REFERENCES

The following normative documents contain provisions which, through reference in this text, constitute provisions of this International Standard. For dated references, subsequent amendments to, or revisions of, any of these publications do not apply. However, parties to agreements based on this International Standard are encouraged to investigate the possibility of applying the most recent editions of the normative documents indicated below. For undated references, the latest edition of the normative document referred to applies. Members of ISO and IEC maintain registers of currently valid International Standards.

ISO 9001:1994, Quality systems — Model for quality assurance in design, development, production, installation and servicing.

ISO 9002:1994, Quality systems — Model for quality assurance in production, installation and servicing.

ISO/IEC Guide 2, General terms and their definitions concerning standardization and related activities.

VIM, International vocabulary of basic and general terms in metrology, issued by BIPM, IEC, IFCC, ISO, IUPAC, IUPAP and OIML.

NOTE 1   Further related standards, guides, etc. on subjects included in this International Standard are given in the bibliography.

NOTE 2   It should be noted that when this International Standard was being developed, the revisions of ISO 9001 and ISO 9002 were anticipated to be published in late 2000 as a merged ISO 9001:2000. This is no longer the case.

## 3 TERMS AND DEFINITIONS

For the purposes of this International Standard, the relevant terms and definitions given in ISO/IEC Guide 2 and VIM apply.

NOTE   General definitions related to quality are given in ISO 8402, whereas ISO/IEC Guide 2 gives definitions specifically related to standardization, certification and laboratory accreditation. Where different definitions are given in ISO 8402, the definitions in ISO/IEC Guide 2 and VIM are preferred.

## 4 MANAGEMENT REQUIREMENTS

### 4.1 ORGANIZATION

**4.1.1**   The laboratory or the organization of which it is part shall be an entity that can be held legally responsible.

**4.1.2**   It is the responsibility of the laboratory to carry out its testing and calibration activities in such a way as to meet the requirements of this International Standard and to satisfy the needs of the client, the regulatory authorities or organizations providing recognition.

**4.1.3**   The laboratory management system shall cover work carried out in the laboratory's permanent facilities, at sites away from its permanent facilities, or in associated temporary or mobile facilities.

**4.1.4**   If the laboratory is part of an organization performing activities other than testing and/or calibration, the responsibilities of key personnel in the organization that have an involvement or influence on the testing and/or calibration activities of the laboratory shall be defined in order to identify potential conflicts of interest.

NOTE 1   Where a laboratory is part of a larger organization, the organizational arrangements should be such that departments having conflicting interests, such as production, commercial marketing or financing do not adversely influence the laboratory's compliance with the requirements of this International Standard.

NOTE 2   If the laboratory wishes to be recognized as a third-party laboratory, it should be able to demonstrate that it is impartial and that it and its personnel are free from any undue commercial, financial and other pressures which might influence their technical judgement. The third-party testing or calibration laboratory should not engage in any activities that may endanger the trust in its independence of judgement and integrity in relation to its testing or calibration activities.

**4.1.5**   The laboratory shall

a) have managerial and technical personnel with the authority and resources needed to carry out their duties and to identify the occurrence of departures from the quality system or from the procedures for performing tests and/or calibrations, and to initiate actions to prevent or minimize such departures (see also 5.2);

b) have arrangements to ensure that its management and personnel are free from any undue internal and external commercial, financial and other pressures and influences that may adversely affect the quality of their work;

c) have policies and procedures to ensure the protection of its clients' confidential information and proprietary rights, including procedures for protecting the electronic storage and transmission of results;

d) have policies and procedures to avoid involvement in any activities that would diminish confidence in its competence, impartiality, judgement or operational integrity;

e) define the organization and management structure of the laboratory, its place in any parent organization, and the relationships between quality management, technical operations and support services;

f) specify the responsibility, authority and interrelationships of all personnel who manage, perform or verify work affecting the quality of the tests and/or calibrations;

g) provide adequate supervision of testing and calibration staff, including trainees, by persons familiar with methods and procedures, purpose of each test and/or calibration, and with the assessment of the test or calibration results;

h) have technical management which has overall responsibility for the technical operations and the provision of the resources needed to ensure the required quality of laboratory operations;

i) appoint a member of staff as quality manager (however named) who, irrespective of other duties and responsibilities, shall have defined responsibility and authority for ensuring that the quality system is implemented and followed at all times; the quality manager shall have direct access to the highest level of management at which decisions are made on laboratory policy or resources;

j. appoint deputies for key managerial personnel (see note).

NOTE   Individuals may have more than one function and it maybe impractical to appoint deputies for every function.

## 4.2 Quality System

**4.2.1**   The laboratory shall establish, implement and maintain a quality system appropriate to the scope of its activities. The laboratory shall document its policies, systems, programmes, procedures and instructions to the extent necessary to assure the quality of the test and/or calibration results. The system's documentation shall be communicated to, understood by, available to, and implemented by the appropriate personnel.

**4.2.2**   The laboratory's quality system policies and objectives shall be defined in a quality manual (however named). The overall objectives shall be documented in a quality policy statement. The quality policy statement shall be issued under the authority of the chief executive. It shall include at least the following:

a) the laboratory management's commitment to good professional practice and to the quality of its testing and calibration in servicing its clients;

b) the management's statement of the laboratory's standard of service;

c) the objectives of the quality system;

d) a requirement that all personnel concerned with testing and calibration activities within the laboratory familiarize themselves with the quality documentation and implement the policies and procedures in their work; and

e) the laboratory management's commitment to compliance with this International Standard.

NOTE   The quality policy statement should be concise and may include the requirement that tests and/or calibrations shall always be carried out in accordance with stated methods and clients' requirements. When the test and/or calibration laboratory is part of a larger organization, some quality policy elements may be in other documents.

**4.2.3**   The quality manual shall include or make reference to the supporting procedures including technical procedures. It shall outline the structure of the documentation used in the quality system.

**4.2.4**   The roles and responsibilities of technical management and the quality manager, including their responsibility for ensuring compliance with this International Standard, shall be defined in the quality manual.

## 4.3 Document Control

### 4.3.1 General

The laboratory shall establish and maintain procedures to control all documents that form part of its quality system (internally generated or from external sources), such as regulations, standards, other normative documents, test and/or calibration methods, as well as drawings, software, specifications, instructions and manuals.

NOTE 1   In this context "document" could be policy statements, procedures, specifications, calibration tables, charts, text books, posters, notices, memoranda, software, drawings, plans, etc. These may be on various media, whether hard copy or electronic, and they may be digital, analog, photographic or written.

NOTE 2   The control of data related to testing and calibration is covered in 5.4.7. The control of records is covered in 4.12.

### 4.3.2 Document Approval and Issue

**4.3.2.1**   All documents issued to personnel in the laboratory as part of the quality system shall be reviewed and approved for use by authorized personnel prior to issue. A master list or an equivalent document control procedure identifying the current revision status and distribution of documents in the quality system shall be established and be readily available to preclude the use of invalid and/or obsolete documents.

**4.3.2.2**   The procedure(s) adopted shall ensure that:

a) authorized editions of appropriate documents are available at all locations where operations essential to the effective functioning of the laboratory are performed;

b) documents are periodically reviewed and, where necessary, revised to ensure continuing suitability and compliance with applicable requirements;

c) invalid or obsolete documents are promptly removed from all points of issue or use, or otherwise assured against unintended use;

d) obsolete documents retained for either legal or knowledge preservation purposes are suitably marked.

**4.3.2.3**    Quality system documents generated by the laboratory shall be uniquely identified. Such identification shall include the date of issue and/or revision identification, page numbering, the total number of pages or a mark to signify the end of the document, and the issuing authority(ies).

### 4.3.3 Document Changes

**4.3.3.1**    Changes to documents shall be reviewed and approved by the same function that performed the original review unless specifically designated otherwise. The designated personnel shall have access to pertinent background information upon which to base their review and approval.

**4.3.3.2**    Where practicable, the altered or new text shall be identified in the document or the appropriate attachments.

**4.3.3.3**    If the laboratory's documentation control system allows for the amendment of documents by hand pending the re-issue of the documents, the procedures and authorities for such amendments shall be defined. Amendments shall be clearly marked, initialled and dated. A revised document shall be formally re-issued as soon as practicable.

**4.3.3.4**    Procedures shall be established to describe how changes in documents maintained in computerized systems are made and controlled.

### 4.4 REVIEW OF REQUESTS, TENDERS AND CONTRACTS

**4.4.1**    The laboratory shall establish and maintain procedures for the review of requests, tenders and contracts. The policies and procedures for these reviews leading to a contract for testing and/or calibration shall ensure that:

a) the requirements, including the methods to be used, are adequately defined, documented and understood (see 5.4.2);

b) the laboratory has the capability and resources to meet the requirements;

c) the appropriate test and/or calibration method is selected and capable of meeting the clients' requirements (see 5.4.2).

Any differences between the request or tender and the contract shall be resolved before any work commences. Each contract shall be acceptable both to the laboratory and the client.

NOTE 1    The request, tender and contract review should be conducted in a practical and efficient manner, and the effect of financial, legal and time schedule aspects should be taken into account. For internal clients, reviews of requests, tenders and contracts can be performed in a simplified way.

NOTE 2    The review of capability should establish that the laboratory possesses the necessary physical, personnel and information resources, and that the laboratory's personnel have the skills and expertise necessary for the performance of the tests and/or calibrations in question. The review may also encompass results of earlier participation in interlaboratory comparisons or proficiency testing and/or the running of trial test or calibration programmes using samples or items of known value in order to determine uncertainties of measurement, limits of detection, confidence limits, etc.

NOTE 3   A contract may be any written or oral agreement to provide a client with testing and/or calibration services.

**4.4.2**   Records of reviews, including any significant changes, shall be maintained. Records shall also be maintained of pertinent discussions with a client relating to the client's requirements or the results of the work during the period of execution of the contract.

NOTE   For review of routine and other simple tasks, the date and the identification (e.g. the initials) of the person in the laboratory responsible for carrying out the contracted work are considered adequate. For repetitive routine tasks, the review need be made only at the initial enquiry stage or on granting of the contract for on-going routine work performed under a general agreement with the client, provided that the client's requirements remain unchanged. For new, complex or advanced testing and/or calibration tasks, a more comprehensive record should be maintained.

**4.4.3**   The review shall also cover any work that is subcontracted by the laboratory.

**4.4.4**   The client shall be informed of any deviation from the contract.

**4.4.5**   If a contract needs to be amended after work has commenced, the same contract review process shall be repeated and any amendments shall be communicated to all affected personnel.

## 4.5 Subcontracting of Tests and Calibrations

**4.5.1**   When a laboratory subcontracts work whether because of unforeseen reasons (e.g. workload, need for further expertise or temporary incapacity) or on a continuing basis (e.g. through permanent subcontracting, agency or franchising arrangements), this work shall be placed with a competent subcontractor. A competent subcontractor is one that, for example, complies with this International Standard for the work in question.

**4.5.2**   The laboratory shall advise the client of the arrangement in writing and, when appropriate, gain the approval of the client, preferably in writing.

**4.5.3**   The laboratory is responsible to the client for the subcontractor's work, except in the case where the client or a regulatory authority specifies which subcontractor is to be used.

**4.5.4**   The laboratory shall maintain a register of all subcontractors that it uses for tests and/or calibrations and a record of the evidence of compliance with this International Standard for the work in question.

## 4.6 Purchasing Services and Supplies

**4.6.1**   The laboratory shall have a policy and procedure(s) for the selection and purchasing of services and supplies it uses that affect the quality of the tests and/or calibrations. Procedures shall exist for the purchase, reception and storage of reagents and laboratory consumable materials relevant for the tests and calibrations.

**4.6.2**   The laboratory shall ensure that purchased supplies and reagents and consumable materials that affect the quality of tests and/or calibrations are not used until they have been inspected or otherwise verified as complying with standard specifications or requirements defined in the methods for the tests and/or calibrations concerned. These services and supplies used shall comply with specified requirements. Records of actions taken to check compliance shall be maintained.

**4.6.3** Purchasing documents for items affecting the quality of laboratory output shall contain data describing the services and supplies ordered. These purchasing documents shall be reviewed and approved for technical content prior to release.

NOTE    The description may include type, class, grade, precise identification, specifications, drawings, inspection instructions, other technical data including approval of test results, the quality required and the quality system standard under which they were made.

**4.6.4**    The laboratory shall evaluate suppliers of critical consumables, supplies and services which affect the quality of testing and calibration, and shall maintain records of these evaluations and list those approved.

## 4.7 Service to the Client

The laboratory shall afford clients or their representatives cooperation to clarify the client's request and to monitor the laboratory's performance in relation to the work performed, provided that the laboratory ensures confidentiality to other clients.

NOTE 1    Such cooperation may include:

a) providing the client or the client's representative reasonable access to relevant areas of the laboratory for the witnessing of tests and/or calibrations performed for the client;
b) preparation, packaging, and dispatch of test and/or calibration items needed by the client for verification purposes.

NOTE 2    Clients value the maintenance of good communication, advice and guidance in technical matters, and opinions and interpretations based on results. Communication with the client, especially in large assignments, should be maintained throughout the work. The laboratory should inform the client of any delays or major deviations in the performance of the tests and/or calibrations.

NOTE 3    Laboratories are encouraged to obtain other feedback, both positive and negative, from their clients (e.g. client surveys). The feedback should be used to improve the quality system, testing and calibration activities and client service.

## 4.8 Complaints

The laboratory shall have a policy and procedure for the resolution of complaints received from clients or other parties. Records shall be maintained of all complaints and of the investigations and corrective actions taken by the laboratory (see also 4.10).

## 4.9 Control of Nonconforming Testing and/or Calibration Work

**4.9.1**    The laboratory shall have a policy and procedures that shall be implemented when any aspect of its testing and/or calibration work, or the results of this work, do not conform to its own procedures or the agreed requirements of the client. The policy and procedures shall ensure that:

a) the responsibilities and authorities for the management of nonconforming work are designated and actions (including halting of work and withholding of test reports and calibration certificates, as necessary) are defined and taken when nonconforming work is identified;

b) an evaluation of the significance of the nonconforming work is made;

c) corrective actions are taken immediately, together with any decision about the acceptability of the nonconforming work;

d) where necessary, the client is notified and work is recalled;

e) the responsibility for authorizing the resumption of work is defined.

NOTE   Identification of nonconforming work or problems with the quality system or with testing and/or calibration activities can occur at various places within the quality system and technical operations. Examples are customer complaints, quality control, instrument calibration, checking of consumable materials, staff observations or supervision, test report and calibration certificate checking, management reviews and internal or external audits.

**4.9.2**   Where the evaluation indicates that the nonconforming work could recur or that there is doubt about the compliance of the laboratory's operations with its own policies and procedures, the corrective action procedures given in 4.10 shall be promptly followed.

## 4.10 CORRECTIVE ACTION

### 4.10.1 General

The laboratory shall establish a policy and procedure and shall designate appropriate authorities for implementing corrective action when nonconforming work or departures from the policies and procedures in the quality system or technical operations have been identified.

NOTE   A problem with the quality system or with the technical operations of the laboratory may be identified through a variety of activities, such as control of nonconforming work, internal or external audits, management reviews, feedback from clients or staff observations.

### 4.10.2 Cause Analysis

The procedure for corrective action shall start with an investigation to determine the root cause(s) of the problem.

NOTE   Cause analysis is the key and sometimes the most difficult part in the corrective action procedure. Often the root cause is not obvious and thus a careful analysis of all potential causes of the problem is required. Potential causes could include client requirements, the samples, sample specifications, methods and procedures, staff skills and training, consumables, or equipment and its calibration.

### 4.10.3 Selection and Implementation of Corrective Actions

Where corrective action is needed, the laboratory shall identify potential corrective actions. It shall select and implement the action(s) most likely to eliminate the problem and to prevent recurrence.

Corrective actions shall be to a degree appropriate to the magnitude and the risk of the problem.

The laboratory shall document and implement any required changes resulting from corrective action investigations.

### 4.10.4 Monitoring of Corrective Actions

The laboratory shall monitor the results to ensure that the corrective actions taken have been effective.

### 4.10.5 Additional Audits

Where the identification of nonconformances or departures casts doubts on the laboratory's compliance with its own policies and procedures, or on its compliance with this International Standard, the laboratory shall ensure that the appropriate areas of activity are audited in accordance with 4.13 as soon as possible.

NOTE    Such additional audits often follow the implementation of the corrective actions to confirm their effectiveness. An additional audit should be necessary only when a serious issue or risk to the business is identified.

### 4.11 PREVENTIVE ACTION

**4.11.1** Needed improvements and potential sources of nonconformances, either technical or concerning the quality system, shall be identified. If preventive action is required, action plans shall be developed, implemented and monitored to reduce the likelihood of the occurrence of such nonconformances and to take advantage of the opportunities for improvement.

**4.11.2** Procedures for preventive actions shall include the initiation of such actions and application of controls to ensure that they are effective.

NOTE 1    Preventive action is a pro-active process to identify opportunities for improvement rather than a reaction to the identification of problems or complaints.

NOTE 2    Apart from the review of the operational procedures, the preventive action might involve analysis of data, including trend and risk analyses and proficiency-testing results.

### 4.12 CONTROL OF RECORDS

### 4.12.1 General

**4.12.1.1**    The laboratory shall establish and maintain procedures for identification, collection, indexing, access, filing, storage, maintenance and disposal of quality and technical records. Quality records shall include reports from internal audits and management reviews as well as records of corrective and preventive actions.

**4.12.1.2**    All records shall be legible and shall be stored and retained in such a way that they are readily retrievable in facilities that provide a suitable environment to prevent damage or deterioration and to prevent loss. Retention times of records shall be established.

NOTE    Records may be in any media, such as hard copy or electronic media.

**4.12.1.3**    All records shall be held secure and in confidence.

**4.12.1.4**    The laboratory shall have procedures to protect and back-up records stored electronically and to prevent unauthorized access to or amendment of these records.

## 4.12.2 Technical Records

**4.12.2.1**  The laboratory shall retain records of original observations, derived data and sufficient information to establish an audit trail, calibration records, staff records and a copy of each test report or calibration certificate issued, for a defined period. The records for each test or calibration shall contain sufficient information to facilitate, if possible, identification of factors affecting the uncertainty and to enable the test or calibration to be repeated under conditions as close as possible to the original. The records shall include the identity of personnel responsible for the sampling, performance of each test and/or calibration and checking of results.

NOTE 1   In certain fields it may be impossible or impractical to retain records of all original observations.

NOTE 2   Technical records are accumulations of data (see 5.4.7) and information which result from carrying out tests and/or calibrations and which indicate whether specified quality or process parameters are achieved. They may include forms, contracts, work sheets, work books, check sheets, work notes, control graphs, external and internal test reports and calibration certificates, clients' notes, papers and feedback.

**4.12.2.2**  Observations, data and calculations shall be recorded at the time they are made and shall be identifiable to the specific task.

**4.12.2.3**  When mistakes occur in records, each mistake shall be crossed out, not erased, made illegible or deleted, and the correct value entered alongside. All such alterations to records shall be signed or initialled by the person making the correction. In the case of records stored electronically, equivalent measures shall be taken to avoid loss or change of original data.

## 4.13 INTERNAL AUDITS

**4.13.1** The laboratory shall periodically, and in accordance with a predetermined schedule and procedure, conduct internal audits of its activities to verify that its operations continue to comply with the requirements of the quality system and this International Standard. The internal audit programme shall address all elements of the quality system, including the testing and/or calibration activities. It is the responsibility of the quality manager to plan and organize audits as required by the schedule and requested by management. Such audits shall be carried out by trained and qualified personnel who are, wherever resources permit, independent of the activity to be audited.

NOTE   The cycle for internal auditing should normally be completed in one year.

**4.13.2**  When audit findings cast doubt on the effectiveness of the operations or on the correctness or validity of the laboratory's test or calibration results, the laboratory shall take timely corrective action, and shall notify clients in writing if investigations show that the laboratory results may have been affected.

**4.13.3**  The area of activity audited, the audit findings and corrective actions that arise from them shall be recorded.

**4.13.4**  Follow-up audit activities shall verify and record the implementation and effectiveness of the corrective action taken.

## 4.14 MANAGEMENT REVIEWS

**4.14.1** In accordance with a predetermined schedule and procedure, the laboratory's executive management shall periodically conduct a review of the laboratory's quality system and testing and/or calibration activities to ensure their continuing suitability and effectiveness, and to introduce necessary changes or improvements. The review shall take account of:

- the suitability of policies and procedures;
- reports from managerial and supervisory personnel;
- the outcome of recent internal audits;
- corrective and preventive actions;
- assessments by external bodies;
- the results of interlaboratory comparisons or proficiency tests;
- changes in the volume and type of the work;
- client feedback;
- complaints;
- other relevant factors, such as quality control activities, resources and staff training.

NOTE 1    A typical period for conducting a management review is once every 12 months.

NOTE 2    Results should feed into the laboratory planning system and should include the goals, objectives and action plans for the coming year.

NOTE 3    A management review includes consideration of related subjects at regular management meetings.

**4.14.2** Findings from management reviews and the actions that arise from them shall be recorded. The management shall ensure that those actions are carried out within an appropriate and agreed timescale.

## 5 TECHNICAL REQUIREMENTS

### 5.1 GENERAL

**5.1.1** Many factors determine the correctness and reliability of the tests and/or calibrations performed by a laboratory. These factors include contributions from:

- human factors (5.2);
- accommodation and environmental conditions (5.3);
- test and calibration methods and method validation (5.4);
- equipment (5.5);
- measurement traceability (5.6);
- sampling (5.7);
- the handling of test and calibration items (5.8).

**5.1.2** The extent to which the factors contribute to the total uncertainty of measurement differs considerably between (types of) tests and between (types of) calibrations. The laboratory shall take account of these factors in developing test and calibration methods and procedures, in the training and qualification of personnel, and in the selection and calibration of the equipment it uses.

## 5.2 PERSONNEL

**5.2.1**   The laboratory management shall ensure the competence of all who operate specific equipment, perform tests and/or calibrations, evaluate results, and sign test reports and calibration certificates. When using staff who are undergoing training, appropriate supervision shall be provided. Personnel performing specific tasks shall be qualified on the basis of appropriate education, training, experience and/or demonstrated skills, as required.

NOTE 1   In some technical areas (e.g. non-destructive testing) it may be required that the personnel performing certain tasks hold personnel certification. The laboratory is responsible for fulfilling specified personnel certification requirements. The requirements for personnel certification might be regulatory, included in the standards for the specific technical field, or required by the client.

NOTE 2   The personnel responsible for the opinions and interpretation included in test reports should, in addition to the appropriate qualifications, training, experience and satisfactory knowledge of the testing carried out, also have:

- relevant knowledge of the technology used for the manufacturing of the items, materials, products, etc. tested, or the way they are used or intended to be used, and of the defects or degradations which may occur during or in service;
- knowledge of the general requirements expressed in the legislation and standards; and
- an understanding of the significance of deviations found with regard to the normal use of the items, materials, products, etc. concerned.

**5.2.2**   The management of the laboratory shall formulate the goals with respect to the education, training and skills of the laboratory personnel. The laboratory shall have a policy and procedures for identifying training needs and providing training of personnel. The training programme shall be relevant to the present and anticipated tasks of the laboratory.

**5.2.3**   The laboratory shall use personnel who are employed by, or under contract to, the laboratory. Where contracted and additional technical and key support personnel are used, the laboratory shall ensure that such personnel are supervised and competent and that they work in accordance with the laboratory's quality system.

**5.2.4**   The laboratory shall maintain current job descriptions for managerial, technical and key support personnel involved in tests and/or calibrations.

NOTE   Job descriptions can be defined in many ways. As a minimum, the following should be defined:

- the responsibilities with respect to performing tests and/or calibrations;
- the responsibilities with respect to the planning of tests and/or calibrations and evaluation of results;
- the responsibilities for reporting opinions and interpretations;
- the responsibilities with respect to method modification and development and validation of new methods;
- expertise and experience required;
- qualifications and training programmes;
- managerial duties.

**5.2.5** The management shall authorize specific personnel to perform particular types of sampling, test and/or calibration, to issue test reports and calibration certificates, to give opinions and interpretations and to operate particular types of equipment. The laboratory shall maintain records of the relevant authorization (s), competence, educational and professional qualifications, training, skills and experience of all technical personnel, including contracted personnel. This information shall be readily available and shall include the date on which authorization and/or competence is confirmed.

## 5.3 Accommodation and Environmental Conditions

**5.3.1** Laboratory facilities for testing and/or calibration, including but not limited to energy sources, lighting and environmental conditions, shall be such as to facilitate correct performance of the tests and/or calibrations.

The laboratory shall ensure that the environmental conditions do not invalidate the results or adversely affect the required quality of any measurement. Particular care shall be taken when sampling and tests and/or calibrations are undertaken at sites other than a permanent laboratory facility. The technical requirements for accommodation and environmental conditions that can affect the results of tests and calibrations shall be documented.

**5.3.2** The laboratory shall monitor, control and record environmental conditions as required by the relevant specifications, methods and procedures or where they influence the quality of the results. Due attention shall be paid, for example, to biological sterility, dust, electromagnetic disturbances, radiation, humidity, electrical supply, temperature, and sound and vibration levels, as appropriate to the technical activities concerned. Tests and calibrations shall be stopped when the environmental conditions jeopardize the results of the tests and/or calibrations.

**5.3.3** There shall be effective separation between neighbouring areas in which there are incompatible activities. Measures shall be taken to prevent cross-contamination.

**5.3.4** Access to and use of areas affecting the quality of the tests and/or calibrations shall be controlled. The laboratory shall determine the extent of control based on its particular circumstances.

**5.3.5** Measures shall be taken to ensure good housekeeping in the laboratory. Special procedures shall be prepared where necessary.

## 5.4 Test and Calibration Methods and Method Validation

### 5.4.1 General

The laboratory shall use appropriate methods and procedures for all tests and/or calibrations within its scope. These include sampling, handling, transport, storage and preparation of items to be tested and/or calibrated, and, where appropriate, an estimation of the measurement uncertainty as well as statistical techniques for analysis of test and/or calibration data.

The laboratory shall have instructions on the use and operation of all relevant equipment, and on the handling and preparation of items for testing and/or calibration, or both, where the absence of such instructions could jeopardize the results of tests and/or calibrations. All instructions, standards, manuals and reference data relevant to the work of the laboratory shall be kept up to date and shall be made readily available to personnel (see 4.3). Deviation from test and calibration methods shall occur only if the deviation has been documented, technically justified, authorized, and accepted by the client.

NOTE   International, regional or national standards or other recognized specifications that contain sufficient and concise information on how to perform the tests and/or calibrations do not need to be supplemented or rewritten as internal procedures if these standards are written in a way that they can be used as published by the operating staff in a laboratory. It may be necessary to provide additional documentation for optional steps in the method or additional details.

## 5.4.2 Selection of Methods

The laboratory shall use test and/or calibration methods, including methods for sampling, which meet the needs of the client and which are appropriate for the tests and/or calibrations it undertakes. Methods published in international, regional or national standards shall preferably be used. The laboratory shall ensure that it uses the latest valid edition of a standard unless it is not appropriate or possible to do so. When necessary, the standard shall be supplemented with additional details to ensure consistent application.

When the client does not specify the method to be used, the laboratory shall select appropriate methods that have been published either in international, regional or national standards, or by reputable technical organizations, or in relevant scientific texts or journals, or as specified by the manufacturer of the equipment. Laboratory-developed methods or methods adopted by the laboratory may also be used if they are appropriate for the intended use and if they are validated. The client shall be informed as to the method chosen. The laboratory shall confirm that it can properly operate standard methods before introducing the tests or calibrations. If the standard method changes, the confirmation shall be repeated.

The laboratory shall inform the client when the method proposed by the client is considered to be inappropriate or out of date.

## 5.4.3 Laboratory-Developed Methods

The introduction of test and calibration methods developed by the laboratory for its own use shall be a planned activity and shall be assigned to qualified personnel equipped with adequate resources.

Plans shall be updated as development proceeds and effective communication amongst all personnel involved shall be ensured.

## 5.4.4 Non-Standard Methods

When it is necessary to use methods not covered by standard methods, these shall be subject to agreement with the client and shall include a clear specification of the client's requirements and the purpose of the test and/or calibration. The method developed shall have been validated appropriately before use.

NOTE   For new test and/or calibration methods, procedures should be developed prior to the tests and/or calibrations being performed and should contain at least the following information:

a) appropriate identification;
b) scope;
c) description of the type of item to be tested or calibrated;
d) parameters or quantities and ranges to be determined;
e) apparatus and equipment, including technical performance requirements;

f) reference standards and reference materials required;

g) environmental conditions required and any stabilization period needed;

h) description of the procedure, including

- affixing of identification marks, handling, transporting, storing and preparation of items,
- checks to be made before the work is started,
- checks that the equipment is working properly and, where required, calibration and adjustment of the equipment before each use,
- the method of recording the observations and results,
- any safety measures to be observed;

i) criteria and/or requirements for approval/rejection;

j) data to be recorded and method of analysis and presentation;

k) the uncertainty or the procedure for estimating uncertainty.

## 5.4.5 Validation of Methods

**5.4.5.1**   Validation is the confirmation by examination and the provision of objective evidence that the particular requirements for a specific intended use are fulfilled.

**5.4.5.2**   The laboratory shall validate non-standard methods, laboratory-designed/developed methods, standard methods used outside their intended scope, and amplifications and modifications of standard methods to confirm that the methods are fit for the intended use. The validation shall be as extensive as is necessary to meet the needs of the given application or field of application. The laboratory shall record the results obtained, the procedure used for the validation, and a statement as to whether the method is fit for the intended use.

NOTE 1    Validation may include procedures for sampling, handling and transportation.

NOTE 2    The techniques used for the determination of the performance of a method should be one of, or a combination of, the following:

- calibration using reference standards or reference materials;
- comparison of results achieved with other methods;
- interlaboratory comparisons;
- systematic assessment of the factors influencing the result;
- assessment of the uncertainty of the results based on scientific understanding of the theoretical principles of the method and practical experience.

NOTE 3    When some changes are made in the validated non-standard methods, the influence of such changes should be documented and, if appropriate, a new validation should be carried out.

**5.4.5.3**   The range and accuracy of the values obtainable from validated methods (e.g. the uncertainty of the results, detection limit, selectivity of the method, linearity, limit of repeatability and/or reproducibility, robustness against external influences and/or cross-sensitivity against interference from the matrix of the sample/test object), as assessed for the intended use, shall be relevant to the clients' needs.

NOTE 1    Validation includes specification of the requirements, determination of the characteristics of the methods, a check that the requirements can be fulfilled by using the method, and a statement on the validity.

NOTE 2    As method-development proceeds, regular review should be carried out to verify that the needs of the client are still being fulfilled. Any change in requirements requiring modifications to the development plan should be approved and authorized.

NOTE 3    Validation is always a balance between costs, risks and technical possibilities. There are many cases in which the range and uncertainty of the values (e.g. accuracy, detection limit, selectivity, linearity, repeatability, reproducibility, robustness and cross-sensitivity) can only be given in a simplified way due to lack of information.

### 5.4.6 Estimation of Uncertainty of Measurement

**5.4.6.1**    A calibration laboratory, or a testing laboratory performing its own calibrations, shall have and shall apply a procedure to estimate the uncertainty of measurement for all calibrations and types of calibrations.

**5.4.6.2**    Testing laboratories shall have and shall apply procedures for estimating uncertainty of measurement. In certain cases the nature of the test method may preclude rigorous, metrologically and statistically valid, calculation of uncertainty of measurement. In these cases the laboratory shall at least attempt to identify all the components of uncertainty and make a reasonable estimation, and shall ensure that the form of reporting of the result does not give a wrong impression of the uncertainty. Reasonable estimation shall be based on knowledge of the performance of the method and on the measurement scope and shall make use of, for example, previous experience and validation data.

NOTE 1    The degree of rigor needed in an estimation of uncertainty of measurement depends on factors such as:

- the requirements of the test method;
- the requirements of the client;
- the existence of narrow limits on which decisions on conformance to a specification are based.

NOTE 2    In those cases where a well-recognized test method specifies limits to the values of the major sources of uncertainty of measurement and specifies the form of presentation of calculated results, the laboratory is considered to have satisfied this clause by following the test method and reporting instructions (see 5.10).

**5.4.6.3**    When estimating the uncertainty of measurement, all uncertainty components which are of importance in the given situation shall be taken into account using appropriate methods of analysis.

NOTE 1    Sources contributing to the uncertainty include, but are not necessarily limited to, the reference standards and reference materials used, methods and equipment used, environmental conditions, properties and condition of the item being tested or calibrated, and the operator.

NOTE 2    The predicted long-term behaviour of the tested and/or calibrated item is not normally taken into account when estimating the measurement uncertainty.

NOTE 3    For further information, see ISO 5725 and the Guide to the Expression of Uncertainty in Measurement (see bibliography).

### 5.4.7 Control of Data

**5.4.7.1** Calculations and data transfers shall be subject to appropriate checks in a systematic manner.

**5.4.7.2** When computers or automated equipment are used for the acquisition, processing, recording, reporting, storage or retrieval of test or calibration data, the laboratory shall ensure that:

a) computer software developed by the user is documented in sufficient detail and is suitably validated as being adequate for use;
b) procedures are established and implemented for protecting the data; such procedures shall include, but not be limited to, integrity and confidentiality of data entry or collection, data storage, data transmission and data processing;
c) computers and automated equipment are maintained to ensure proper functioning and are provided with the environmental and operating conditions necessary to maintain the integrity of test and calibration data.

NOTE   Commercial off-the-shelf software (e.g. wordprocessing, database and statistical programmes) in general use within their designed application range may be considered to be sufficiently validated. However, laboratory software configuration/modifications should be validated as in 5.4.7.2a).

### 5.5 EQUIPMENT

**5.5.1** The laboratory shall be furnished with all items of sampling, measurement and test equipment required for the correct performance of the tests and/or calibrations (including sampling, preparation of test and/or calibration items, processing and analysis of test and/or calibration data). In those cases where the laboratory needs to use equipment outside its permanent control, it shall ensure that the requirements of this International Standard are met.

**5.5.2** Equipment and its software used for testing, calibration and sampling shall be capable of achieving the accuracy required and shall comply with specifications relevant to the tests and/or calibrations concerned. Calibration programmes shall be established for key quantities or values of the instruments where these properties have a significant effect on the results. Before being placed into service, equipment (including that used for sampling) shall be calibrated or checked to establish that it meets the laboratory's specification requirements and complies with the relevant standard specifications. It shall be checked and/or calibrated before use (see 5.6).

**5.5.3** Equipment shall be operated by authorized personnel. Up-to-date instructions on the use and maintenance of equipment (including any relevant manuals provided by the manufacturer of the equipment) shall be readily available for use by the appropriate laboratory personnel.

**5.5.4** Each item of equipment and its software used for testing and calibration and significant to the result shall, when practicable, be uniquely identified.

**5.5.5** Records shall be maintained of each item of equipment and its software significant to the tests and/or calibrations performed. The records shall include at least the following:

a) the identity of the item of equipment and its software;
b) the manufacturer's name, type identification, and serial number or other unique identification;

c) checks that equipment complies with the specification (see 5.5.2);

d) the current location, where appropriate;

e) the manufacturer's instructions, if available, or reference to their location;

f) dates, results and copies of reports and certificates of all calibrations, adjustments, acceptance criteria, and the due date of next calibration;

g) the maintenance plan, where appropriate, and maintenance carried out to date;

h) any damage, malfunction, modification or repair to the equipment.

**5.5.6**   The laboratory shall have procedures for safe handling, transport, storage, use and planned maintenance of measuring equipment to ensure proper functioning and in order to prevent contamination or deterioration.

NOTE   Additional procedures may be necessary when measuring equipment is used outside the permanent laboratory for tests, calibrations or sampling.

**5.5.7**   Equipment that has been subjected to overloading or mishandling, gives suspect results, or has been shown to be defective or outside specified limits, shall be taken out of service. It shall be isolated to prevent its use or clearly labelled or marked as being out of service until it has been repaired and shown by calibration or test to perform correctly. The laboratory shall examine the effect of the defect or departure from specified limits on previous tests and/or calibrations and shall institute the "Control of nonconforming work" procedure (see 4.9).

**5.5.8**   Whenever practicable, all equipment under the control of the laboratory and requiring calibration shall be labelled, coded or otherwise identified to indicate the status of calibration, including the date when last calibrated and the date or expiration criteria when recalibration is due.

**5.5.9**   When, for whatever reason, equipment goes outside the direct control of the laboratory, the laboratory shall ensure that the function and calibration status of the equipment are checked and shown to be satisfactory before the equipment is returned to service.

**5.5.10**   When intermediate checks are needed to maintain confidence in the calibration status of the equipment, these checks shall be carried out according to a defined procedure.

**5.5.11**   Where calibrations give rise to a set of correction factors, the laboratory shall have procedures to ensure that copies (e.g. in computer software) are correctly updated.

**5.5.12**   Test and calibration equipment, including both hardware and software, shall be safeguarded from adjustments which would invalidate the test and/or calibration results.

## 5.6 Measurement Traceability

### 5.6.1 General

All equipment used for tests and/or calibrations, including equipment for subsidiary measurements (e.g. for environmental conditions) having a significant effect on the accuracy or validity of the result of the test, calibration or sampling shall be calibrated before being put into service. The laboratory shall have an established programme and procedure for the calibration of its equipment.

NOTE   Such a programme should include a system for selecting, using, calibrating, checking, controlling and maintaining measurement standards, reference materials used as measurement standards, and measuring and test equipment used to perform tests and calibrations.

## 5.6.2 Specific Requirements

### 5.6.2.1 Calibration

**5.6.2.1.1**　For calibration laboratories, the programme for calibration of equipment shall be designed and operated so as to ensure that calibrations and measurements made by the laboratory are traceable to the International System of Units (SI) (*Systeme international d'unités*).

A calibration laboratory establishes traceability of its own measurement standards and measuring instruments to the SI by means of an unbroken chain of calibrations or comparisons linking them to relevant primary standards of the SI units of measurement. The link to SI units may be achieved by reference to national measurement standards. National measurement standards may be primary standards, which are primary realizations of the SI units or agreed representations of SI units based on fundamental physical constants, or they may be secondary standards which are standards calibrated by another national metrology institute. When using external calibration services, traceability of measurement shall be assured by the use of calibration services from laboratories that can demonstrate competence, measurement capability and traceability. The calibration certificates issued by these laboratories shall contain the measurement results, including the measurement uncertainty and/or a statement of compliance with an identified metrological specification (see also 5.10.4.2).

NOTE 1　Calibration laboratories fulfilling the requirements of this International Standard are considered to be competent. A calibration certificate bearing an accreditation body logo from a calibration laboratory accredited to this International Standard, for the calibration concerned, is sufficient evidence of traceability of the calibration data reported.

NOTE 2　Traceability to SI units of measurement may be achieved by reference to an appropriate primary standard (see VIM:1993, 6.4) or by reference to a natural constant, the value of which in terms of the relevant SI unit is known and recommended by the General Conference of Weights and Measures (CGPM) and the International Committee for Weights and Measures (CIPM).

NOTE 3　Calibration laboratories that maintain their own primary standard or representation of SI units based on fundamental physical constants can claim traceability to the SI system only after these standards have been compared, directly or indirectly, with other similar standards of a national metrology institute.

NOTE 4　The term "identified metrological specification" means that it must be clear from the calibration certificate which specification the measurements have been compared with, by including the specification or by giving an unambiguous reference to the specification.

NOTE 5　When the terms "international standard" or "national standard" are used in connection with traceability, it is assumed that these standards fulfil the properties of primary standards for the realization of SI units.

NOTE 6　Traceability to national measurement standards does not necessarily require the use of the national metrology institute of the country in which the laboratory is located.

NOTE 7　If a calibration laboratory wishes or needs to obtain traceability from a national metrology institute other than in its own country, this laboratory should select a national metrology institute that actively participates in the activities of BIPM either directly or through regional groups.

NOTE 8   The unbroken chain of calibrations or comparisons may be achieved in several steps carried out by different laboratories that can demonstrate traceability.

**5.6.2.1.2**   There are certain calibrations that currently cannot be strictly made in SI units. In these cases calibration shall provide confidence in measurements by establishing traceability to appropriate measurement standards such as:

- the use of certified reference materials provided by a competent supplier to give a reliable physical or chemical characterization of a material;
- the use of specified methods and/or consensus standards that are clearly described and agreed by all parties concerned.

Participation in a suitable programme of interlaboratory comparisons is required where possible.

### 5.6.2.2 Testing

**5.6.2.2.1**   For testing laboratories, the requirements given in 5.6.2.1 apply for measuring and test equipment with measuring functions used, unless it has been established that the associated contribution from the calibration contributes little to the total uncertainty of the test result. When this situation arises, the laboratory shall ensure that the equipment used can provide the uncertainty of measurement needed.

NOTE   The extent to which the requirements in 5.6.2.1 should be followed depends on the relative contribution of the calibration uncertainty to the total uncertainty. If calibration is the dominant factor, the requirements should be strictly followed.

**5.6.2.2.2**   Where traceability of measurements to SI units is not possible and/or not relevant, the same requirements for traceability to, for example, certified reference materials, agreed methods and/or consensus standards, are required as for calibration laboratories (see 5.6.2.1.2).

## 5.6.3 Reference Standards and Reference Materials

### 5.6.3.1 Reference Standards

The laboratory shall have a programme and procedure for the calibration of its reference standards. Reference standards shall be calibrated by a body that can provide traceability as described in 5.6.2.1. Such reference standards of measurement held by the laboratory shall be used for calibration only and for no other purpose, unless it can be shown that their performance as reference standards would not be invalidated. Reference standards shall be calibrated before and after any adjustment.

### 5.6.3.2 Reference Materials

Reference materials shall, where possible, be traceable to SI units of measurement, or to certified reference materials. Internal reference materials shall be checked as far as is technically and economically practicable.

### 5.6.3.3 Intermediate Checks

Checks needed to maintain confidence in the calibration status of reference, primary, transfer or working standards and reference materials shall be carried out according to defined procedures and schedules.

### 5.6.3.4 Transport and Storage

The laboratory shall have procedures for safe handling, transport, storage and use of reference standards and reference materials in order to prevent contamination or deterioration and in order to protect their integrity.

NOTE     Additional procedures may be necessary when reference standards and reference materials are used outside the permanent laboratory for tests, calibrations or sampling.

## 5.7 SAMPLING

**5.7.1**   The laboratory shall have a sampling plan and procedures for sampling when it carries out sampling of substances, materials or products for subsequent testing or calibration. The sampling plan as well as the sampling procedure shall be available at the location where sampling is undertaken. Sampling plans shall, whenever reasonable, be based on appropriate statistical methods. The sampling process shall address the factors to be controlled to ensure the validity of the test and calibration results.

NOTE 1   Sampling is a defined procedure whereby a part of a substance, material or product is taken to provide for testing or calibration of a representative sample of the whole. Sampling may also be required by the appropriate specification for which the substance, material or product is to be tested or calibrated. In certain cases (e.g. forensic analysis), the sample may not be representative but is determined by availability.

NOTE 2   Sampling procedures should describe the selection, sampling plan, withdrawal and preparation of a sample or samples from a substance, material or product to yield the required information.

**5.7.2**   Where the client requires deviations, additions or exclusions from the documented sampling procedure, these shall be recorded in detail with the appropriate sampling data and shall be included in all documents containing test and/or calibration results, and shall be communicated to the appropriate personnel.

**5.7.3**   The laboratory shall have procedures for recording relevant data and operations relating to sampling that forms part of the testing or calibration that is undertaken. These records shall include the sampling procedure used, the identification of the sampler, environmental conditions (if relevant) and diagrams or other equivalent means to identify the sampling location as necessary and, if appropriate, the statistics the sampling procedures are based upon.

## 5.8 HANDLING OF TEST AND CALIBRATION ITEMS

**5.8.1**   The laboratory shall have procedures for the transportation, receipt, handling, protection, storage, retention and/or disposal of test and/or calibration items, including all provisions necessary to protect the integrity of the test or calibration item, and to protect the interests of the laboratory and the client.

**5.8.2**   The laboratory shall have a system for identifying test and/or calibration items. The identification shall be retained throughout the life of the item in the laboratory. The system shall be designed and operated so as to ensure that items cannot be confused physically or when referred to in records or other documents. The system shall, if appropriate, accommodate a sub-division of groups of items and the transfer of items within and from the laboratory.

**5.8.3**  Upon receipt of the test or calibration item, abnormalities or departures from normal or specified conditions, as described in the test or calibration method, shall be recorded. When there is doubt as to the suitability of an item for test or calibration, or when an item does not conform to the description provided, or the test or calibration required is not specified in sufficient detail, the laboratory shall consult the client for further instructions before proceeding and shall record the discussion.

**5.8.4**  The laboratory shall have procedures and appropriate facilities for avoiding deterioration, loss or damage to the test or calibration item during storage, handling and preparation. Handling instructions provided with the item shall be followed. When items have to be stored or conditioned under specified environmental conditions, these conditions shall be maintained, monitored and recorded. Where a test or calibration item or a portion of an item is to be held secure, the laboratory shall have arrangements for storage and security that protect the condition and integrity of the secured items or portions concerned.

NOTE 1    Where test items are to be returned into service after testing, special care is required to ensure that they are not damaged or injured during the handling, testing or storing/waiting processes.

NOTE 2    A sampling procedure and information on storage and transport of samples, including information on sampling factors influencing the test or calibration result, should be provided to those responsible for taking and transporting the samples.

NOTE 3    Reasons for keeping a test or calibration item secure can be for reasons of record, safety or value, or to enable complementary tests and/or calibrations to be performed later.

## 5.9 Assuring the Quality of Test and Calibration Results

The laboratory shall have quality control procedures for monitoring the validity of tests and calibrations undertaken. The resulting data shall be recorded in such a way that trends are detectable and, where practicable, statistical techniques shall be applied to the reviewing of the results. This monitoring shall be planned and reviewed and may include, but not be limited to, the following:

   a) regular use of certified reference materials and/or internal quality control using secondary reference materials;
   b) participation in interlaboratory comparison or proficiency-testing programmes;
   c) replicate tests or calibrations using the same or different methods;
   d) retesting or recalibration of retained items;
   e) correlation of results for different characteristics of an item.

NOTE    The selected methods should be appropriate for the type and volume of the work undertaken.

## 5.10 Reporting the Results

### 5.10.1 General

The results of each test, calibration, or series of tests or calibrations carried out by the laboratory shall be reported accurately, clearly, unambiguously and objectively, and in accordance with any specific instructions in the test or calibration methods.

The results shall be reported, usually in a test report or a calibration certificate (see note 1), and shall include all the information requested by the client and necessary for the interpretation of the test or calibration results and all information required by the method used. This information is normally that required by 5.10.2, and 5.10.3 or 5.10.4.

In the case of tests or calibrations performed for internal clients, or in the case of a written agreement with the client, the results may be reported in a simplified way. Any information listed in 5.10.2 to 5.10.4 which is not reported to the client shall be readily available in the laboratory which carried out the tests and/or calibrations.

NOTE 1   Test reports and calibration certificates are sometimes called test certificates and calibration reports, respectively.

NOTE 2   The test reports or calibration certificates may be issued as hard copy or by electronic data transfer provided that the requirements of this International Standard are met.

### 5.10.2 Test Reports and Calibration Certificates

Each test report or calibration certificate shall include at least the following information, unless the laboratory has valid reasons for not doing so:

a) a title (e.g. "Test Report" or "Calibration Certificate");
b) the name and address of the laboratory, and the location where the tests and/or calibrations were carried out, if different from the address of the laboratory;
c) unique identification of the test report or calibration certificate (such as the serial number), and on each page an identification in order to ensure that the page is recognized as a part of the test report or calibration certificate, and a clear identification of the end of the test report or calibration certificate;
d) the name and address of the client;
e) identification of the method used;
f) a description of, the condition of, and unambiguous identification of the item(s) tested or calibrated;
g) the date of receipt of the test or calibration item(s) where this is critical to the validity and application of the results, and the date(s) of performance of the test or calibration;
h) reference to the sampling plan and procedures used by the laboratory or other bodies where these are relevant to the validity or application of the results;
i) the test or calibration results with, where appropriate, the units of measurement;
j) the name(s), function(s) and signature(s) or equivalent identification of person(s) authorizing the test report or calibration certificate;
k) where relevant, a statement to the effect that the results relate only to the items tested or calibrated.

NOTE 1   Hard copies of test reports and calibration certificates should also include the page number and total number of pages.

NOTE 2   It is recommended that laboratories include a statement specifying that the test report or calibration certificate shall not be reproduced except in full, without written approval of the laboratory.

### 5.10.3 Test Reports

**5.10.3.1**   In addition to the requirements listed in 5.10.2, test reports shall, where necessary for the interpretation of the test results, include the following:

a) deviations from, additions to, or exclusions from the test method, and information on specific test conditions, such as environmental conditions;
b) where relevant, a statement of compliance/non-compliance with requirements and/or specifications;
c) where applicable, a statement on the estimated uncertainty of measurement; information on uncertainty is needed in test reports when it is relevant to the validity or application of the test results, when a client's instruction so requires, or when the uncertainty affects compliance to a specification limit;
d) where appropriate and needed, opinions and interpretations (see 5.10.5);
e) additional information which may be required by specific methods, clients or groups of clients.

**5.10.3.2**   In addition to the requirements listed in 5.10.2 and 5.10.3.1, test reports containing the results of sampling shall include the following, where necessary for the interpretation of test results:

a) the date of sampling;
b) unambiguous identification of the substance, material or product sampled (including the name of the manufacturer, the model or type of designation and serial numbers as appropriate);
c) the location of sampling, including any diagrams, sketches or photographs;
d) a reference to the sampling plan and procedures used;
e) details of any environmental conditions during sampling that may affect the interpretation of the test results;
f) any standard or other specification for the sampling method or procedure, and deviations, additions to or exclusions from the specification concerned.

### 5.10.4 Calibration Certificates

**5.10.4.1**   In addition to the requirements listed in 5.10.2, calibration certificates shall include the following, where necessary for the interpretation of calibration results:

a) the conditions (e.g. environmental) under which the calibrations were made that have an influence on the measurement results;
b) the uncertainty of measurement and/or a statement of compliance with an identified metrological specification or clauses thereof;
c) evidence that the measurements are traceable (see note 2 in 5.6.2.1.1).

**5.10.4.2**   The calibration certificate shall relate only to quantities and the results of functional tests. If a statement of compliance with a specification is made, this shall identify which clauses of the specification are met or not met.

When a statement of compliance with a specification is made omitting the measurement results and associated uncertainties, the laboratory shall record those results and maintain them for possible future reference.

When statements of compliance are made, the uncertainty of measurement shall be taken into account.

**5.10.4.3** When an instrument for calibration has been adjusted or repaired, the calibration results before and after adjustment or repair, if available, shall be reported.

**5.10.4.4** A calibration certificate (or calibration label) shall not contain any recommendation on the calibration interval except where this has been agreed with the client. This requirement may be superseded by legal regulations.

## 5.10.5 Opinions and Interpretations

When opinions and interpretations are included, the laboratory shall document the basis upon which the opinions and interpretations have been made. Opinions and interpretations shall be clearly marked as such in a test report.

NOTE 1    Opinions and interpretations should not be confused with inspections and product certifications as intended in ISO/IEC 17020 and ISO/IEC Guide 65.

NOTE 2    Opinions and interpretations included in a test report may comprise, but not be limited to, the following:

- an opinion on the statement of compliance/noncompliance of the results with requirements;
- fulfilment of contractual requirements;
- recommendations on how to use the results;
- guidance to be used for improvements.

NOTE 3    In many cases it might be appropriate to communicate the opinions and interpretations by direct dialogue with the client. Such dialogue should be written down.

## 5.10.6 Testing and Calibration Results Obtained from Subcontractors

When the test report contains results of tests performed by subcontractors, these results shall be clearly identified. The subcontractor shall report the results in writing or electronically.

When a calibration has been subcontracted, the laboratory performing the work shall issue the calibration certificate to the contracting laboratory.

## 5.10.7 Electronic Transmission of Results

In the case of transmission of test or calibration results by telephone, telex, facsimile or other electronic or electromagnetic means, the requirements of this International Standard shall be met (see also 5.4.7).

## 5.10.8 Format of Reports and Certificates

The format shall be designed to accommodate each type of test or calibration carried out and to minimize the possibility of misunderstanding or misuse.

NOTE 1    Attention should be given to the lay-out of the test report or calibration certificate, especially with regard to the presentation of the test or calibration data and ease of assimilation by the reader.

NOTE 2   The headings should be standardized as far as possible.

## 5.10.9 Amendments to Test Reports and Calibration Certificates

Material amendments to a test report or calibration certificate after issue shall be made only in the form of a further document, or data transfer, which includes the statement:

"Supplement to Test Report [or Calibration Certificate], serial number... [or as otherwise identified],"

or an equivalent form of wording.

Such amendments shall meet all the requirements of this International Standard.

When it is necessary to issue a complete new test report or calibration certificate, this shall be uniquely identified and shall contain a reference to the original that it replaces.

# Annex A (Informative)

# Nominal Cross-References to ISO 9001:1994 and ISO 9002:1994

**TABLE A.1**
**Nominal cross-references to ISO 9001:1994 and ISO 9002:1994**

| ISO 9001:1994 | ISO 9002:1994 | ISO/IEC 17025 |
|---|---|---|
| Clause 1 | Clause 1 | Clause 1 |
| Clause 2 | Clause 2 | Clause 2 |
| Clause 3 | Clause 3 | Clause 3 |
| 4.1.1 | 4.1.1 | 4.1.3, 4.2.2 |
| 4.1.2.1 | 4.1.2.1 | 4.1.5 a), f), h); 4.2.4; 4.9.1 a); 4.10.1 and 5.2.5 |
| 4.1.2.2 | 4.1.2.2 | 4.1.5 a), g), h) and 5.5.1 |
| 4.1.2.3 | 4.1.2.3 | 4.1.5 i) |
| 4.1.3 | 4.1.3 | 4.14 |
| 4.2.1 and 4.2.2 | 4.2.1 and 4.2.2 | 4.2.1, 4.2.2, 4.2.3 |
| 4.2.3 | 4.2.3 | 4.2.1, 4.2.2 and 4.14 |
| 4.3 | 4.3 | 4.4 |
| 4.4 | 4.4 (n. a.) | 1.5, 5.4.2, 5.4.3, 5.4.4, 5.4.5 |
| 4.5 | 4.5 | 4.3, 5.4.7, 5.5.11 |
| 4.6.1 | 4.6.1 | 4.6, 5.5, 5.6.1, 5.6.2.1, 5.6.2.2 |
| 4.6.2 | 4.6.2 | 4.5, 4.6 |
| 4.6.3 | 4.6.3 | 4.6 |
| 4.6.4 | 4.6.4 | 4.5, 4.6.4, 4.7, 5.5.2 |
| 4.7 | 4.7 | 5.8, 5.10.6 |
| 4.8 | 4.8 | 5.5.4, 5.8 |
| 4.9 | 4.9 | 4.12, 5.3, 5.4, 5.5, 5.8, 5.9 |
| 4.10.1 | 4.10.1 | 5.4 |
| 4.10.2 | 4.10.2 | 4.5, 4.6, 5.5.2, 5.8 |
| 4.10.3 | 4.10.3 | 4.9, 5.5.9, 5.8.3, 5.8.4, 5.9 |
| 4.10.4 | 4.10.4 | 5.4.7, 5.9, 5.10.1 |
| 4.10.5 | 4.10.5 | 4.12.2 |
| 4.11.1 | 4.11.1 | 5.4, 5.5, 5.6 |
| 4.11.2 | 4.11.2 | 5.3, 5.4.1, 5.4.5, 5.5, 5.6 |
| 4.12 | 4.12 | 5.5.12, 5.8, 5.9.2 |
| 4.13 | 4.13 | 4.9 |
| 4.14 | 4.14.1 | 4.10 and 4.11 |
| 4.15 | 4.15 | 5.9 |
| 4.16 | 4.16 | 4.12 |

| ISO 9001:1994 | ISO 9002:1994 | ISO/IEC 17025 |
|---|---|---|
| 4.17 | 4.17 | 4.10.5, 4.13 (4.12) |
| 4.18 | 4.18 | 5.2, 5.5.3 |
| 4.19 | 4.19 | 4.7, 5.2.1, 5.10.5 |
| 4.20 | 4.20 | 5.9 |

n.a. = not applicable

ISO/IEC 17025 covers several technical competence requirements that are not covered by ISO 9001:1994 and ISO 9002:1994.

# Annex B (Informative)

# Guidelines for Establishing Applications for Specific Fields

**B.1**   The requirements specified in this International Standard are stated in general terms and, while they are applicable to all test and calibration laboratories, explanations might be needed. Such explanations on applications are herein referred to as applications. Applications should not include additional general requirements not included in this International Standard.

**B.2**   Applications can be thought of as an elaboration of the generally stated criteria (requirements) of this International Standard for specified fields of test and calibration, test technologies, products, materials or specific tests or calibrations. Accordingly, applications should be established by persons having appropriate technical knowledge and experience, and should address items that are essential or most important for the proper conduct of a test or calibration.

**B.3**   Depending on the application at hand, it may be necessary to establish applications for the technical requirements of this International Standard. Establishing applications may be accomplished by simply providing detail or adding extra information to the already generally stated requirements in each of the clauses (e.g. specific limitations to the temperature and humidity in the laboratory). In some cases the applications will be quite limited, applying only to a given test or calibration method or to a group of calibration or test methods. In other cases the applications may be quite broad, applying to the testing or calibration of various products or items or to entire fields of testing or calibration.

**B.4**   If the applications apply to a group of test or calibration methods in an entire technical field, common wording should be used for all of the methods. Alternatively, it may be necessary to develop a separate document of applications to supplement this International Standard for specific types or groups of tests or calibrations, products, materials or technical fields of tests or calibrations. Such a document should provide only the necessary supplementary information, while maintaining this International Standard as the governing document through reference. Applications which are too specific should be avoided in order to limit the proliferation of detailed documents.

**B.5**   The guidance in this annex should be used by accreditation bodies and other types of evaluation bodies when they develop applications for their own purposes (e.g. accreditation in specific areas).

# BIBLIOGRAPHY

[1] ISO 5725-1, Accuracy (trueness and precision) of measurement methods and results — Part 1: General principles and definitions.

[2] ISO 5725-2, Accuracy (trueness and precision) of measurement methods and results — Part 2: Basic method for the determination of repeatability and reproducibility of a standard measurement method.

[3] ISO 5725-3, Accuracy (trueness and precision) of measurement methods and results — Part 3: Intermediate measures of the precision of a standard measurement method.

[4] ISO 5725-4, Accuracy (trueness and precision) of measurement methods and results — Part 4: Basic methods for the determination of the trueness of a standard measurement method.

[5] ISO 5725-6, Accuracy (trueness and precision) of measurement methods and results — Part 6: Use in practice of accuracy values.

[6] ISO 8402, Quality management and quality assurance — Vocabulary.

[7] ISO 9000-1:1994, Quality management and quality assurance standards — Part 1: Guidelines for selection and use.

[8] ISO 9000-3: 1997, Quality management and quality assurance standards — Part 3: Guidelines for the application of ISO 9001:1994 to the development, supply, installation and maintenance of computer software.

[9] ISO 9004-1:1994, Quality management and quality system elements — Part 1: Guidelines.

[10] ISO 9004-4:1993, Quality management and quality system elements — Part 4: Guidelines for quality improvement.

[11] ISO 10011-1, Guidelines for auditing quality systems — Part 1: Auditing.

[12] ISO 10011-2, Guidelines for auditing quality systems — Part 2: Qualification criteria for quality system auditors.

[13] ISO 10011-3, Guidelines for auditing quality systems — Part 3: Management of audit programmes.

[14] ISO 10012-1, Quality assurance requirements for measuring equipment — Part 1: Metrological confirmation for measuring equipment.

[15] ISO 10012-2, Quality assurance requirements for measuring equipment — Part 2: Guidelines for control of measurement processes.

[16] ISO/IEC 17020, General criteria for the operation of various types of bodies performing inspection.

[17] ISO Guide 30, Terms and definitions used in connection with reference materials.

[18] ISO Guide 31, Contents of certificates of reference materials.

[19] ISO Guide 32, Calibration in analytical chemistry and use of certified reference materials.

[20] ISO Guide 33, Uses of certified reference materials.

[21] ISO Guide 34, General requirements for the competence of reference material producers.

[22] ISO Guide 35, Certification of reference materials — General and statistical principles.

[23] ISO/IEC Guide 43-1, Proficiency testing by interlaboratory comparisons — Part 1: Development and operation of proficiency testing schemes.

[24] ISO/IEC Guide 43-2, Proficiency testing by interlaboratory comparisons — Part 2: Selection and use of proficiency testing schemes by laboratory accreditation bodies.

[25] ISO/IEC Guide 58:1993, Calibration and testing laboratory accreditation systems — General requirements for operation and recognition.

[26] ISO/IEC Guide 65, General requirements for bodies operating product certification systems.

[27] Guide to the Expression of Uncertainty in Measurement, issued by BIPM, IEC, IFCC, ISO, IUPAC, IUPAP and OIML.

[28] Information and documents on laboratory accreditation can be found on the ILAC (International Laboratory Accreditation Cooperation): www.ilac.org.

# INTERNATIONAL STANDARD ISO 19011*

**GUIDELINES FOR QUALITY AND/OR ENVIRONMENTAL MANAGEMENT SYSTEMS AUDITING****

## CONTENTS

---

\* First edition 2002-10-01
\*\* Reference number ISO 19011:2002(E)

## FOREWORD

ISO (the International Organization for Standardization) is a worldwide federation of national standards bodies (ISO member bodies). The work of preparing International Standards is normally carried out through ISO technical committees. Each member body interested in a subject for which a technical committee has been established has the right to be represented on that committee. International organizations, governmental and non-governmental, in liaison with ISO, also take part in the work. ISO collaborates closely with the International Electrotechnical Commission (IEC) on all matters of electrotechnical standardization.

International standards are drafted in accordance with the rules given in the ISOIIEC Directives, part 3.

The main task of technical committees is to prepare International Standards. Draft International Standards accepted by the technical committees are circulated to the member bodies for voting. Publication as an International Standard requires approval by at least 75% of the members casting a vote.

Attention is drawn to the possibility that some of the elements of this International Standard may be the subject of patent rights. ISO shall not be held responsible for identifying any or all such patent rights.

ISO 19011 was prepared jointly by Technical Committee ISOITC 176, Quality management and quality assurance, Subcommittee SC 3, Supporting technologies, and Technical Committee ISOITC 207, Environmental management, Subcommittee SC 2, Environmental auditing and related environmental investigations.

This first edition of ISO 19011 cancels and replaces ISO 10011-1:1990, ISO 10011-2:1991, ISO 10011-3:1991, ISO 14010:1996, ISO 14011:1996 and ISO 14012:1996.

## INTRODUCTION

The ISO 9000 and ISO 14000 series of International Standards emphasize the importance of audits as a management tool for monitoring and verifying the effective implementation of an organization's quality and/or environmental policy. Audits are also an essential part of conformity assessment activities such as external certification/registration and of supply chain evaluation and surveillance.

This International Standard provides guidance on the management of audit programs, the conduct of internal or external audits of quality and/or environmental management systems, as well as on the competence and evaluation of auditors. It is intended to apply to a broad range of potential users, including auditors, organizations implementing quality and/or environmental management systems, organizations needing to conduct audits of quality and/or environmental management systems for contractual reasons, and organizations involved in auditor certification or training, in certification/registration of management systems, in accreditation or in standardization in the area of conformity assessment.

The guidance in this International Standard is intended to be flexible. As indicated at various points in the text, the use of these guidelines can differ according to the size, nature and complexity of the organizations to be audited, as well as the objectives and scopes of the audits to be conducted. Throughout this International Standard, supplementary guidance or examples on specific topics are provided in the form of practical help in boxed text. In some instances, this is intended to support the use of this International Standard in small organizations.

Clause 4 describes the principles of auditing. These principles help the user to appreciate the essential nature of auditing and they are a necessary prelude to clauses 5, 6 and 7.

Clause 5 provides guidance on managing audit programs and covers such issues as assigning responsibility for managing audit programs, establishing the audit program objectives, coordinating auditing activities and providing sufficient audit team resources.

Clause 6 provides guidance on conducting audits of quality and/or environmental management systems, including the selection of audit teams.

Clause 7 provides guidance on the competence needed by an auditor and describes a process for evaluating auditors.

Where quality and environmental management systems are implemented together, it is at the discretion of the user of this International Standard as to whether the quality management system and environmental management system audits are conducted separately or together.

Although this International Standard is applicable to the auditing of quality and/or environmental management systems, the user can consider adapting or extending the guidance provided herein to apply to other types of audits, including other management system audits.

This International Standard provides only guidance; however, users can apply this to develop their own audit-related requirements.

In addition, any other individual or organization with an interest in monitoring conformance to requirements, such as product specifications or laws and regulations, may find the guidance in this International Standard useful.

# 1 SCOPE

This International Standard provides guidance on the principles of auditing, managing audit programs, conducting quality management system audits and environmental management system audits, as well as guidance on the competence of quality and environmental management system auditors.

It is applicable to all organizations needing to conduct internal or external audits of quality and/or environmental management systems or to manage an audit program.

The application of this International Standard to other types of audit is possible in principle, provided that special consideration is paid to identifying the competence needed by the audit team members in such cases.

# 2 NORMATIVE REFERENCES

The following normative documents contain provisions which, through references in this text, constitute provisions of this International Standard. For dated references, subsequent amendments to, or revisions of, any of these publications do not apply. However, parties to agreements based on this International Standard are encouraged to investigate the possibility of applying the most recent edition of the normative documents indicated below. For undated references, the latest edition of the normative document referred to apply. Members of ISO and IEC maintain registers of currently valid International Standards.

ISO 9000:2000, *Quality management systems — Fundamentals and vocabulary*
ISO 14050:2002, *Environmental management — Vocabulary*

# 3 TERMS AND DEFINITIONS

For the purposes of this International Standard, the terms and definitions given in ISO 9000 and ISO 14050 apply, unless superseded by the terms and definitions given below.

A term in a definition or note which is defined elsewhere in this clause is indicated by boldface followed by its entry number in parentheses. Such a boldface term may be replaced in the definition by its complete definition.

## 3.1

### Audit

systematic, independent and documented process for obtaining **audit evidence** (3.3) and evaluating it objectively to determine the extent to which the **audit criteria** (3.2) are fulfilled

NOTE 1 Internal audits, sometimes called first-party audits, are conducted by, or on behalf of, the organization itself for management review and other internal purposes, and may form the basis for an organization's self-declaration of conformity. In many cases, particularly in smaller organizations, independence can be demonstrated by the freedom from responsibility for the activity being audited.
NOTE 2 External audits include those generally termed second- and third-party audits. Second-party audits are conducted by parties having an interest in the organization, such as customers, or by other persons on their behalf. Third-party audits are conducted by external, independent auditing organizations, such as those providing registration or certification of conformity to the requirements of ISO 9001 or ISO 14001.
NOTE 3 When a quality management system and an environmental management system are audited together, this is termed a combined audit.
NOTE 4 When two or more auditing organizations cooperate to audit a single **auditee** (3.7), this is termed a joint audit.

## 3.2

### Audit criteria

set of policies, procedures or requirements

NOTE     Audit criteria are used as a reference against which **audit evidence** (3.3) is compared.

## 3.3

### Audit evidence

records, statements of fact or other information, which are relevant to the **audit criteria** (3.2) and verifiable

NOTE     Audit evidence may be qualitative or quantitative.

## 3.4

### Audit findings

results of the evaluation of the collected **audit evidence** (3.3) against **audit criteria** (3.2)

NOTE     Audit findings can indicate either conformity or nonconformity with audit criteria or opportunities for improvement.

## 3.5

### Audit conclusion

outcome of an **audit** (3.1), provided by the **audit team** (3.9) after consideration of the audit objectives and all **audit findings** (3.4)

## 3.6

**Audit client**
organization or person requesting an **audit** (3.1)

NOTE   The audit client may be the **auditee** (3.7) or any other organization which has the regulatory or contractual right to request an audit.

## 3.7

**Auditee**
organization being audited

## 3.8

**Auditor**
person with the **competence** (3.14) to conduct an **audit** (3.1)

## 3.9

**Audit team**
one or more **auditors** (3.8) conducting an **audit** (3.1), supported if needed by **technical experts** (3.10)

NOTE 1   One auditor of the audit team is appointed as the audit team leader.
NOTE 2   The audit team may include auditors-in-training.

## 3.10

**Technical expert**
person who provides specific knowledge or expertise to the **audit team** (3.9)

NOTE 1   Specific knowledge or expertise is that which relates to the organization, the process or activity to be audited, or language or culture.
NOTE 2   A technical expert does not act as an **auditor** (3.8) in the audit team.

## 3.11

**Audit program**
set of one or more **audits** (3.1) planned for a specific time frame and directed towards a specific purpose

NOTE   An audit program includes all activities necessary for planning, organizing and conducting the audits.

## 3.12

**Audit plan**
description of the activities and arrangements for an **audit** (3.1)

## 3.13

**Audit scope**
extent and boundaries of an **audit** (3.1)

NOTE   The audit scope generally includes a description of the physical locations, organizational units, activities and processes, as well as the time period covered.

### 3.14

**Competence**
demonstrated personal attributes and demonstrated ability to apply knowledge and skills

## 4   PRINCIPLES OF AUDITING

Auditing is characterized by reliance on a number of principles. These make the audit an effective and reliable tool in support of management policies and controls, providing information on which an organization can act to improve its performance. Adherence to these principles is a prerequisite for providing audit conclusions that are relevant and sufficient and for enabling auditors working independently from one another to reach similar conclusions in similar circumstances.

The following principles relate to auditors.

a) Ethical conduct: the foundation of professionalism
   Trust, integrity, confidentiality and discretion are essential to auditing.
b) Fair presentation: the obligation to report truthfully and accurately
   Audit findings, audit conclusions and audit reports reflect truthfully and accurately the audit activities. Significant obstacles encountered during the audit and unresolved diverging opinions between the audit team and the auditee are reported.
c) Due professional care: the application of diligence and judgment in auditing
   Auditors exercise care in accordance with the importance of the task they perform and the confidence placed in them by audit clients and other interested parties. Having the necessary competence is an important factor.

Further principles relate to the audit, which is by definition independent and systematic.

d) Independence: the basis for the impartiality of the audit and objectivity of the audit conclusions
   Auditors are independent of the activity being audited and are free from bias and conflict of interest. Auditors maintain an objective state of mind throughout the audit process to ensure that the audit findings and conclusions will be based only on the audit evidence.
e) Evidence-based approach: the rational method for reaching reliable and reproducible audit conclusions in a systematic audit process
   Audit evidence is verifiable. It is based on samples of the information available, since an audit is conducted during a finite period of time and with finite resources. The appropriate use of sampling is closely related to the confidence that can be placed in the audit conclusions.

The guidance given in the remaining clauses of this International Standard is based on the principles set out above.

## 5   MANAGING AN AUDIT PROGRAM

### 5.1   GENERAL

An audit program may include one or more audits, depending upon the size, nature and complexity of the organization to be audited. These audits may have a variety of objectives and may also include joint or combined audits (see Notes 3 and 4 to the definition of audit in 3.1).

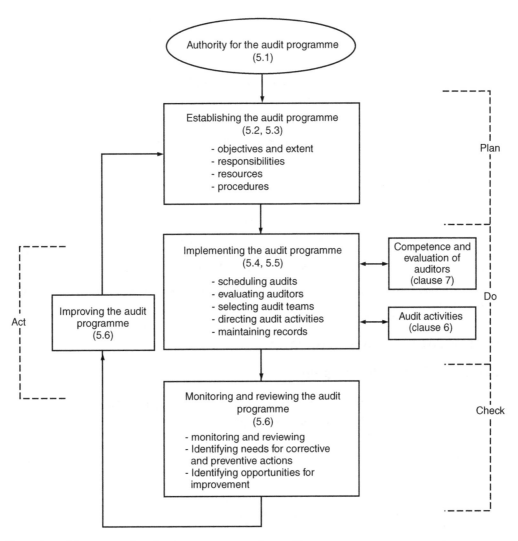

**FIGURE 1** Illustration of the process flow for the management of an audit program.

An audit program also includes all activities necessary for planning and organizing the types and number of audits, and for providing resources to conduct them effectively and efficiently within the specified time frames.

An organization may establish more than one audit program.

The organization's top management should grant the authority for managing the audit program.

Those assigned the responsibility for managing the audit program should

a) establish, implement, monitor, review and improve the audit program, and

b) identify the necessary resources and ensure they are provided.

Figure 1 illustrates the process flow for the management of an audit program.

NOTE 1    Figure 1 also illustrates the application of the Plan-Do-Check-Act methodology in this International Standard.

NOTE 2    The numbers in this and all subsequent figures refer to the relevant clauses of this International Standard.

If an organization to be audited operates both quality management and environmental management systems, combined audits may be included in the audit program. In such a case, special attention should be paid to the competence of the audit team.

Two or more auditing organizations may cooperate, as part of their audit programs, to conduct a joint audit. In such a case, special attention should be paid to the division of responsibilities, the provision of any additional resources, the competence of the audit team and the appropriate procedures. Agreement on these should be reached before the audit commences.

---

**Practical help — Examples of audit programs**

Examples of audit programs include the following:

a) a series of internal audits covering an organization-wide quality management system for the current year;
b) second-party management system audits of potential suppliers of critical products to be conducted within 6 months;
c) certification/registration and surveillance audits conducted by a third-party certification/registration body on an environmental management system within a time period agreed contractually between the certification body and the client.

An audit program also includes appropriate planning, the provision of resources and the establishment of procedures to conduct audits within the program.

---

## 5.2   AUDIT PROGRAM OBJECTIVES AND EXTENT

### 5.2.1   Objectives of an Audit Program

Objectives should be established for an audit program, to direct the planning and conduct of audits.
These objectives can be based on consideration of

a) management priorities,
b) commercial intentions,
c) management system requirements,
d) statutory, regulatory and contractual requirements,
e) need for supplier evaluation,
f) customer requirements,
g) needs of other interested parties, and
h) risks to the organization.

---

**Practical help — Examples of audit program objectives**

Examples of audit program objectives include the following:

a) to meet requirements for certification to a management system standard;
b) to verify conformance with contractual requirements;
c) to obtain and maintain confidence in the capability of a supplier;
d) to contribute to the improvement of the management system.

---

### 5.2.2   Extent of an Audit Program

The extent of an audit program can vary and will be influenced by the size, nature and complexity of the organization to be audited, as well as by the following:

a) the scope, objective and duration of each audit to be conducted;
b) the frequency of audits to be conducted;
c) the number, importance, complexity, similarity and locations of the activities to be audited;

d) standards, statutory, regulatory and contractual requirements and other audit criteria;
e) the need for accreditation or registration/certification;
f) conclusions of previous audits or results of a previous audit program review;
g) any language, cultural and social issues;
h) the concerns of interested parties;
i) significant changes to an organization or its operations.

## 5.3 AUDIT PROGRAM RESPONSIBILITIES, RESOURCES AND PROCEDURES

### 5.3.1 Audit Program Responsibilities

The responsibility for managing an audit program should be assigned to one or more individuals with a general understanding of audit principles, of the competence of auditors and the application of audit techniques. They should have management skills as well as technical and business understanding relevant to the activities to be audited.

Those assigned the responsibility for managing the audit program should

a) establish the objectives and extent of the audit program,
b) establish the responsibilities and procedures, and ensure resources are provided,
c) ensure the implementation of the audit program,
d) ensure that appropriate audit program records are maintained, and
e) monitor, review and improve the audit program.

### 5.3.2 Audit Program Resources

When identifying resources for the audit program, consideration should be given to

a) financial resources necessary to develop, implement, manage and improve audit activities,
b) audit techniques,
c) processes to achieve and maintain the competence of auditors, and to improve auditor performance,
d) the availability of auditors and technical experts having competence appropriate to the particular audit program objectives,
e) the extent of the audit program, and
f) travelling time, accommodation and other auditing needs.

### 5.3.3 Audit Program Procedures

Audit program procedures should address the following:

a) planning and scheduling audits;
b) assuring the competence of auditors and audit team leaders;
c) selecting appropriate audit teams and assigning their roles and responsibilities;
d) conducting audits;
e) conducting audit follow-up, if applicable;
f) maintaining audit program records;
g) monitoring the performance and effectiveness of the audit program;
h) reporting to top management on the overall achievements of the audit program.

For smaller organizations, the activities above can be addressed in a single procedure

## 5.4   AUDIT PROGRAM IMPLEMENTATION

The implementation of an audit program should address the following:

a) communicating the audit program to relevant parties;
b) coordinating and scheduling audits and other activities relevant to the audit program;
c) establishing and maintaining a process for the evaluation of the auditors and their continual professional development, in accordance with respectively 7.6 and 7.5;
d) ensuring the selection of audit teams;
e) providing necessary resources to the audit teams;
f) ensuring the conduct of audits according to the audit program;
g) ensuring the control of records of the audit activities;
h) ensuring review and approval of audit reports, and ensuring their distribution to the audit client and other specified parties;
i) ensuring audit follow-up, if applicable.

## 5.5   AUDIT PROGRAM RECORDS

Records should be maintained to demonstrate the implementation of the audit program and should include the following:

a) records related to individual audits, such as
   audit plans,
   audit reports,
   nonconformity reports,
   corrective and preventive action reports, and
   audit follow-up reports, if applicable;
b) results of audit program review;
c) records related to audit personnel covering subjects such as
   auditor competence and performance evaluation,
   audit team selection, and
   maintenance and improvement of competence.

   Records should be retained and suitably safeguarded.

## 5.6   AUDIT PROGRAM MONITORING AND REVIEWING

The implementation of the audit program should be monitored and, at appropriate intervals, reviewed to assess whether its objectives have been met and to identify opportunities for improvement. The results should be reported to top management.
   Performance indicators should be used to monitor characteristics such as

   the ability of the audit teams to implement the audit plan,
   conformity with audit programs and schedules, and
   feedback from audit clients, auditees and auditors.

   The audit program review should consider, for example,

a) results and trends from monitoring,
b) conformity with procedures,
c) evolving needs and expectations of interested parties,
d) audit program records,
e) alternative or new auditing practices, and
f) consistency in performance between audit teams in similar situations.

Results of audit program reviews can lead to corrective and preventive actions and the improvement of the audit program.

# 6 AUDIT ACTIVITIES

## 6.1 GENERAL

This clause contains guidance on planning and conducting audit activities as part of an audit program. Figure 2 provides an overview of typical audit activities. The extent to which the provisions of this clause are applicable depends on the scope and complexity of the specific audit and the intended use of the audit conclusions.

NOTE   The dotted lines indicate that any audit follow-up actions are usually not considered to be part of the audit.

## 6.2 INITIATING THE AUDIT

### 6.2.1 Appointing the Audit Team Leader

Those assigned the responsibility for managing the audit program should appoint the audit team leader for the specific audit.

Where a joint audit is conducted, it is important to reach agreement among the auditing organizations before the audit commences on the specific responsibilities of each organization, particularly with regard to the authority of the team leader appointed for the audit.

### 6.2.2 Defining Audit Objectives, Scope and Criteria

Within the overall objectives of an audit program, an individual audit should be based on documented objectives, scope and criteria.

The audit objectives define what is to be accomplished by the audit and may include the following:

a) determination of the extent of conformity of the auditee's management system, or parts of it, with audit criteria;
b) evaluation of the capability of the management system to ensure compliance with statutory, regulatory and contractual requirements;
c) evaluatation of the effectiveness of the management system in meeting its specified objectives;
d) identification of areas for potential improvement of the management system.

The audit scope describes the extent and boundaries of the audit, such as physical locations, organizational units, activities and processes to be audited, as well as the time period covered by the audit.

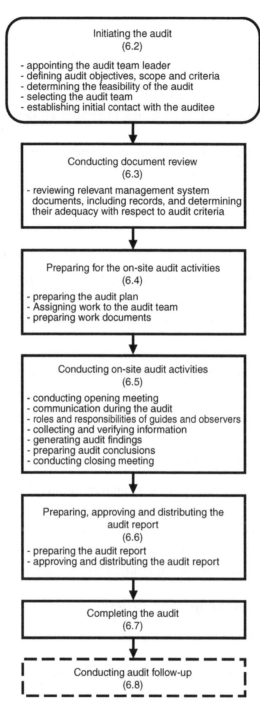

NOTE    The dotted lines indicate that any audit follow-up actions are usually not considered to be part of the audit.

**FIGURE 2**  Overview of typical audit activities.

The audit criteria are used as a reference against which conformity is determined and may include applicable policies, procedures, standards, laws and regulations, management system requirements, contractual requirements or industry/business sector codes of conduct.

The audit objectives should be defined by the audit client. The audit scope and criteria should be defined between the audit client and the audit team leader in accordance with audit program procedures. Any changes to the audit objectives, scope or criteria should be agreed to by the same parties.

Where a combined audit is to be conducted, it is important that the audit team leader ensures that the audit objectives, scope and criteria are appropriate to the nature of the combined audit.

### 6.2.3 Determining the Feasibility of the Audit

The feasibility of the audit should be determined, taking into consideration such factors as the availability of

sufficient and appropriate information for planning the audit,
adequate cooperation from the auditee, and
adequate time and resources.

Where the audit is not feasible, an alternative should be proposed to the audit client, in consultation with the auditee.

### 6.2.4 Selecting the Audit Team

When the audit has been declared feasible, an audit team should be selected, taking into account the competence needed to achieve the objectives of the audit. If there is only one auditor, the auditor should perform all applicable duties of an audit team leader. Clause 7 contains guidance on determining the competence needed and describes processes for evaluating auditors.

In deciding the size and composition of the audit team, consideration should be given to the following:

a) audit objectives, scope, criteria and estimated duration of the audit;
b) whether the audit is a combined or joint audit;
c) the overall competence of the audit team needed to achieve the objectives of the audit;
d) statutory, regulatory, contractual and accreditation/certification requirements, as applicable;
e) the need to ensure the independence of the audit team from the activities to be audited and to avoid conflict of interest;
f) the ability of the audit team members to interact effectively with the auditee and to work together;
g) the language of the audit, and an understanding of the auditee's particular social and cultural characteristics; these issues may be addressed either by the auditor's own skills or through the support of a technical expert.

The process of assuring the overall competence of the audit team should include the following steps:

identification of the knowledge and skills needed to achieve the objectives of the audit;
selection of the audit team members such that all of the necessary knowledge and skills are present in the audit team.

If not fully covered by the auditors in the audit team, the necessary knowledge and skills may be satisfied by including technical experts. Technical experts should operate under the direction of an auditor.

Auditors-in-training may be included in the audit team, but should not audit without direction or guidance.

Both the audit client and the auditee can request the replacement of particular audit team members on reasonable grounds based on the principles of auditing described in clause 4. Examples of reasonable grounds include conflict of interest situations (such as an audit team member having been a former employee of the auditee or having provided consultancy services to the auditee) and previous unethical behavior. Such grounds should be communicated to the audit team leader and to those assigned responsibility for managing the audit program, who should resolve the issue with the audit client and auditee before making any decisions on replacing audit team members.

### 6.2.5   Establishing Initial Contact with the Auditee

The initial contact for the audit with the auditee may be informal or formal, but should be made by those assigned responsibility for managing the audit program or the audit team leader. The purpose of the initial contact is

a) to establish communication channels with the auditee's representative,
b) to confirm the authority to conduct the audit,
c) to provide information on the proposed timing and audit team composition,
d) to request access to relevant documents, including records,
e) to determine applicable site safety rules,
f) to make arrangements for the audit, and
g) to agree on the attendance of observers and the need for guides for the audit team.

## 6.3   CONDUCTING DOCUMENT REVIEW

Prior to the on-site audit activities the auditee's documentation should be reviewed to determine the conformity of the system, as documented, with audit criteria. The documentation may include relevant management system documents and records, and previous audit reports. The review should take into account the size, nature and complexity of the organization, and the objectives and scope of the audit. In some situations, this review may be deferred until the on-site activities commence, if this is not detrimental to the effectiveness of the conduct of the audit. In other situations, a preliminary site visit may be conducted to obtain a suitable overview of available information.

If the documentation is found to be inadequate, the audit team leader should inform the audit client, those assigned responsibility for managing the audit program, and the auditee. A decision should be made as to whether the audit should be continued or suspended until documentation concerns are resolved.

## 6.4   PREPARING FOR THE ON-SITE AUDIT ACTIVITIES

### 6.4.1   Preparing the Audit Plan

The audit team leader should prepare an audit plan to provide the basis for the agreement among the audit client, audit team and the auditee regarding the conduct of the audit. The plan should facilitate scheduling and coordination of the audit activities.

The amount of detail provided in the audit plan should reflect the scope and complexity of the audit. The details may differ, for example, between initial and subsequent audits and also between internal and external audits. The audit plan should be sufficiently flexible to permit changes, such as changes in the audit scope, which can become necessary as the on-site audit activities progress.

The audit plan should cover the following:

a) the audit objectives;
b) the audit criteria and any reference documents;
c) the audit scope, including identification of the organizational and functional units and processes to be audited;
d) the dates and places where the on-site audit activities are to be conducted;
e) the expected time and duration of on-site audit activities, including meetings with the auditee's management and audit team meetings;
f) the roles and responsibilities of the audit team members and accompanying persons;
g) the allocation of appropriate resources to critical areas of the audit.

The audit plan should also cover the following, as appropriate:

h) identification of the auditee's representative for the audit;
i) the working and reporting language of the audit where this is different from the language of the auditor and/or the auditee;
j) the audit report topics;
k) logistic arrangements (travel, on-site facilities, etc.);
l) matters related to confidentiality;
m) any audit follow-up actions.

The plan should be reviewed and accepted by the audit client, and presented to the auditee, before the on-site audit activities begin.

Any objections by the auditee should be resolved between the audit team leader, the auditee and the audit client. Any revised audit plan should be agreed among the parties concerned before continuing the audit.

## 6.4.2 Assigning Work to the Audit Team

The audit team leader, in consultation with the audit team, should assign to each team member responsibility for auditing specific processes, functions, sites, areas or activities. Such assignments should take into account the need for the independence and competence of auditors and the effective use of resources, as well as different roles and responsibilities of auditors, auditors-in-training and technical experts. Changes to the work assignments may be made as the audit progresses to ensure the achievement of the audit objectives.

## 6.4.3 Preparing Work Documents

The audit team members should review the information relevant to their audit assignments and prepare work documents as necessary for reference and for recording audit proceedings. Such work documents may include

checklists and audit sampling plans, and
forms for recording information, such as supporting evidence, audit findings and records of meetings.

The use of checklists and forms should not restrict the extent of audit activities, which can change as a result of information collected during the audit.

Work documents, including records resulting from their use, should be retained at least until audit completion. Retention of documents after audit completion is described in 6.7. Those documents involving confidential or proprietary information should be suitably safeguarded at all times by the audit team members.

## 6.5 Conducting On-Site Audit Activities

### 6.5.1 Conducting the Opening Meeting

An opening meeting should be held with the auditee's management or, where appropriate, those responsible for the functions or processes to be audited. The purpose of an opening meeting is

- a) to confirm the audit plan,
- b) to provide a short summary of how the audit activities will be undertaken,
- c) to confirm communication channels, and
- d) to provide an opportunity for the auditee to ask questions.

---

**Practical help — Opening the meeting**

In many instances, for example internal audits in a small organization, the opening meeting may simply consist of communicating that an audit is being conducted and explaining the nature of the audit.

For other audit situations, the meeting should be formal and records of the attendance should be kept. The meeting should be chaired by the audit team leader, and the following items should be considered, as appropriate:

a) introduction of the participants, including an outline of their roles;
b) confirmation of the audit objectives, scope and criteria;
c) confirmation of the audit timetable and other relevant arrangements with the auditee, such as the date and time for the closing meeting, any interim meetings between the audit team and the auditee's management, and any late changes;
d) methods and procedures to be used to conduct the audit, including advising the auditee that the audit evidence will only be based on a sample of the information available and that therefore there is an element of uncertainty in auditing;
e) confirmation of formal communication channels between the audit team and the auditee;
f) confirmation of the language to be used during the audit;
g) confirmation that, during the audit, the auditee will be kept informed of audit progress;
h) confirmation that the resources and facilities needed by the audit team are available;
i) confirmation of matters relating to confidentiality;
j) confirmation of relevant work safety, emergency and security procedures for the audit team;
k) confirmation of the availability, roles and identities of any guides;
l) the method of reporting, including any grading of nonconformities;
m) information about conditions under which the audit may be terminated;
n) information about any appeal system on the conduct or conclusions of the audit.

---

### 6.5.2 Communication during the Audit

Depending upon the scope and complexity of the audit, it can be necessary to make formal arrangements for communication within the audit team and with the auditee during the audit.

The audit team should confer periodically to exchange information, assess audit progress, and to reassign work between the audit team members as needed.

During the audit, the audit team leader should periodically communicate the progress of the audit and any concerns to the auditee and audit client, as appropriate. Evidence collected during the audit that suggests an immediate and significant risk (e.g. safety, environmental or quality) should be reported without delay to the auditee and, as appropriate, to the audit client. Any concern about an issue outside the audit scope should be noted and reported to the audit team leader, for possible communication to the audit client and auditee.

Where the available audit evidence indicates that the audit objectives are unattainable, the audit team leader should report the reasons to the audit client and the auditee to determine appropriate action. Such action may include reconfirmation or modification of the audit plan, changes to the audit objectives or audit scope, or termination of the audit.

Any need for changes to the audit scope which can become apparent as on-site auditing activities progress should be reviewed with and approved by the audit client and, as appropriate, the auditee.

### 6.5.3    Roles and Responsibilities of Guides and Observers

Guides and observers may accompany the audit team but are not a part of it. They should not influence or interfere with the conduct of the audit.

When guides are appointed by the auditee, they should assist the audit team and act on the request of the audit team leader. Their responsibilities may include the following:

a) establishing contacts and timing for interviews;
b) arranging visits to specific parts of the site or organization;
c) ensuring that rules concerning site safety and security procedures are known and respected by the audit team members;
d) witnessing the audit on behalf of the auditee;
e) providing clarification or assisting in collecting information.

### 6.5.4    Collecting and Verifying Information

During the audit, information relevant to the audit objectives, scope and criteria, including information relating to interfaces between functions, activities and processes, should be collected by appropriate sampling and should be verified. Only information that is verifiable may be audit evidence. Audit evidence should be recorded.

The audit evidence is based on samples of the available information. Therefore there is an element of uncertainty in auditing, and those acting upon the audit conclusions should be aware of this uncertainty.

Figure 3 provides an overview of the process, from collecting information to reaching audit conclusions.

Methods to collect information include

interviews,
observation of activities, and
review of documents.

---

**Practical help — Sources of information**

The sources of information chosen may vary according to the scope and complexity of the audit and may include the following:

a) interviews with employees and other persons;
b) observations of activities and the surrounding work environment and conditions;
c) documents, such as policy, objectives, plans, procedures, standards, instructions; licences and permits, specifications, drawings, contracts and orders,
d) records, such as inspection records, minutes of meetings, audit reports, records of monitoring programs and the results of measurements;
e) data summaries, analyses and performance indicators;
f) information on the auditee's sampling programs and on procedures for the control of sampling and measurement processes;
g) reports from other sources, for example, customer feedback, other relevant information from external parties and supplier ratings;
h) computerized databases and web sites.

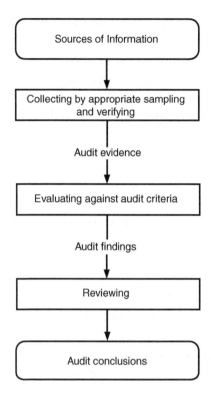

**FIGURE 3** Overview of the process from collecting information to reaching audit conclusions.

---

**Practical help — Conducting interviews**

Interviews are one of the important means of collecting information and should be carried out in a manner adapted to the situation and the person interviewed. However, the auditor should consider the following:

a) interviews should be held with persons from appropriate levels and functions performing activities or tasks within the scope of the audit;

b) interviews should be conducted during the normal working hours and, where practical, at the normal workplace of the person being interviewed;

c) every attempt should be made to put the person being interviewed at ease prior to and during the interview;

d) the reason for the interview and any note taking should be explained;

e) interviews can be initiated by asking the persons to describe their work;

f) questions that bias the answers (i.e. leading questions) should be avoided;

g) the results from the interview should be summarized and reviewed with the interviewed person;

h) the interviewed persons should be thanked for their participation and cooperation.

---

## 6.5.5   Generating Audit Findings

Audit evidence should be evaluated against the audit criteria to generate the audit findings. Audit findings can indicate either conformity or nonconformity with audit criteria. When specified by the audit objectives, audit findings can identify an opportunity for improvement.

The audit team should meet as needed to review the audit findings at appropriate stages during the audit.

Conformity with audit criteria should be summarized to indicate locations, functions or processes that were audited. If included in the audit plan, individual audit findings of conformity and their supporting evidence should also be recorded.

Nonconformities and their supporting audit evidence should be recorded. Nonconformities may be graded. They should be reviewed with the auditee to obtain acknowledgement that the audit evidence is accurate, and that the nonconformities are understood. Every attempt should be made to resolve any diverging opinions concerning the audit evidence and/or findings, and unresolved points should be recorded.

## 6.5.6 Preparing Audit Conclusions

The audit team should confer prior to the closing meeting

a) to review the audit findings, and any other appropriate information collected during the audit, against the audit objectives,

b) to agree on the audit conclusions, taking into account the uncertainty inherent in the audit process,

c) to prepare recommendations, if specified by the audit objectives, and

d) to discuss audit follow-up, if included in the audit plan.

---

**Practical help — Audit conclusions**

Audit conclusions can address issues such as

a) the extent of conformity of the management system with the audit criteria,

b) the effective implementation, maintenance and improvement of the management system, and

c) the capability of the management review process to ensure the continuing suitability, adequacy, effectiveness and improvement of the management system.

If specified by the audit objectives, audit conclusions can lead to recommendations regarding improvements, business relationships, certification/registration or future auditing activities.

---

## 6.5.7 Conducting the Closing Meeting

A closing meeting, chaired by the audit team leader, should be held to present the audit findings and conclusions in such a manner that they are understood and acknowledged by the auditee, and to agree, if appropriate, on the timeframe for the auditee to present a corrective and preventive action plan. Participants in the closing meeting should include the auditee, and may also include the audit client and other parties. If necessary, the audit team leader should advise the auditee of situations encountered during the audit that may decrease the reliance that can be placed on the audit conclusions.

In many instances, for example internal audits in a small organization, the closing meeting may consist of just communicating the audit findings and conclusions.

For other audit situations, the meeting should be formal and minutes, including records of attendance, should be kept.

Any diverging opinions regarding the audit findings and/or conclusions between the audit team and the auditee should be discussed and if possible resolved. If not resolved, all opinions should be recorded.

If specified by the audit objectives, recommendations for improvements should be presented. It should be emphasized that recommendations are not binding.

## 6.6 PREPARING, APPROVING AND DISTRIBUTING THE AUDIT REPORT

### 6.6.1 Preparing the Audit Report

The audit team leader should be responsible for the preparation and contents of the audit report.

The audit report should provide a complete, accurate, concise and clear record of the audit, and should include or refer to the following:

a) the audit objectives;
b) the audit scope, particularly identification of the organizational and functional units or processes audited and the time period covered;
c) identification of the audit client;
d) identification of audit team leader and members;
e) the dates and places where the on-site audit activities were conducted;
f) the audit criteria;
g) the audit findings;
h) the audit conclusions.

The audit report may also include or refer to the following, as appropriate:

i) the audit plan;
j) list of auditee representatives;
k) a summary of the audit process, including the uncertainty and/or any obstacles encountered that could decrease the reliability of the audit conclusions;
l) confirmation that the audit objectives have been accomplished within the audit scope in accordance with the audit plan;
m) any areas not covered, although within the audit scope;
n) any unresolved diverging opinions between the audit team and the auditee;
o) recommendations for improvement, if specified in the audit objectives;
p) agreed follow-up action plans, if any;
q) a statement of the confidential nature of the contents;
r) the distribution list for the audit report.

### 6.6.2 Approving and Distributing the Audit Report

The audit report should be issued within the agreed time period. If this is not possible, the reasons for the delay should be communicated to the audit client and a new issue date should be agreed.

The audit report should be dated, reviewed and approved in accordance with audit program procedures.

The approved audit report should then be distributed to recipients designated by the audit client.

The audit report is the property of the audit client. The audit team members and all report recipients should respect and maintain the confidentiality of the report.

## 6.7 COMPLETING THE AUDIT

The audit is completed when all activities described in the audit plan have been carried out and the approved audit report has been distributed.

Documents pertaining to the audit should be retained or destroyed by agreement between the participating parties and in accordance with audit program procedures and applicable statutory, regulatory and contractual requirements.

Unless required by law, the audit team and those responsible for managing the audit program should not disclose the contents of documents, any other information obtained during the audit, or the audit report, to any other party without the explicit approval of the audit client and, where appropriate, the approval of the auditee. If disclosure of the contents of an audit document is required, the audit client and auditee should be informed as soon as possible.

## 6.8  Conducting Audit Follow-Up

The conclusions of the audit may indicate the need for corrective, preventive or improvement actions, as applicable. Such actions are usually decided and undertaken by the auditee within an agreed timeframe and are not considered to be part of the audit. The auditee should keep the audit client informed of the status of these actions.

The completion and effectiveness of corrective action should be verified. This verification may be part of a subsequent audit.

The audit program may specify follow-up by members of the audit team, which adds value by using their expertise. In such cases, care should be taken to maintain independence in subsequent audit activities.

## 7  COMPETENCE AND EVALUATION OF AUDITORS

### 7.1  General

Confidence and reliance in the audit process depends on the competence of those conducting the audit. This competence is based on the demonstration of

the personal attributes described in 7.2, and
the ability to apply the knowledge and skills described in 7.3 gained through the education, work experience, auditor training and audit experience described in 7.4.

This concept of competence of auditors is illustrated in Figure 4. Some of the knowledge and skills described in 7.3 are common to auditors of quality and environmental management systems, and some are specific to auditors of the individual disciplines.

Auditors develop, maintain and improve their competence through continual professional development and regular participation in audits (see 7.5).

A process for evaluating auditors and audit team leaders is described in 7.6.

### 7.2  Personal Attributes

Auditors should possess personal attributes to enable them to act in accordance with the principles of auditing described in clause 4.

An auditor should be:

a) ethical, i.e. fair, truthful, sincere, honest and discreet;
b) open-minded, i.e. willing to consider alternative ideas or points of view;

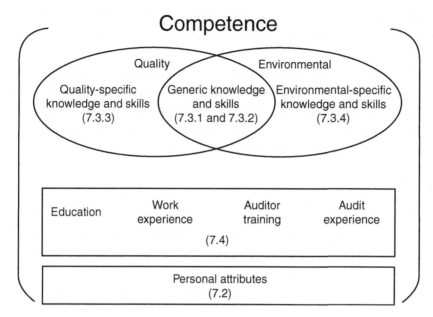

**FIGURE 4** Concept of competence.

c) diplomatic, i.e. tactful in dealing with people;
d) observant, i.e. actively aware of physical surroundings and activities;
e) perceptive, i.e. instinctively aware of and able to understand situations;
f) versatile, i.e. adjusts readily to different situations;
g) tenacious, i.e. persistent, focused on achieving objectives;
h) decisive, i.e. reaches timely conclusions based on logical reasoning and analysis; and
i) self-reliant, i.e. acts and functions independently while interacting effectively with others.

## 7.3    Knowledge and Skills

### 7.3.1    Generic Knowledge and Skills of Quality Management System and Environmental Management System Auditors

Auditors should have knowledge and skills in the following areas.

a) Audit principles, procedures and techniques: to enable the auditor to apply those appropriate to different audits and ensure that audits are conducted in a consistent and systematic manner. An auditor should be able
to apply audit principles, procedures and techniques,
to plan and organize the work effectively,
to conduct the audit within the agreed time schedule,
to prioritize and focus on matters of significance,
to collect information through effective interviewing, listening, observing and reviewing documents, records and data,
to understand the appropriateness and consequences of using sampling techniques for auditing,
to verify the accuracy of collected information,
to confirm the sufficiency and appropriateness of audit evidence to support audit findings and conclusions,

to assess those factors that can affect the reliability of the audit findings and conclusions,

to use work documents to record audit activities,

to prepare audit reports,

to maintain the confidentiality and security of information, and

to communicate effectively, either through personal linguistic skills or through an interpreter.

b) Management system and reference documents: to enable the auditor to comprehend the scope of the audit and apply audit criteria. Knowledge and skills in this area should cover

the application of management systems to different organizations,

interaction between the components of the management system,

quality or environmental management system standards, applicable procedures or other management system documents used as audit criteria,

recognizing differences between and priority of the reference documents,

application of the reference documents to different audit situations, and

information systems and technology for, authorization, security, distribution and control of documents, data and records.

c) Organizational situations: to enable the auditor to comprehend the organization's operational context. Knowledge and skills in this area should cover

organizational size, structure, functions and relationships,

general business processes and related terminology, and

cultural and social customs of the auditee.

d) Applicable laws, regulations and other requirements relevant to the discipline: to enable the auditor to work within, and be aware of, the requirements that apply to the organization being audited. Knowledge and skills in this area should cover

local, regional and national codes, laws and regulations,

contracts and agreements,

international treaties and conventions, and

other requirements to which the organization subscribes.

### 7.3.2 Generic Knowledge and Skills of Audit Team Leaders

Audit team leaders should have additional knowledge and skills in audit leadership to facilitate the efficient and effective conduct of the audit. An audit team leader should be able

to plan the audit and make effective use of resources during the audit,

to represent the audit team in communications with the audit client and auditee,

to organize and direct audit team members,

to provide direction and guidance to auditors-in-training,

to lead the audit team to reach the audit conclusions,

to prevent and resolve conflicts, and

to prepare and complete the audit report.

### 7.3.3 Specific Knowledge and Skills of Quality Management System Auditors

Quality management system auditors should have knowledge and skills in the following areas.

a) Quality-related methods and techniques: to enable the auditor to examine quality management systems and to generate appropriate audit findings and conclusions. Knowledge and skills in this area should cover quality terminology,

quality management principles and their application, and

quality management tools and their application (for example statistical process control, failure mode and effect analysis, etc.).

b) Processes and products, including services: to enable the auditor to comprehend the technological context in which the audit is being conducted. Knowledge and skills in this area should cover

sector-specific terminology,

technical characteristics of processes and products, including services, and

sector-specific processes and practices.

### 7.3.4  Specific Knowledge and Skills of Environmental Management System Auditors

Environmental management system auditors should have knowledge and skills in the following areas.

a) Environmental management methods and techniques: to enable the auditor to examine environmental management systems and to generate appropriate audit findings and conclusions. Knowledge and skills in this area should cover

environmental terminology,

environmental management principles and their application, and

environmental management tools (such as environmental aspect/impact evaluation, life cycle assessment, environmental performance evaluation, etc.).

b) Environmental science and technology: to enable the auditor to comprehend the fundamental relationships between human activities and the environment. Knowledge and skills in this area should cover

the impact of human activities on the environment,

interaction of ecosystems,

environmental media (e.g. air, water, land),

management of natural resources (e.g. fossil fuels, water, flora and fauna), and

general methods of environmental protection.

c) Technical and environmental aspects of operations: to enable the auditor to comprehend the interaction of the auditee's activities, products, services and operations with the environment. Knowledge and skills in this area should cover

sector-specific terminology,

environmental aspects and impacts,

methods for evaluating the significance of environmental aspects,

critical characteristics of operational processes, products and services,

monitoring and measurement techniques, and

technologies for the prevention of pollution.

### 7.4  EDUCATION, WORK EXPERIENCE, AUDITOR TRAINING AND AUDIT EXPERIENCE

### 7.4.1  Auditors

Auditors should have the following education, work experience, auditor training and audit experience.

a) They should have completed an education sufficient to acquire the knowledge and skills described in 7.3.

b) They should have work experience that contributes to the development of the knowledge and skills described in 7.3.3 and 7.3.4. This work experience should be in a technical, managerial or professional position involving the exercise of judgement, problem solving, and communication with other managerial or professional personnel, peers, customers and/or other interested parties.

Part of the work experience should be in a position where the activities undertaken contribute to the development of knowledge and skills in

the quality management field for quality management system auditors, and

the environmental management field for environmental management system auditors.

c) They should have completed auditor training that contributes to the development of the knowledge and skills described in 7.3.1 as well as in 7.3.3 and 7.3.4. This training may be provided by the person's own organization or by an external organization.

d) They should have audit experience in the activities described in clause 6. This experience should have been gained under the direction and guidance of an auditor who is competent as an audit team leader in the same discipline.

NOTE  The extent of direction and guidance (here and in 7.4.2, 7.4.3 and Table 1) needed during an audit is at the discretion of those assigned the responsibility for managing the audit program and the audit team leader. The provision of direction and guidance does not imply constant supervision and does not require someone to be assigned solely to the task.

### 7.4.2  Audit Team Leaders

An audit team leader should have acquired additional audit experience to develop the knowledge and skills described in 7.3.2. This additional experience should have been gained while acting in the role of an audit team leader under the direction and guidance of another auditor who is competent as an audit team leader.

### 7.4.3  Auditors who Audit Both Quality and Environmental Management Systems

Quality management system or environmental management system auditors who wish to become auditors in the second discipline

a) should have the training and work experience needed to acquire the knowledge and skills for the second discipline, and

b) should have conducted audits covering the management system in the second discipline under the direction and guidance of an auditor who is competent as an audit team leader in the second discipline.

An audit team leader in one discipline should meet the above recommendations to become an audit team leader in the second discipline.

### 7.4.4  Levels of Education, Work Experience, Auditor Training and Audit Experience

Organizations should establish the levels of the education, work experience, auditor training and audit experience an auditor needs to gain the knowledge and skills appropriate to the audit program by applying Steps 1 and 2 of the evaluation process described in 7.6.2.

**TABLE 1**
**Example of levels of education, work experience, auditor training and audit experience for auditors conducting certification or similar audits**

| Parameter | Auditor | Auditor in both disciplines | Audit team leader |
|---|---|---|---|
| Education | Secondary education (see Note 1) | Same as for auditor | Same as for auditor |
| Total work experience | 5 years (see Note 2) | Same as for auditor | Same as for auditor |
| Work experience in quality or environmental management field | At least 2 years of the total 5 years | 2 years in the second discipline (see Note 3) | Same as for auditor |
| Auditor training | 40 h of audit training | 24 h of training in the second discipline (see Note 4) | Same as for auditor |
| Audit experience | Four complete audits for a total of at least 20 days of audit experience as an auditor-in-training under the direction and guidance of an auditor competent as an audit team leader (see Note 5). The audits should be completed within the last three consecutive years | Three complete audits for a total of at least 15 days of audit experience in the second discipline under the direction and guidance of an auditor competent as an audit team leader in the second discipline (see Note 5). The audits should be completed within the last two consecutive years | Three complete audits for a total of at least 15 days of audit experience acting in the role of an audit team leader under the direction and guidance of an auditor competent as an audit team leader (see Note 5). The audits should be completed within the last two consecutive years |

NOTE 1    Secondary education is that part of the national educational system that comes after the primary or elementary stage, but that is completed prior to entrance to a university or similar educational institution.

NOTE 2    The number of years of work experience may be reduced by 1 year if the person has completed appropriate post-secondary education.

NOTE 3    The work experience in the second discipline may be concurrent with the work experience in the first discipline.

NOTE 4    The training in the second discipline is to acquire knowledge of the relevant standards, laws, regulations, principles, methods and techniques.

NOTE 5    A complete audit is an audit covering all of the steps described in 6.3 to 6.6. The overall audit experience should cover the entire management system standard.

Experience has shown that the levels given in Table 1 are appropriate for auditors conducting certification or similar audits. Depending on the audit program, higher or lower levels may be appropriate.

## 7.5    MAINTENANCE AND IMPROVEMENT OF COMPETENCE

### 7.5.1    Continual Professional Development

Continual professional development is concerned with the maintenance and improvement of knowledge, skills and personal attributes. This can be achieved through means such as additional work experience, training, private study, coaching, attendance at meetings, seminars and conferences or other relevant activities. Auditors should demonstrate their continual professional development.

The continual professional development activities should take into account changes in the needs of the individual and the organization, the practice of auditing, standards and other requirements.

### 7.5.2    Maintenance of Auditing Ability

Auditors should maintain and demonstrate their auditing ability through regular participation in audits of quality and/or environmental management systems.

## 7.6    Auditor Evaluation

### 7.6.1    General

The evaluation of auditors and audit team leaders should be planned, implemented and recorded in accordance with audit program procedures to provide an outcome that is objective, consistent, fair and reliable. The evaluation process should identify training and other skill enhancement needs.

The evaluation of auditors occurs at the following different stages:

the initial evaluation of persons who wish to become auditors;

the evaluation of the auditors as part of the audit team selection process described in 6.2.4;

the continual evaluation of auditor performance to identify needs for maintenance and improvement of knowledge and skills.

Figure 5 illustrates the relationship between these stages of evaluation.

The process steps described in 7.6.2 may be used in each of these stages of evaluation.

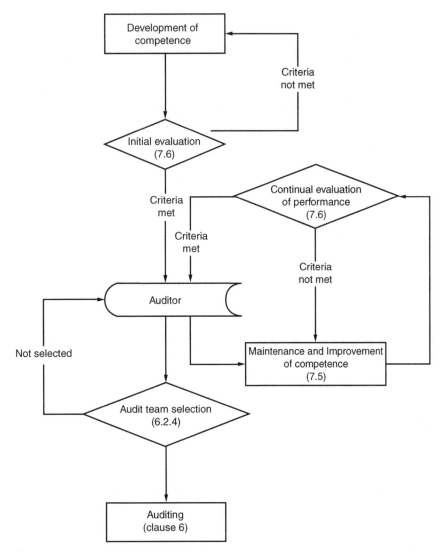

**FIGURE 5**  Relationship between the stages of evaluation.

**TABLE 2**
**Evaluation methods**

| Evaluation method | Objectives | Examples |
|---|---|---|
| Review of records | To verify the background of the auditor | Analysis of records of education, training, employment and audit experience |
| Positive and negative feedback | To provide information about how the performance of the auditor is perceived | Surveys, questionnaires, personal references, testimonials, complaints, performance evaluation, peer review |
| Interview | To evaluate personal attributes and communication skills, to verify information and test knowledge and to acquire additional information | Face-to-face and telephone interviews |
| Observation | To evaluate personal attributes and the ability to apply knowledge and skills | Role playing, witnessed audits, on the-job performance |
| Testing | To evaluate personal attributes and knowledge and skills and their application | Oral and written exams, psychometric testing |
| Post-audit review | To provide information where direct observation may not be possible or appropriate | Review of the audit report and discussion with the audit client, auditee, colleagues and with the auditor |

## 7.6.2   Evaluation Process

The evaluation process involves four main steps.

**Step 1** — Identify the personal attributes, and the knowledge and skills to meet the needs of the audit program. In deciding the appropriate knowledge and skills, the following should be considered:

the size, nature and complexity of the organization to be audited;
the objectives and extent of the audit program;
certification/registration and accreditation requirements;
the role of the audit process in the management of the organization to be audited;
the level of confidence required in the audit program;
the complexity of the management system to be audited.

**Step 2** — Set the evaluation criteria. The criteria may be quantitative (such as the years of work experience and education, number of audits conducted, hours of audit training) or qualitative (such as having demonstrated personal attributes, knowledge or the performance of the skills, in training or in the workplace).

**Step 3** — Select the appropriate evaluation method. Evaluation should be undertaken by a person or a panel using one or more of the methods selected from those in Table 2. In using Table 2, the following should be noted:

the methods outlined represent a range of options and may not apply in all situations;
the various methods outlined can differ in their reliability;
typically, a combination of methods should be used to ensure an outcome that is objective, consistent, fair and reliable.

**Step 4** — Conduct the evaluation. In this step the information collected about the person is compared against the criteria set in Step 2. Where a person does not meet the criteria, additional training, work and/or audit experience are required, following which there should be a re-evaluation.

An example of how the steps of the evaluation process might be applied and documented for a hypothetical internal audit program is illustrated in Table 3.

## TABLE 3
## Application of the evaluation process for an auditor in a hypothetical internal audit program

| Areas of competence | Step 1 Personal attributes, and knowledge and skills | Step 2 Evaluation criteria | Step 3 Evaluation methods |
|---|---|---|---|
| Personal attributes | Ethical, open-minded, diplomatic, observant, perceptive, versatile, tenacious, decisive, self-reliant. | Satisfactory performance in the workplace. | Performance evaluation |
| **Generic knowledge and skills** | | | |
| Audit principles, procedures and techniques | Ability to conduct an audit according to in-house procedures, communicating with known workplace colleagues. | Completed an internal auditor training course. Performed three audits as a member of an internal audit team. | Review of training records. Observation. Peer review |
| Management system and reference documents | Ability to apply the relevant parts of the Management System Manual and related procedures. | Read and understood the procedures in the Management System Manual relevant to the audit objectives, scope and criteria. | Review of training records. Testing. Interview |
| Organizational situations | Ability to operate effectively within the organization's culture and organizational and reporting structure. | Worked for the organization for at least one year in a supervisory role | Review of employment records |
| Applicable laws, regulations and other requirements | Ability to identify and understand the application of the relevant laws and regulations related to the processes, products and/or discharges to the environment. | Completed a training course on the laws relevant to the activities and processes to be audited. | Review of training records |
| **Quality-specific knowledge and skills** | | | |
| Quality-related methods and techniques | Ability to describe the in-house quality control methods. Ability to differentiate between requirements for in-process and final testing. | Completed training in the application of quality control methods. Demonstrated work place use of in-process and final testing procedures. | Review of training records. Observation |
| Processes and products, including services | Ability to identify the products, their manufacturing process, specifications and end-use. | Worked in the production planning as process planning clerk. Worked in the service department. | Review of employment records |
| **Environmental-specific knowledge and skills** | | | |
| Environmental management methods and techniques | Ability to understand methods for evaluating environmental performance. | Completed training in environmental performance evaluation. | Review of training records |
| Environmental science and technology | Ability to understand how the pollution prevention and control methods used by the organization address the organization's significant environmental aspects. | Six months of work experience in pollution prevention and control in a similar manufacturing environment. | Review of employment records |
| Technical and environmental aspects of operations | Ability to recognize the organization's environmental aspects and their impacts (e.g. materials, their reactions with one another and potential impact on the environment in the event of spillage or release). Ability to assess the emergency response procedures applicable to environmental incidents. | Completed an in-house training course on materials storage, mixing, use, disposal and their environmental impacts. Completed training in the Emergency Response | Plan and experience as a member of the emergency response team.<br><br>Review of training records, course content and results. Review of training and employment records |

# IPEC GMP for Bulk Excipients

We were unable to obtain distribution rights for the IPEC GMP Guide for Bulk Pharmaceutical Excipients from the International Pharmaceutical Excipients Council.

You may obtain a copy of the document directly by contacting the IPEC-Americas Headquarters Office:

Telephone:  703-875-2127
Fax:        703-525-5157
e-mail:     info@ipecamericas.org

IPEC GMP for Bulk Excipients

T - #0019 - 011124 - C0 - 276/219/26 - PB - 9780367393328 - Gloss Lamination